The Missing Lemur Link

An Ancestral Step in the Evolution of Human Behaviour

Lemurs share a common distant ancestor with humans. Following their own evolutionary pathway, lemurs provide the ideal model to shed light on the behavioural traits of primates, including conflict management, communication strategies and society building, and how these aspects of social living relate to those found in the anthropoid primates.

Adopting a comparative approach throughout, lemur behaviour is cross-examined with that of monkeys, apes and humans. This book reviews and expands upon the newest fields of research in lemur behavioural biology, including recent analytical approaches that have so far been limited to studies of haplorrhine primates. Different methodological approaches are harmonised in this volume to break conceptual walls between both primate taxa and different disciplines.

Through a focus on the methodologies behind lemur behaviour and social interactions, future primate researchers will be encouraged to produce directly comparable results.

Ivan Norscia carries out research at the Natural History Museum, University of Pisa, Italy. He started investigating the behavioural ecology of lemurs in dry and wet forests of Madagascar and through his research has contributed to the redefinition of a lemur species (*Avahi meridionalis*). His research later expanded to the behaviour of monkeys, apes and humans.

Elisabetta Palagi is a department member of the Natural History Museum, University of Pisa, Italy. Her research centres upon lemur individual recognition and multi-modal signalling, which has expanded to other primate and non-primate animals. Most recently, in conjunction with Ivan Norscia, she has adopted a cross-species comparison approach to shed light on the biological foundation of human behaviour.

Cambridge Studies in Biological and Evolutionary Anthropology

Consulting editors

C. G. Nicholas Mascie-Taylor, *University of Cambridge*
Robert A. Foley, *University of Cambridge*

Series editors

Agustín Fuentes, *University of Notre Dame*
Sir Peter Gluckman, *The Liggins Institute, The University of Auckland*
Nina G. Jablonski, *Pennsylvania State University*
Clark Spencer Larsen, *The Ohio State University*
Michael P. Muehlenbein, *Indiana University, Bloomington*
Dennis H. O'Rourke, *The University of Utah*
Karen B. Strier, *University of Wisconsin*
David P. Watts, *Yale University*

Also available in the series

53. *Technique and Application in Dental Anthropology* Joel D. Irish & Greg C. Nelson (eds.) 978 0 521 87061 0
54. *Western Diseases: An Evolutionary Perspective* Tessa M. Pollard 978 0 521 61737 6
55. *Spider Monkeys: The Biology, Behavior and Ecology of the Genus Ateles* Christina J. Campbell 978 0 521 86750 4
56. *Between Biology and Culture* Holger Schutkowski (ed.) 978 0 521 85936 3
57. *Primate Parasite Ecology: The Dynamics and Study of Host-Parasite Relationships* Michael A. Huffman & Colin A. Chapman (eds.) 978 0 521 87246 1
58. *The Evolutionary Biology of Human Body Fatness: Thrift and Control* Jonathan C. K. Wells 978 0 521 88420 4
59. *Reproduction and Adaptation: Topics in Human Reproductive Ecology* C. G. Nicholas Mascie-Taylor & Lyliane Rosetta (eds.) 978 0 521 50963 3
60. *Monkeys on the Edge: Ecology and Management of Long-Tailed Macaques and their Interface with Humans* Michael D. Gumert, Agustín Fuentes & Lisa Jones-Engel (eds.) 978 0 521 76433 9
61. *The Monkeys of Stormy Mountain: 60 Years of Primatological Research on the Japanese Macaques of Arashiyama* Jean-Baptiste Leca, Michael A. Huffman & Paul L. Vasey (eds.) 978 0 521 76185 7
62. *African Genesis: Perspectives on Hominin Evolution* Sally C. Reynolds & Andrew Gallagher (eds.) 978 1 107 01995 9
63. *Consanguinity in Context* Alan H. Bittles 978 0 521 78186 2
64. *Evolving Human Nutrition: Implications for Public Health* Stanley Ulijaszek, Neil Mann & Sarah Elton (eds.) 978 0 521 86916 4
65. *Evolutionary Biology and Conservation of Titis, Sakis and Uacaris* Liza M. Veiga, Adrian A. Barnett, Stephen F. Ferrari & Marilyn A. Norconk (eds.) 978 0 521 88158 6
66. *Anthropological Perspectives on Tooth Morphology: Genetics, Evolution, Variation* G. Richard Scott & Joel D. Irish (eds.) 978 1 107 01145 8

67. *Bioarchaeological and Forensic Perspectives on Violence: How Violent Death is Interpreted from Skeletal Remains* Debra L. Martin & Cheryl P. Anderson (eds.) 978 1 107 04544 6
68. *The Foragers of Point Hope: The Biology and Archaeology of Humans on the Edge of the Alaskan Arctic* Charles E. Hilton, Benjamin M. Auerbach & Libby W. Cowgill (eds.) 978 1 107 02250 8
69. *Bioarchaeology: Interpreting Behavior from the Human Skeleton, 2nd Ed.* Clark Spencer Larsen 978 0 521 83869 6 & 978 0 521 54748 2
70. *Fossil Primates* Susan Cachel 978 1 107 00530 3
71. *Demography and Evolutionary Ecology of Hadza Hunter-Gatherers* Nicholas Blurton Jones 978 1 107 06982 4
72. *Skeletal Biology of the Ancient Rapanui (Easter Islanders)* Vincent H. Stefan & George W. Gill (eds.) 978 1 107 02366 6
73. *The Dwarf and Mouse Lemurs of Madagascar: Biology, Behavior and Conservation Biogeography of the Cheirogaleidae* Shawn M. Lehman, Ute Radespiel & Elke Zimmermann (eds.) 978 1 107 07559 7

The Missing Lemur Link

An Ancestral Step in the Evolution of Human Behaviour

IVAN NORSCIA
University of Pisa, Italy

and

ELISABETTA PALAGI
University of Pisa, Italy

CAMBRIDGE
UNIVERSITY PRESS

CAMBRIDGE
UNIVERSITY PRESS

University Printing House, Cambridge CB2 8BS, United Kingdom

Cambridge University Press is part of the University of Cambridge.

It furthers the University's mission by disseminating knowledge in the pursuit of education, learning and research at the highest international levels of excellence.

www.cambridge.org
Information on this title: www.cambridge.org/9781107016088

© Ivan Norscia & Elisabetta Palagi, 2016

This publication is in copyright. Subject to statutory exception
and to the provisions of relevant collective licensing agreements,
no reproduction of any part may take place without the written
permission of Cambridge University Press.

First published 2016

Printed in the United Kingdom by TJ International Ltd., Padstow, Cornwall

A catalogue record for this publication is available from the British Library

ISBN 978-1-107-01608-8 Hardback

Cambridge University Press has no responsibility for the persistence or accuracy of URLs for external or third-party internet websites referred to in this publication, and does not guarantee that any content on such websites is, or will remain, accurate or appropriate.

In memory of primatologist and mentor Alison Jolly,
and beloved friend Letizia Pantani

Contents

List of contributors	page xi
A message from Jane Goodall	xiii
Foreword by Alison Jolly and Ian Tattersall	xv
Preface by the authors	xviii
Preface box by Judith Masters and Fabien Génin – Overcoming the Scala Naturae: why strepsirrhines are not primitive	xxvi
Acknowledgements	xxxii

Part I Communication: from sociality to society — 1

1 Who are you? How lemurs recognise each other in a smell-centred world — 3

Box 1.1 by Paolo Pelosi – Speaking of which: breaking the olfactory code — 4

Box 1.2 by Eckhard W. Heymann – Speaking of which: odour communication in tamarins — 9

Box 1.3 by Ipek G. Kulahci and Asif A. Ghazanfar – Speaking of which: a multimodal approach to individual recognition — 20

2 What do you mean? Multimodal communication for a better signal transmission — 30

Box 2.1 by Bridget M. Waller and Katie E. Slocombe – Speaking of which: lipsmacking as an example of multimodal signalling in monkeys — 34

Box 2.2 by James P. Higham – Speaking of which: the complex scenarios of multimodal communication — 39

3 A vertical living: sexual selection strategies and upright locomotion — 54

Box 3.1 by David A. Puts – Speaking of which: vertical locomotion and its impact on the evolution of human sexual behaviour — 60

Part II How conflicts shape societies — 75

4 Bossing around the forest: power asymmetry and hierarchy — 77

Box 4.1 by Takeshi Furuichi – Speaking of which: female social status of wild bonobos — 83

Box 4.2 by Bernard Thierry – Speaking of which: cross-species comparisons in social styles — 94

5 Something to make peace for: conflict management and resolution — 112

Box 5.1 by Cary J. Roseth – Speaking of which: peacemaking in children — 116
Box 5.2 by Giada Cordoni – Speaking of which: conflict management in non-primate mammals — 121
Box 5.3 by Peter Verbeek – Speaking of which: the contribution of peace ethology to life science — 134

6 Anxiety…from scratch: emotional response to tense situations — 146

Box 6.1 by Filippo Aureli – Speaking of which: anxiety-related behaviour — 150
Box 6.2 by Lisa Gould – Speaking of which: sex and hormones in wild lemurs — 161

Part III Why lemurs keep in touch — 183

7 Playing lemurs: why primates have been playing for a long time — 185

Box 7.1 by Marc Bekoff – Speaking of which: why studying play is so fascinating — 190
Box 7.2 by Sergio M. Pellis – Speaking of which: the neurobiology of play — 202

8 Sex is not on discount: mating market and lemurs — 219

Box 8.1 by Karen B. Strier – Speaking of which: reflections on primate sexual behaviour with special reference to muriquis — 227
Box 8.2 by Rebecca J. Lewis – Speaking of which: grooming, service exchange and dominance in the Verreaux's sifaka — 234

Part IV Closing remarks — 245

9 Understanding lemurs: future directions in lemur cognition — 247

Box 9.1 by Evan L. MacLean and Brian Hare – Speaking of which: lemur cognition and sociality — 255
Box 9.2 by Elsa Addessi and Elisabetta Visalberghi – Speaking of which: choice-making and capuchin monkeys — 264

Looking back to the future – Michael Huffman — 280
Index — 285

List of Contributors

Elsa Addessi
Unit of Cognitive Primatology and Primate Centre Istituto di Scienze e Tecnologie della Cognizione Consiglio Nazionale delle Ricerche via Aldrovandi, Rome, Italy

Filippo Aureli
Instituto de Neuroetologia, Universidad Veracruzana, Xalapa, Veracruz, Mexico & Research Centre in Evolutionary Anthropology and Palaeoecology, Liverpool John Moores University, Liverpool, United Kingdom

Marc Bekoff
Ecology and Evolutionary Biology, University of Colorado, Boulder, Colorado

Giada Cordoni
Natural History Museum (Calci, Pisa), University of Pisa, Italy

Takeshi Furuichi
Primate Research Institute, Kyoto University Inuyama, Aichi, Japan

Fabien Génin
Department of Zoology & Entomology, University of Fort Hare, South Africa

Asif A. Ghazanfar
Princeton Neuroscience Institute, Psychology Department, Princeton University, Princeton, NJ, USA

Lisa Gould
Dept. of Anthropology, University of Victoria, Victoria, BC

Brian Hare
Department of Evolutionary Anthropology, Duke University, Durham, NC, USA

Eckhard W. Heymann
Abteilung Verhaltensökologie & Soziobiologie - Deutsches Primatenzentrum, Kellnerweg, Göttingen, Germany

James P. Higham
Dept. of Anthropology, New York University, New York, NY, USA

Ipek G. Kulahci
Princeton Neuroscience Institute, Psychology Department, Princeton University, Princeton, NJ, USA

Evan L. MacLean
Department of Evolutionary Anthropology, Duke University, Durham, NC, USA

Judith C. Masters
Department of Zoology & Entomology, University of Fort Hare, South Africa

Sergio M. Pellis
Department of Neuroscience, University of Lethbridge, Lethbridge, Alberta, Canada

Paolo Pelosi
State Key Laboratory for Biology of Plant Diseases and Insect Pests, Institute of Plant Protection, Chinese Academy of Agricultural Sciences, Beijing, China

David A. Puts
Department of Anthropology, Carpenter Building, Pennsylvania State University, University Park, PA

Cary J. Roseth
Department of Counseling, Educational Psychology and Special Education, Michigan State University, East Lansing, MI, USA

Katie E. Slocombe
Department of Psychology, University of York, York, UK

Karen B. Strier
Department of Anthropology, University of Wisconsin-Madison, Madison, Wisconsin, USA

Bernard Thierry
Département Ecologie, Physiologie et Ethologie, Institut Pluridisciplinaire Hubert Curien, Université de Strasbourg, Centre National de la Recherche Scientifique, Strasbourg, France

Peter Verbeek
Dept. of Comparative Culture, Miyazaki International College, Miyazaki, Miyazaki Prefecture, Japan

Elisabetta Visalberghi
Unit of Cognitive Primatology and Primate Centre Istituto di Scienze e Tecnologie della Cognizione Consiglio Nazionale delle Ricerche via Aldrovandi, Rome, Italy

Bridget M. Waller
Department of Psychology, University of Portsmouth, Portsmouth, UK

A message from Jane Goodall

In 1960 Dr Louis Leakey, palaeontologist and anthropologist, suggested that I should travel to the Gombe National Park, in what is now Tanzania, to study the behaviour of the chimpanzees there. He believed that apes and humans shared a common ancestor some 6 or 7 million years ago and that an understanding of chimpanzee behaviour in the wild might help him better understand how our prehistoric ancestors may have behaved. Subsequently, when I had been in the field for some 18 months, he arranged for me to go to Cambridge University to work for a PhD (although I had never been to college). In the early sixties ethologists tended to be reductionists, searching for simple explanations for very complex behaviours. When I talked about the Gombe chimpanzees having different personalities, complex cognitive abilities, minds, and emotions such as pleasure and sadness, anger, frustration and so on, I was accused of anthropomorphism as those attributes, I was told, were unique to the human animal. Some scientists were even shocked that I referred to the chimpanzees by names rather than numbers. Fortunately, I had learned from my dog, as a child, that none of that was true.

Since those days the science of animal behaviour has come a long way and there have been many studies of animal cognition, animal emotions and even personality. In other words the sharp line once believed to separate humans from the rest of the animal kingdom was blurred. There was a difference of degree between the behaviour of humans and chimpanzees, but not of kind. Since then other studies have shown that many of the complex behaviours shared by the great apes and humans are present also in monkeys. And during the last decade the barrier between monkeys and lemurs has also been found to be increasingly fragile. Lemurs – the spirits of our ancestors according to some Malagasy traditions – exchange services, suffer and manage anxiety, reconcile, and use play to increase tolerance towards unfamiliar individuals.

Many species of lemurs are extremely endangered, and an understanding of their behaviour, in addition to helping us to understand the origins of some of our own behaviour, is also crucial for determining the best conservation strategies. We need to understand how the lemurs interact with each other and with their forest environment. We need to understand the complexity of the natural world in order to save biodiversity. And we need to inform and educate people so that they understand how the different species of animals and plants in a given ecosystem are interdependent on each other and why it is important to save this complex environment not only to protect wildlife but also for their own future well-being. This vision is in line with the core values of the Jane Goodall Institute that I founded in 1977 (www.janegoodall.org).

Respect, nourish and protect all living things; people, animals and the environment are all interconnected. Knowledge leads to understanding, and understanding

Chimpanzee juvenile at Gombe Stream National Park, Tanzania. Photo: Ivan Norscia.

will encourage us to take action. Every individual has the ability to make a positive difference. Flexibility and open-mindedness are essential to enable us to respond to a changing world. Integrity and compassion are required in all that we do and say.

The primates of Madagascar, as well as the primates of the rest of the world, are our closest relatives and deserve our attention and care. This book helps us to better understand lemurs and their place in the evolutionary path that our own species has followed.

Foreword

We human beings are distinctly unusual in the way we behave, and most particularly in the way in which we process information about the world around us. The final finishing touches to the distinctive manner in which we think and interact seem to have been acquired rather recently. But they were based upon, and made possible by, a complex biological edifice that had been under construction for a very long time indeed. If anything that happened along the winding road of primate and human evolution had failed to happen, for whatever reason, we would not be the creatures we are today. As a result, we are not only who we are, but we are also who we were. Arguably, that long history stretches back to the very beginnings of life on planet Earth; but probably the most useful place to start any rational account of it is with the origin and early evolution of our own unique mammalian order, Primates.

The primate – and ultimately human – edifice has been built, incrementally, over the past sixty million years or so. The foundations and lower floors of that edifice were furnished by the early primates, whose fossils we know from North American, Eurasian and African rocks dating between about sixty and forty million years ago. But these forms are long vanished; and the closest thing we have to living, observable, analogies for those ancient relatives are the strepsirrhine primates of the tropical Old World. Also including a handful of nocturnal species from tropical Africa and Asia, the strepsirrhines are incomparably well represented by the lemurs of the mini-continent of Madagascar. Isolated in their island domain for upward of fifty million years, the lemurs have radiated into an astonishing array of adaptive types: large and small, nocturnal and diurnal, suspensory and cursorial, leaf-eating and fruit-eating, highly arboreal and at least partly terrestrial, eurytopes and stenotopes, gregarious and solitary. Sadly, the larger-bodied representatives of this amazingly diverse and apparently monophyletic primate group are recently extinct, at least partly due to the depredations of humankind. But the surviving species, reckoned to number between fifty and a hundred, depending on the criteria you use to recognise them, still present us with a mind-boggling adaptive diversity.

Nonetheless, despite their undeniable diversity in phylogeny and form, in certain fundamental – and particularly cognitive – respects, all of the Malagasy primates represent variations on a single basic theme: a theme apparently close to what was also exemplified by those early primates from which we, too, ultimately sprang. Accordingly, if we can contrive to fathom the ways in which lemurs interact with each other and with the world around them, we will have made a major stride towards understanding the biological underpinnings on which the yet more complex cognitive modes of human beings and other 'higher' primates are based.

Strepsirrhine primates have wet muzzles, the better to smell with, and possess functional Jacobson's organs that amplify the sense of smell. Their cousins the haplorrhines, namely monkeys, apes and humans, have dry noses and upper lips. Correlated with their olfactory prowess, many strepsirrhines have specialised scent glands, which they use to mark branches, their own tails, or even each other. Most groom themselves and each other with a tooth-comb composed of the lower canines and incisors, and not with their hands like monkeys and apes. Their upper canines, however, are fiercely sharp, honing against a lower premolar. The specialised organs of both aggression and of mutual trust or affection are right there at the front of the face. Lemurs use their hands mostly for locomotion, or for handling food. The only highly manipulative strepsirrhine is the weird and wonderful aye-aye, with its specialised finger for tapping to test hollow objects and probing into the hollow, once interesting contents are found. For many years, other strepsirrhines were thought stupid because they did not pass hand-oriented psychological tests, and this book deals with newer research that gives them problems to solve according to their own inclinations.

One of the biggest outstanding questions is to what extent the social relations of strepsirrhines resemble the complex relations of monkeys and apes, and in this domain one major problem that has so far resisted complete solution is why, in most species of Malagasy lemurs, females are clearly dominant over the males. Female dominance sets the lemurs apart from the huge majority of other mammals, including almost all monkeys and apes. Meanwhile, for those species that lead 'solitary' lives, sociality turns out to be largely a question of keeping track of neighbours, male and female, in a widespread community. This book offers new insights about the relationship of this complex of strepsirrhine abilities to their social behaviours. In some cases these insights challenge previous thinking: for instance, about reconciliation after aggression, and about the social roles of play and grooming. There are also some unnoticed parallels, such as why both sifakas and human beings carry sexual signals on their (and our) chests.

The Missing Lemur Link results from a unique collaboration between a laboratory scientist who has spent many years closely observing lemur behaviours in controlled situations, and an inveterate field worker who deeply understands the complex and unpredictable natural world in which his subjects live. Both authors are widely versed in studies of other primates as well as of the lemurs, and this enables them to make considerable progress in these pages towards a synthesis of the vast amount of information on communication and other forms of primate interaction that has by now accumulated. This aspect of their work makes this book a truly a path-breaking contribution because, largely as a result of the lemurs' apparently rather elementary forms of face-to-face social interaction and of their high dependence on olfaction, researchers have tended to compartmentalise them away from the other primates. Yet there are clearly cognitive continuities between the strepsirrhine and haplorrhine primate suborders, and characterising these commonalities will provide an essential key to understanding how modern anthropoids emerged from an ultimately quite strepsirrhine-like ancestral condition. In these pages Elisabetta Palagi and Ivan Norscia have performed an enormous service to primatology by focusing on methodology in a way that will encourage future researchers on primates of all groups to produce directly comparable results.

Most gratifyingly of all, this excellent new book shows us just how much has been achieved in the study of the lemurs in recent years, and firmly points the way ahead.

Alison Jolly & Ian Tattersall

Preface

Diversity organises unity which organises

Edgar Morin, 1977

Theoretical discussion and debate are open on which part of social behaviour is really exclusive of *Homo sapiens* and other apes, and what, instead, is rooted in the common ground of primate origins. Since the time of our common ancestors, lemurs have been following their own evolutionary pathway to become 'modern primates'. This process, enhanced by Madagascar's isolation from the rest of the world, which started about eighty million years ago, has led lemurs to possess a puzzling combination of features, such as a small brain and communication highly based on smell combined with biological peculiarities, such as female dominance, lack of sexual dimorphism and strict seasonal breeding. On the other hand, group-living lemurs share basic features with social monkeys and apes such as cohesive multimale-multifemale societies, female philopatry and individual recognition. Lemurs are the ideal model to shed light on the 'primate behavioural potential' in terms of conflict management, communication strategies and society building, and on how much these crucial aspects of social living are similar to or different from those found in monkeys, apes and humans. Part of the perceived gap between strepsirrhines and haplorrhines – also mentioned by Alison Jolly and Ian Tattersall in their foreword – probably lies in the divergent research methodologies applied to these two groups. The aim of this book is to review and expand upon the newest fields of research in lemur behavioural biology, including recent analytical approaches that, so far, have been restricted to haplorrhines.

Breaking the wall

Are lemurs primates? This question may sound rhetorical, but unfortunately the answer (yes, they are) cannot be taken for granted.

The idea for this book came to our minds after a few influential Italian anthropologists had questioned the relevance of lemur behavioural studies to the anthropological domain and, therefore, to the understanding of the evolution of human behaviour. Monkeys were discriminated against as well, because – according to these scholars – they did not fit in the discipline, unless direct comparisons with humans were drawn. While the significance of lemurs and monkeys to human evolution was debated (and no consensus reached), apes were automatically included in the anthropological domain, thankfully. Of course the discrimination of strepsirrhines and monkeys was instrumental in favouring other research strains. And, of course, the comparative approach can be applied only when the subjects of the comparison have been sufficiently and equally investigated. So, here we are. In this book the comparison is served.

It is not possible to draw the line between different primate taxa (e.g., strepsirrhines versus monkeys, monkeys versus apes) and determine where (or when) a behavioural or cognitive ability ends and another starts. On one side, the division of organisms into categories is conceptually comforting because differences are more easily detectable than similarities and more functional to describe the diversity of the biological world. From Aristotle's *De Anima* (fourth century BC) to Linnaeus' *Scala Naturae* (eighteenth century AD) and the work of modern taxonomists, organisms have been categorised and organised according to different criteria reflecting the knowledge of each historical period. On the other side, such a 'separatist' (but probably adaptive) approach turns the spotlight on diversity more than on similarity. As a result, organisms are often described in the light of their divergences (wolves are carnivores and chimpanzees are primates, e.g., based on their anatomy) rather than on the basis of their common features (e.g., wolves and chimpanzees both possess a social brain). In this respect, 'behavioural taxonomy' is different from any other taxonomy. The challenging work of a researcher that approaches a new phenomenon is recognising the diversity and placing it into the evolutionary continuity that links organisms to one another. De Waal and Ferrari (2010) pointed out that such continuity is acknowledged in anatomy, genetics and neuroscience but not in cognition. According to these authors, comparative cognition is dominated by a top-down approach that focuses on the expression of complex skills and not on their most basal building blocks. For example, we can say that humans are unique because they use articulate language to communicate. Thus, humans are different from great apes who *do not* possess this ability (top-down approach). Yet, great apes and humans use a combination of gestures, vocalisations and facial expressions to convey information (the building blocks of complex and sophisticated communication). Thus, *both* great apes *and* humans use the same (visual and auditory) modalities to communicate, which constitute the biological bases of the human language, which in turn goes beyond the ability to articulate sounds to make a speech (bottom-up approach). The question is to what extent the abilities of other primates inform on the evolution of human complex cognitive skills.

Cognition is a proxy of behaviour, and both cognition and behaviour are the immaterial expression of a complicated network of interactions within and between multiple biological structures (e.g., from genotype to physiology and morpho-anatomy) and environmental stimuli. As occurs with the cognitive domain, also for the behavioural domain it is not always easy to assess when continuity ends and diversity begins.

As well described by Alison Jolly and Ian Tattersall in the foreword, strepsirrhines have their own peculiarities, which make them unquestionably different from other primates. Yet, as the two authors also emphasise, 'there are clearly cognitive continuities between strepsirrhine and haplorrhine suborders'. Understanding these continuities leads to the understanding of the biological bases of human behaviour. And having emerged early from pre-primate mammalian ancestors, strepsirrhines offer not only the possibility to explore the connection between human and non-human primates but also the connection to their mammalian predecessors, as Michael Huffman argues in his afterword referring to our 'humble beginnings'.

The scientific theories framing the evolutionary linkage between humans and other primates (and animals) date back to the schools of thought of the nineteenth century (represented by Darwin, Huxley, Wallace, etc.), which eventually resulted in the formulation of the modern theory of evolution, also influenced by the 'transformational thinking' explained in Judith Masters and Fabien Génin's box.

Only recently, modern physical anthropology has turned its gaze onto the study of non-human primate behaviour. The physical anthropologist Louis Leakey promoted the study of the behaviour of the apes as the portal to understand the behaviour of our hominid ancestors, just because 'behaviour does not fossilise'. Jane Goodall started studying chimpanzees because of the very belief that knowing chimpanzees would help humans to understand humans. As she explains in the message opening this book, 'in the early sixties ethologists tended to be reductionists, searching for simple explanations for very complex behaviours'. The understanding of complex behaviours was prevented not only by the principle of reduction but also by the principle of disjunction, another pillar of classical science. One of the consequences of the application of the disjunction principle is the separation between disciplines, with the risk that they become hermetic from each other. Instead, it is now clear that a multidisciplinary and comparative approach is necessary to shed light on the biological foundation of human behaviour, which should start with the assessment of the behaviour of the animals with which we share the closest common ancestor: the other primates. There is no agreement over the extent to which extant strepsirrhines and haplorrhines can actually resemble their extinct ancestors (see Masters and Génin's box). However, Leakey's point on the understanding of 'ancestral behaviour' by using other primates as a reference remains valid. As postulated by the sociologist and philosopher Edgar Morin (1990), complex systems, such as behaviour, cannot be explained by simply disaggregating them into their single components because this would deny the nature of complexity itself. The apparent circularity of this reasoning underlies the difficulty of giving a linear definition to complexity. The process of distinguishing applied by classical science should be used not to separate but to reunite aspects that are often proposed as antithetic, such as genes and memes, individuals and societies and – why not? – strepsirrhines and haplorrhines.

This book tries to break the conceptual walls not only between primate taxa but also between the different disciplines such as anthropology, zoology, evolutionary biology, comparative psychology, behavioural ecology, ethology, sociology, neuroscience and so on. Testimonials of this continuity are first and foremost the authors who contributed to this volume through their opening and closing remarks. Alison Jolly and Ian Tattersall, who wrote the foreword, were pioneers in dealing with both lemurs and other primates from the perspectives of physical and behavioural anthropology. Indeed, Alison Jolly's *The Evolution of Primate Behavior* (1985) and *Lucy's Legacy: Sex and Intelligence in Human Evolution* (2001) and Ian Tattersall's

The Primates of Madagascar (1982) and *The Monkey in the Mirror: Essays on the Science of What Makes Us Human* (2003) are the flagships of the (re)connection between strepsirrhines and haplorrhines, including humans. Michael Huffman, who offers interesting insights in his Afterword, during his career has bridged monkeys and apes by cross-comparing their behavioural traditions, social learning, self-medication, and their cooperation and competition with humans. The primate bridge is completed by Jane Goodall, who provided early evidence of the behavioural connection between humans and other apes. Alison Jolly, who in 1962 greatly appreciated the talk that Jane Goodall gave at a symposium of the London Zoological Society, would later provide similar evidence of the behavioural connection between strepsirrhines and other primates. Alison Jolly's ideas are clearly reported in her 1966 paper published in *Science*. In sum, Jane Goodall, Alison Jolly, Ian Tattersall and Michael Huffman by opening and closing this book also open and close the uninterrupted bridge that connects humans, apes, monkeys and the lemurs of Madagascar.

What lemurs can do (or book plan)

This book is focused *on* lemurs but is *not just about* lemurs. Lemurs are the pivot to explore different behavioural phenomena within the entire primate order. Other than reviewing and reinterpreting the existing literature, we enriched the text with unpublished results coming from wild and captive studies in order to make the framework one of the most comprehensive (but still incomplete!) to date. Our original contribution to this volume was made possible thanks to the help of three Italian zoos: Zoo di Pistoia, Parco Zoo Falconara and Parco Punta Verde, which opened their facilities to our studies in captivity and offered financial support for our investigation in the wild, which we concentrated in the Berenty forest. This study site, in southern Madagascar, is ideal to conduct proper observational studies because it offers excellent observational conditions. This is in order to collect good quality data ensuring reliable results and to opportunistically set up quasi-experimental protocols to properly test behavioural hypotheses in the wild.

The holistic, unifying vision we propose in this book is complemented by the extremely valuable contributions that many eminent scholars, women and men from different disciplines accepted to include in this volume in the form of thematic boxes. The topics are organised into three sections. Hereafter, we briefly present the content of each section without lingering on details (the book is there to be read!). Instead, we prefer providing the rationale of the book and a 'reading key' to appreciate how and why the boxes of our contributors were integrated into each chapter.

In the first section, 'Communication: from sociality to society', we begin by investigating individual recognition, as the foundation to build stable social relationships. We explore the differences between lemurs and other primates and the role of olfaction in orienting lemur–lemur recognition. In his box, Paolo Pelosi shows how biochemistry has been applied (quite recently) to the study of olfaction and explains how scientists are starting to understand the way a chemical message is decoded, after the discovery of odorant-binding proteins (OBPs) and olfactory receptors. Ipek Kulahci and Asif Ghazanfar introduce the readers to the importance of combining different sensory modalities (e.g., olfaction *x* hearing; hearing *x* sight) for individual recognition, not only in humans but also in other animals, including lemurs.

Once we made sure that lemurs can recognise each other, we moved on to how they communicate. While we focus on the different sensory modalities that – associated with olfaction – can be used to convey information to others, Eckhard Heymann explains how and why olfactory communication is also used in non-strepsirrhine primates. In particular, he makes evidence-based speculations on the meaning of scent marking in New World monkeys. James Higham, in his box, brings in the use of olfaction in Old World monkeys – combined with visual cues – and underlines the possible pressures that may have pushed animals towards the recruitment of different sensory modalities to communicate. Bridget Waller and Katie Slocombe focus on the importance of considering multimodality to better comprehend how primates convey and receive information. They use lipsmacking in macaques as a good example of complex and multimodal communication.

In the final part of this section, we draw attention to the fact that not only humans move vertically and use frontal signals. From lemurs to apes, several primate species characterised by upright locomotion converge in the use of face-to-face secondary sexual signals despite the differences in the dominant sex, mating system and socioecological habits. While our chapter stresses how the prominent sexual signals are carried out by the 'chosen sex' (females, in male-dominated societies, and vice versa), David Puts' box focuses on humans and expands the perspective on how other factors, such as male intrasexual competition, have possibly interfered in shaping men's 'ornaments'.

The second section, 'How conflicts shape societies', turns to how the conflict of interest between individuals can build social networks and how individuals manage to coexist in social groups. Conflicts determine the dominance structure of animal groups through aggression, bring forth strategies for their natural resolution, and induce emotional responses to deal with tense situations.

In our journey towards understanding the complexity of lemur society, we firstly demonstrate that the two main features used to label hierarchical relations in lemurs – namely, linearity and female dominance – are miles away from describing their dominance profile. As occurs in non-lemur primates, lemur power relations are more complex within groups and variable between groups (and species) than expected. Similarly, as Bernard Thierry explains in his box, even if all macaques display stable linear dominance rank orders, they differ by a number of social patterns. Thierry also stresses the importance of appropriate methodology and interspecific comparison to detect differences and similarities in the dominance arrangement of monkeys' social groups. By making us reflect on the covariation of social traits, Thierry's box leads us to appreciate the possible divergence between what we may call socio-behavioural taxonomy and classical taxonomy *sensu stricto*. Takeshi Furuichi moves to bonobos, a great ape characterised, as different lemur species, by female dominance. However, as we can read through his box, the analogy ends here. In *Lemur catta*, for example, females form matrilines and inherit territories whereas males migrate. In *Pan paniscus*, females are not the philopatric sex: they migrate from the natal group and their social relationships in the resident group are based on alliances more than on kinship. This twist between taxa allows us to appreciate the different ways in which female power is gained and managed.

After assessing the foundation upon which dominance relationships are built, we explore how the conflicts of interest that inevitably rise in social groups are managed. By using the pieces of evidence that have piled up in the last decade on lemurs, we make an argument to show that the Relational Model proposed by Frans de Waal applies to strepsirrhines, as well as to monkeys and apes. Lemurs can avoid and tolerate each other, but they are also able to actively reconcile with one another after conflicts. Additionally, lemurs engage in pre- and post-conflict strategies by adjusting them to the different contexts. Our contributors' boxes furnish interesting hints to discuss the strategies adopted by lemurs in the light of those adopted by other social species. On one side, Cary Roseth focuses on how children adopt, combine and tune different strategies (coercive and prosocial) – according to different social contexts (e.g., stable/unstable) – to cope with conflicts of interest. On the other side, Giada Cordoni goes through the occurrence of peace-making outside the primate order, helping us to understand if being primates or, more generally, being social explains certain aspects of conflict management. Peter Verbeek resumes this concept and moves forward by placing peacemaking into the wider framework of peaceful behaviour, which includes cooperation, helping, sharing, etc. and which also concerns invertebrate species. In this view, peaceful behaviour is something more than natural conflict resolution and, according to Verbeek, the ethology of peace (a building block of the science of peace) is fundamental to understand life itself.

We conclude the section by explaining how social tension and fearful situations (e.g., predation attempts) can increase anxiety, which is measurable via self-directed behaviours. This field of research, well explored in non-strepsirrhine primates, has been neglected in lemurs. The box of Filippo Aureli extensively explains how the use of self-directed behaviour (mainly scratching) to measure anxiety has been validated over the years, in different haplorrhine species and in different contexts. Self-directed behaviour can be used as an indicator to detect anxiety increase and also anxiety reduction (e.g., consequent to receiving or giving grooming, depending on the species considered). We show that also in lemurs scratching can be linked to a variation in the emotional homeostasis of the subject. Moreover, lemurs can use different social behaviours, such as play, grooming, and/or conciliatory contacts, to buffer their anxiety, which results in the reduction of scratching levels. We point out that anxiety and stress are two related but distinct phenomena because stress responses are expressed via the increase in the levels of glucocorticoid hormones after a perturbing event. While we focus on the use of scratching to measure anxiety, Lisa Gould in her box explains how the measure of hormonal variations has been used in lemurs, for example to assess whether intrasexual competition increases stress levels in males during the mating period.

The last section, 'Why lemurs keep in touch', pivots around social play, grooming and scent marking as doors to reach others and obtain the direct and indirect benefits of establishing new (or re-establishing old) social interactions. We delve into primate social play by focusing on its possible immediate functions to the benefit of social life. In particular, we hone in on facial and body signals, as a means to obtain and maintain symmetrical and balanced playful interactions, with no one player prevailing over the other. Fair sessions are in fact based on the ability of the subjects to fine-tune playful patterns and declare 'good' intentions through signals. Sergio Pellis and Marc Bekoff, in their boxes, expand on social play in non-primate animals and provide examples from their study species such as magpies, rats and dogs. Marc Bekoff explains why social play can be considered as a precursor of social equity and fairness, and does it by making inferences on how canids use the play bow as an honest signal of benign intent. Sergio Pellis stresses the importance of juvenile play in the development of adult social skills and of the brain mechanisms that underlie those skills.

In the second and last chapter of this section, we consider the mechanisms used by animals when a resource cannot be taken through coercion. In particular, we discuss the strategies adopted by male and female primates to increase their reproductive success in the light of some aspects of the biological market theory. Lemur males and females, which are seasonal breeders and live in an extremely seasonal environment, exchange different commodities or services in different periods of the year. During the mating period, males advertise their quality via olfactory tournaments and provide grooming to access females. Outside the mating period grooming is traded for grooming between sexes. The absence of an arena of aggressive encounters, which elicits the need to enact appropriate trading tactics, can be linked – among other factors – to the reduced sexual dimorphism. A similar phenomenon is found in New World monkeys. Karen Strier, in her box, describes how muriqui males, which live in a peaceful society, use sperm competition to sire offspring. Therefore, she opens to other mechanisms adopted by primate males to compete to access females' eggs in a relaxed way. By resuming the overall rationale of the chapter, Rebecca Lewis in her box frames lemur male–female relationships into the broader network of power dynamics. Interestingly, she points the finger towards economic theories and the study of non-primate social institutions as new possible fields of primatological research.

We conclude the book by singling out a topic that, in the last decade, has attracted the attention of several investigators: lemur cognition. As Alison Jolly and Ian Tattersall point out in the foreword, for a long time lemurs have been considered too dumb to pass cognitive tests. In the final chapter, 'Understanding lemurs: future directions in lemur cognition', we penetrate into the understory of the experimental trials used to comprehend what the actual cognitive abilities of lemurs are. Various non-social cognitive tasks have been presented to lemurs to assess their numerical sensitivity, their ability to represent tools, and the factors influencing their choices. Another field of research has directed its effort towards the understanding of lemur social learning and social intelligence. Evan MacLean and Brian Hare, in their box, explore some aspects of the Social Intelligence Hypothesis put forth by Alison Jolly in 1966. They introduce us to the significance of lemurs for identifying those characteristics of cognition that are unique to primates and to the possible use of strepsirrhines as an out-group for studying monkeys and apes. They also consider lemurs as a good 'natural experiment' that can be used to understand how socioecological factors may have favoured cognitive evolution. Based on previous research on corvids they were able to demonstrate, for example, that lemurs with different social organisations differ in their abilities of transitive reasoning. Elsa Addessi and Elisabetta Visalberghi add to the primate cognitive framework, by addressing the cognitive skills of New World monkeys, starting from their ability to use tools both in the wild and in captivity. They show how the experiments on food choice have demonstrated the presence of learning via social facilitation more than imitation in capuchin monkeys. As much as the outcomes of the investigation in platyrrhines have weakened the alleged cognitive barrier between New and Old World monkeys, the early results of the research on lemur cognition already indicate that strepsirrhines possess certain cognitive abilities previously attributed to monkeys only.

We would like to conclude this preface by saying that in this book the readers will find far more questions than we can answer. The study of certain aspects of lemur behaviour is in its infancy and, as each reader will get to see, only a few species of lemurs have been investigated out of the very many lemur species recognised so far. In part this can be blamed on the difficulties of observing lemurs in their natural settings: a precise ethological investigation requires good visibility of subjects and behavioural patterns and some lemur species live in impossible, impenetrable forests or have a cryptic lifestyle. In part there is the necessity to increase the ethological quantitative and standardised studies on strepsirrhines, both in captivity and in the wild (penetrable) settings. With all the limits that may exist when studying strepsirrhines, there is much work that can be done and remains to be done.

In every chapter we face the selected topic starting from humans and moving to other primate and non-primate animals, concluding with lemurs. This is not because we adhere to the above mentioned top-down approach but because we want to show that even starting from the alleged 'top' of the evolutionary building (humans) it is possible to stress the continuity existing between one living form and another. Many (if not all) human behavioural peculiarities could not be considered as such or be expressed in the way they are without the evolutionary scaffold that sustains them. We introduce the main hypotheses, theories and evidence existing in the literature. Of course, all is filtered according to our own ideas. We do not expect the readers to agree upon everything we say. On the contrary, we hope that it does not occur. We believe that the confrontation of ideas gives origin to more structured scientific debate and to the possible solutions that can sort it out.

Overcoming the Scala Naturae: why strepsirrhines are not primitive

Judith Masters and Fabien Génin

Tooth-combed primates first came under intense scientific scrutiny in the late eighteenth and early nineteenth centuries. This was the time of transformational thinking, when scientists like Etienne Geoffroy Saint-Hilaire studied anatomy to identify missing links among body plans, and so reveal the pattern of nature.

> A constant truth for a man who has observed a large number of the productions of the Earth, is that there exists, in all their parts, a great harmony and necessary relationships; it seems that nature has enclosed herself within certain limits, and has formed all living beings on only one unique plan, essentially the same in its principle, but which she has varied in a thousand ways in all its accessory parts. If you consider in particular on class of animals, it is there chiefly that her plan becomes evident to us: we shall find that the diverse forms under which she has pleased herself to make each species all derive one from another: it suffices her to change a few of the proportions of the organs to render them suitable for new functions, or to extend or constrain their uses. ...the forms in each class of animals, however varied they may be, are fundamentally all the result of organs common to all: nature refuses to employ new ones. (Geoffroy Saint-Hilaire, 1796, pp. 20–21, our translation)

Thus, although most contemporary evolutionary biologists view Darwin's (1859) *On the Origin of Species* as the first major publication in evolutionary theory, the rapid acceptance of Darwin's new interpretive framework owed a great deal to the innovative and meticulous work of the comparative anatomists over the preceding one hundred years. Transformationists understood very clearly that taxonomic associations had underlying natural causes: 'This order of things indeed exists: there are some series in which all the species are so similar to each other, that it has inspired everybody to refer to them as natural families...' wrote Geoffroy Saint-Hilaire (1796, p. 21, our translation); and it was to the good fortune of strepsirrhine biology that this leading exponent of tranformationism chose 'lemurs' as his first focal group. The strepsirrhine primates delighted Geoffroy particularly because the diversity in their characteristics allowed him to demonstrate the affinity of tarsiers with primates. Tarsiers had been classified as gerbils and didelphids by earlier researchers, but Geoffroy's identification of the African galagos enabled him to devise a transformation series from tarsiers through galagos to lorises and lemurs, 'which it seems that nature has formed so that there might not be a sudden jump in her work, and to arrive, by an imperceptible gradation, at the form of the tarsier so extraordinary, even unique in creation' (Geoffroy Saint-Hilaire, 1796, p. 44, our translation).

Geoffroy's attempt at systematising the lemur-like primates was undermined by a manifestly artificial taxonomic system published by Karl Wilhelm Illiger (1811), the director of the zoological museum in Berlin. Illiger distributed the tarsiers and strepsirrhines among three families: the Prosimii, containing the indriids, lemurids and lorisids; the Macrotarsi, including the tarsiers and galagids; and the Leptodactyla, comprising the aye-ayes. The term 'prosimian' emerged from this classification, and carried the connotation that these taxa were primitive or archaic primates. Haeckel (1883) took this assumption further, arguing that prosimians be excluded from the order Primates on the grounds that they formed a wastebasket of primitive groups whose closest affinities were to other mammal taxa: 'these curious animals are probably the little changed descendants of the primeval group of Placentalia...' which had in turn evolved from 'Handed or Ape-footed Marsupials'. The aye-ayes were interpreted as transitional between primates and rodents, the Macrotarsi as having given rise to insectivores, and Illiger's Prosimii – renamed the Brachytarsi by Haeckel – as forming the link between ancestral mammals and apes.

Despite the obvious philosophical problems with the *Scala Naturae* approach of arranging living taxa into graded series of increasing 'complexity', it has proved difficult to uproot entirely from primate systematics – and particularly from our understanding of strepsirrhines and their evolution. To many researchers, living prosimians retain the aura of 'archaic taxa' that ceased to evolve sometime in the Cenozoic, and hence present living exemplars of ancestral primates. Simpson's (1945) suborder Prosimii resembled those of Illiger and Haeckel in that it comprised a group of taxa that lacked any defining character, other than that they were non-anthropoid primates. His phylogeny published in 1961 (reproduced here as Figure p.1) illustrates this view graphically. Several primate palaeontologists defended Simpson's unnatural classification system on the grounds that including tarsiers among the anthropoids would compromise the diagnosis of the anthropoids. 'The suborder Prosimiae is meant to include all primates not sufficiently evolved to warrant a separate suborder' wrote Van Valen (1969).

This approach was successfully challenged by developments in phylogenetic philosophy and methodology in the closing quarter of the twentieth century. Recent molecular reconstructions reveal that the Haplorrhini and Strepsirrhini shared a common ancestor in the very late Cretaceous and earliest Palaeocene, not long after the emergence of the primate clade. This common ancestor had a petrosal bulla, a complete postorbital bar, a divergent hallux (and possibly a divergent pollex as well), and nails instead of claws – at least on the hallux; but neither the primate nor the anthropoid ancestor – nor indeed the strepsirrhine ancestor – could have been confused with any living tarsier or strepsirrhine. Despite several statements to the contrary that haunt the literature, the features that characterise the Strepsirrhini are not all plesiomorphic primate traits, neither are they functionally inferior to the character states that evolved in other lineages. Strepsirrhines have had a long and diverse history, and their morphology attests to this fact. A feature like the unfused mandibular symphysis is indeed likely to be the ancestral state for mammals, but this does not mean that prosimians routinely experience mandibular dislocations in the course of their daily lives, as thousands of museum specimens will testify. Where extreme masticatory stresses have demanded a more robust state, as in the case of some Eocene adapiforms as well as some subfossil Malagasy lemurs, mandibular fusion has evolved convergently to resemble the state in the anthropoids. On the other hand, the stapedial basicranial circulation shared by most lemuriform taxa and the Eocene adapiforms was not evolutionarily prior to the promontory pattern observed in Haplorrhini; both circulatory patterns were present among the diverse Eocene euprimate fauna of North America and Eurasia (Rosenberger and Szalay, 1980). Similarly, all variations in the position and shape of the ectotympanic relative to the auditory bulla used to distinguish living primate groups (intrabullar vs extrabullar; ring-like vs tubular) were present in the Eocene, and there is no obvious 'primitive' condition (Martin, 1990).

Strepsirrhines were named for the shape of their external nares, i.e., twisted or aborally slit nostrils (Hofer, 1977), which also occur in several groups of non-primate mammals (and indeed in some anthropoid primates), and are therefore considered plesiomorphic. The strepsirrhine condition in primates includes a naked rhinarium with a median furrow that clefts the free margin of the upper lip. Additionally, the upper lip is tethered to the gum between the medial incisors by bilateral, sulcate labial folds, necessitating a gap between the roots of the upper medial incisors, and allowing a physical continuity between the rhinarium and the vomeronasal organ (Hofer, 1977). Rosenberger and Strasser (1985) presented an interpretation of this anatomical complex as a behavioural–physiological adaptation to facilitate the transfer of water-soluble pheromones from the rhinarium (and the environment) directly to the vomeronasal organ. They suggested further that this structural arrangement may have pre-adapted early strepsirrhines to evolve a tooth-comb as a mechanism to stimulate glandular secretions and to distribute odours through the fur during auto- and allogrooming. Beard (1988) challenged this elegant hypothesis, stating that the continuity between the rhinarium and vomeronasal organ occurs in other, non-primate taxa (like tree shrews), and therefore cannot be considered an adaptive innovation underlying the initial divergence between haplorrhines and strepsirrhines. In our view, however, the fact that other mammals make use of a similar pheromonal pathway does not contradict the Rosenberger–Strasser hypothesis. The oronasal complex of living Strepsirrhini is not homologous to structures seen in tree shrews, tenrecs and hedgehogs, indicating that these systems evolved convergently – and providing good evidence that they are the result of natural selection, and therefore derived.

But the organ system that has most conveniently lent itself to this gradistic interpretation is that comprising the placenta and the foetal membranes. For example, Luckett (1976, p. 260) stated categorically that the non-invasive, diffuse epitheliochorial placenta and the pattern of foetal membrane development evident in living strepsirrhines is a retention of the ancestral condition not just for primates, but for all eutherians. Martin (1990) has published a detailed rebuttal to this interpretation. The idea that a highly invasive haemochorial placenta, in which the chorion is bathed directly by the maternal blood, is necessary for the development of relatively large-brained or precocial offspring is not supported by the data (Martin, 1990). Both epitheliochorial and haemochorial placentation appear to have evolved convergently in diverse mammalian groups (Hübrecht, 1908), and 'there is no good reason for regarding the placentation of any living primate species as representative of an ancestral stage' (Martin, 1990, p. 457). Finally, the more closely strepsirrhine placentas are examined, the more complex they appear. Both mouse lemurs and galagos have been shown to have mixed placentas, with an endothelial labyrinthine centre and an initial invasive attachment, surrounded by an epitheliochorial ring (Reng, 1977).

Strepsirrhines, and particularly Malagasy strepsirrhines, have had a long history facing very different selection pressures from their haplorrhine relatives, and are unlikely to represent valid models for their Palaeogene ancestors. They are limited today in their distribution to the subtropical-tropical regions of Africa and South-east Asia and their surrounding islands. Does this also limit their usefulness to inform us more generally about the evolution of the primate clade and of ourselves? Most decidedly not. Strepsirrhine primates make up at least 50% of standing primate diversity, whether considered at the family, the genus or the species level. No investigation of the extent of the primate radiation or of its ecological adaptability can be complete without a representative sample of living strepsirrhine taxa. Omitting them entirely or even including one or two 'model taxa' (e.g., 'the mouse lemur' or 'the ring-tailed lemur') is akin to trying to understand chemistry using only the elements on the left hand side of the periodic table. The 'big picture' is impossible to understand. It is high time these 'other primates' were brought into the centre of primatological research.

We thank the National Research Foundation, South Africa, for their continued support of our research.

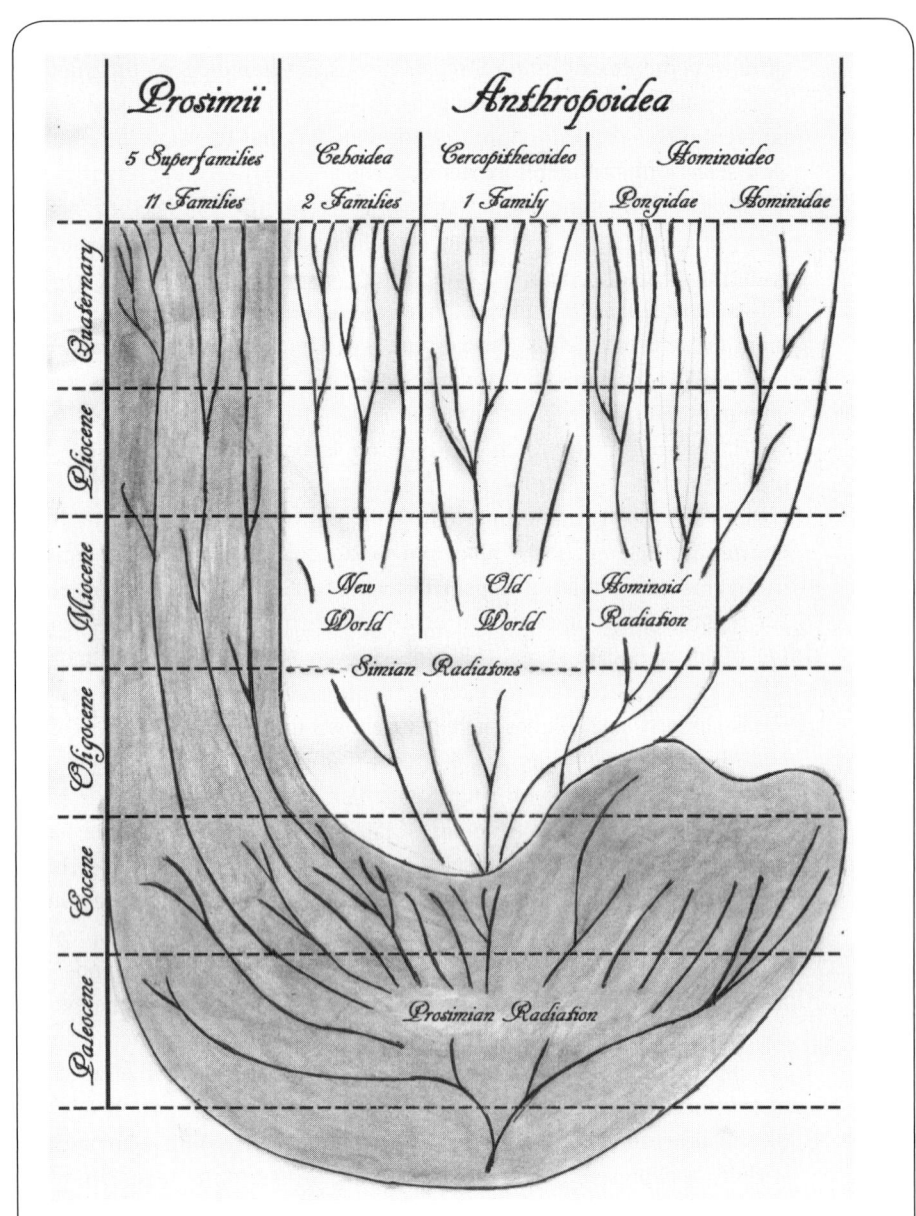

Figure p.1 Diagram of phylogeny and classification of the primates. Drawing by Carmelo Gómez González, based on Simpson (1961).

References

Beard, K. C. (1988). The phylogenetic significance of strepsirhinism in Paleogene primates. *International Journal of Primatology*, 9, 83–95/96.

Darwin, C. R. (1859). *On the Origin of Species by Means of Natural Selection, or the Preservation of Favoured Races in the Struggle for Life*. London: John Murray.

de Waal, F. B. & Ferrari, P. F. (2010). Towards a bottom-up perspective on animal and human cognition. *Trends in Cognitive Sciences*, 14, 201–207.

Geoffroy Saint-Hilaire, E. (1796). Mammifères. Memoire sur les rapports naturels des makis *Lemur*, L. et Description d'une espèce nouvelle de Mammifère. *Magazin Encyclopédique*, 1, 20–50.

Haeckel, E. (1883). *The History of Creation: or the Development of the Earth and its Inhabitants by the Action of Natural Causes*, 3rd ed. translated by E. R. Lankester. London: Kegan Paul, Trench and Co.

Hofer, H. O. (1977). The anatomical relations of the ductus vomeronasalis and the occurrence of taste buds in the papilla palatina of *Nycticebus coucang* (Primates, Prosimiae) with remarks on strepsirrhinism. *Gegenbaurs Morphologisches Jahrbuch*, 123, 836–856.

Hübrecht, A. A. W. (1908). Early ontogenetic phenomena in mammals and their bearing on our interpretation of the phylogeny of vertebrates. *Quarterly Journal of Microscopic Science*, 53, 1–181.

Jolly, A. (1985). *The Evolution of Primate Behavior*, 2nd edition. New York: Macmillan.

Jolly, A. (2001). *Lucy's Legacy: sex and intelligence in human evolution*. Cambridge, Massachusetts: Harvard University Press.

Illiger, C. (1811). *Prodromus Systematis Mammalium et Avium*. Berlin: Sumptibus C. Salfeld.

Luckett, W. P. (1976). Cladistic relationships among primate higher categories: evidence of the fetal membranes and placenta. *Folia Primatologica*, 25, 245–276.

Martin, R. D. (1990). *Primate Origins and Evolution: a Phylogenetic Reconstruction*. Princeton, NJ: Princeton University Press.

Morin, E. (1977). *La Methode. 1. La Nature de la nature*. Paris: Seuil.

Morin, E. (1990). *Introduction à la pensée complexe* (Vol. 96). Paris: Esf.

Reng, R. (1977). Die Placenta von *Microcebus murinus* Miller. *Zeitschrift für Säugetierkunde*, 42, 201–214.

Rosenberger, A. L. & Strasser, E. (1985). Toothcomb origins: support for the grooming hypothesis. *Primates*, 26, 73–84.

Rosenberger, A. L. & Szalay, F. S. (1980). On the tarsiiform origins of Anthropoidea. In: R. L. Ciochon & A. B. Chiarelli, *Evolutionary Biology of the New World Monkeys and Continental Drift*. New York: Plenum Press.

Simpson, G. G. (1945). The principles of classification and a classification of the mammals. *Bulletin of the American Museum of Natural History*, 85, 1–350.

Simpson, G. G. (1961). *Principles of Animal Taxonomy*. New York: Columbia University Press.

Tattersall, I. (1982). *The Primates of Madagascar*. New York: Columbia University Press.

Tattersall, I. (2003). *The Monkey in the Mirror: Essays on the Science of What Makes Us Human*. Harvest Books.

Van Valen, L. (1969). The classification of the primates. *American Journal of Physical Anthropology*, 30, 295–296.

Acknowledgements

Very many people have supported us, in very different ways, during the investigations that we carried out in the past two decades. We remember and we are grateful to each one of them for not leaving us alone while we ventured into the exploration of the deepest biological roots of human behaviour. It would probably take another book to give proper credit to all of them. Therefore, in this section we will mention the people that, directly and indirectly, contributed to the realisation of this book.

Our opening thanks go to Jane Goodall, Alison Jolly, Ian Tattersall and our close friend Michael Huffman, for providing notable and extremely insightful introductory and closing sections to the book. We would also like to thank all the box authors, friends and colleagues, who added a great deal of value to the book with their significant contributions: Elsa Addessi, Filippo Aureli, Marc Bekoff, Giada Cordoni, Takeshi Furuichi, Asif Ghazanfar, Lisa Gould, Brian Hare, Eckhard Heymann, James Higham, Ipek Kulahci, Rebecca Lewis, Evan MacLean, Sergio Pellis, Paolo Pelosi, David Puts, Cary Roseth, Katie Slocombe, Karen Strier, Bernard Thierry, Peter Verbeek, Elisabetta Visalberghi and Bridget Waller. In addition, we wish to thank our family members, friends and colleagues who enriched the book with their photos and drawings: Daniela Antonacci, Carmelo Gómez González, Takeshi Furuichi, Achim Johann, Ipek Kuhlaci, Rebecca Lewis, Ute Radespiel, Tommaso Ragaini, Paola Richard, Laurie Santos and Chiara Scopa. Special thanks go to Riccardo Guasco, for capturing the meaning and sense of the book into the wonderful and truly inspiring image cover.

Much of the information included in this book comes from the data gathered in the past 15 years with the support of field experts, Valentina Carrai and Giuseppe Donati, and with the help of students and field assistants, whom we would like to thank: Chandra Brondi, Viola Caltabiano, Manrica Cresci, Stefania Dall'Olio, Kadoffe Dauphin, Antonella Gregorace, Stefano Kaburu, Jean Lambotsimihampy, Olivier Rahanitriniaina, Seheno Randriamanga, Givet Sambo, Valentina Sclafani, Giulia Spada, Sabrina Telara and Alessandra Zannella. Our research studies on lemurs have been possible because we were allowed access, over the years, to both wild and captive settings. We express our gratitude to our friends (and directors of the Italian parks) Paolo Cavicchio (Giardino Zoologico di Pistoia), Iole Palanca (Parco Zoo Falconara) and Maria Rodeano (Parco Zoo Punta Verde, Lignano Sabbiadoro), who opened the doors of their zoos for our research in captivity and who have been funding our research in the wild since 2006. For the possibility to perform lemur studies in Madagascar since 2001, we also wish to thank: Peter Kappeler and Léon Razafimanantsoa for allowing and facilitating data collection in the field station in the Kirindy forest-CFPF of the German Primate Center (DPZ); Jörg Ganzhorn (Hamburg University, Germany), Manon Vincelette and Jean Baptiste Ramanamanjato (Qit Madagascar Minerals, QMM, Taolagnaro, Madagascar) for granting access to the field stations of Sainte Luce and Mandena; and the De Heaulme family and Alison Jolly for the opportunity to investigate lemur sociobiology in the Berenty forest. On the institutional side, we wish to thank the Commission Tripartite du Département des Eaux et Forêts, and the Parque Botanique et Zoologique de Tsimbasasa.

Last, but actually first, we would like to thank our families – Elisabetta's family: Luca Ragaini, Tommaso Ragaini, Graziano Palagi, Mariella Taddei, Patrizia Taddei; Ivan's family: Clara Mazzanti, Luisa Buonaguidi, Ugo Mazzanti, Carmelo Gómez González, and Novello Conforti – for never-ending support, especially in difficult moments.

Part I

Communication: from sociality to society

1 Who are you? How lemurs recognise each other in a smell-centred world

He approached these faces – even of those near and dear – as if they were abstract puzzles or tests. He did not relate to them, he did not behold. No face was familiar to him, seen as a 'thou', being just identified as a set of features, an 'it'. Thus, there was formal, but no trace of personal, gnosis. And with this went his indifference, or blindness, to expression. A face, to us, is a person looking out – we see, as it were, the person through his persona, his face. But for Dr P. there was no persona in this sense – no outward persona, and no person within... His absurd abstractness of attitude...which rendered him incapable of perceiving identity, or particulars, rendered him incapable of judgment.

Oliver Sacks, 1970 (p. 13)

1.1 Individual recognition: why and how

Human society is founded on individual recognition. The distinction of 'others from others' rules every aspect of a community, from the personal to the social level. In his '*The Man Who Mistook His Wife for a Hat*', Oliver Sacks (1970) stresses how the ability to discern individuals is crucial to make a judgement over them and their true identity. Understanding individuals' uniqueness is pivotal to choose partners and friends, care for family members, vote for a candidate, form political alliances, seal trade agreements, and follow rock icons or religious leaders. Individual recognition is not unique to humans. It is, instead, an ancient cognitive skill shared with other primates and rooted in humans' mammalian history. From lemurs to apes, such ability is critical, for example, to select mates and supporters, form parent–offspring bonds, establish dominance relationships, set up coalitions, exchange or interchange commodities and follow group leaders (Thom and Hurst, 2004). Individual recognition is an excellent example of the cognitive continuity that bridges humans and other primates.

Individuals can be discriminated according to their unique features, or cues. Different from signals, specifically designed to convey information beneficial to the sender (Bradbury and Veherencamp, 1998), cues carry potential information whose relevance depends on the receiver. Paradoxically, the same cue can mean something for one animal and nothing for another which does not read the cue as a cue at all! Consequently, defining a cue is not easy. In general, we can consider as a cue any sensory information (visual, tactile, auditory, olfactory, etc.) that gives rise to a sensory estimate (Ernst and Bülthoff, 2004).

Which characteristics does a cue need to be effective in individual recognition? Firstly, a cue must possess a fixed component, a 'fingerprint' that can be recognised independently of other background variations. For example, a child is able to recognise the voice of their mother regardless of its volume, intonation, and/or prosody. These three elements do not answer to the question: who is she? So, they are not functional to individual recognition itself. They provide additional information of the emotional and physical state of the mother as she speaks. The child identifies a steady property of their mother's voice (auditory cue): the tone (e.g., see Sakkalou and Gattis, 2012).

Secondly, cues must be highly diverse between individuals. For instance, if voice tones in the same family members (e.g., sisters, mother/daughters, etc.) are too similar, they are easy to be mistaken when other cues are missing (e.g., at the phone).

Finally, identity cues should stay temporally consistent or change gradually, thus allowing the receiver to update their sensory information. To stick with the previous examples, a mother can recognise her son's voice even if they have not been in touch for a long time. Of course, it may be difficult to recognise the voice of someone we have not heard from for 50 years!

Renewing identity documents every five to ten years is mandatory in every country because the older the picture on the document gets, the more difficult is for the officer (e.g., at the airport) to match the actual face of the document owner with the old (and maybe black and white!) one. This is why new identification methods have been implemented, such as the iris scanner, based on invariant individual cues.

The presence of identity cues is not enough, per se, to ensure that individual recognition of an object by a subject takes place. The following elements are necessary: encountering the object, elaborating the cues carried by such object, making a mental connection between the object and its cues, and storing such connection in memory in order to 'reload' it when the same object is met again. A full mental representation of the object is built and saved. A second time, a single cue can be enough to recall the full representation of the object by the subject.

The prefix 're' incorporated in the word recognition (from latin *re*=again, and *cognòscere*: to become acquainted) implies a *re*newed identification of a stimulus that has been already encountered. Thus, the memory of a previous experience is crucial.

Box 1.1 | **by Paolo Pelosi**

Speaking of which: breaking the olfactory code

Olfaction is the language of chemistry. Being a chemist, I have always been fascinated by molecular structures, first of all by their beautiful architecture, but also by their diverse properties, from physical appearance, such as colour, odour, consistency and optical properties, to their capacity of interacting with other molecules and giving birth to new entities. But molecules are also the words of a language, a language most animal species use all the time to communicate and exchange information.

Chemical communication can be so complex, as in the case of social insects, that it can acquire all the characters of a spoken language, with its own rules of grammar and syntax. Discovering the architecture of this language and breaking its code is one of the most fascinating and exciting tasks, with challenges similar to those encountered when trying to decipher an ancient scripture. Despite all the information of molecular biology, we are still far from 'breaking the olfactory code', but a large wealth of information, accumulated during the last three decades, has provided a solid ground and indicated clear guidelines for future research. Now the main problem is coping with the extreme complexity of olfaction, notably human olfaction.

My research has been focused for the last three decades on a class of soluble binding proteins for odorants, appropriately named for such property as odorant-binding proteins (OBPs), which I discovered by chance, while searching for olfactory receptors.

When I first approached the study of olfaction, in the late seventies, I followed the main trend, which consisted in investigating relationships between odour and molecular structure. In practice, you present different compounds to human subjects to smell and ask questions about the type and intensity of odours. By comparing structural parameters of the molecules and odour properties, you can draw correlations and formulate hypotheses on how the human nose works. This method proved troublesome and not very efficient, mainly due to the previously unforeseen complexity of the olfactory code. At the same time I wondered why most scientists were afraid of tinkering directly with this sort of 'black box' that was the olfactory system and were addressing indirect questions instead. In other words, this meant applying biochemical tools to olfaction. It was a very risky project, but I had nothing to lose and decided to open this mysterious black box.

The discovery of OBPs in mammals (Pelosi *et al.*, 1982), together with a parallel and almost simultaneous finding of a similar class of proteins in insects (the OBPs of insects) by Richard Vogt (Vogt and Riddiford, 1981), marked the first step of biochemical research in olfaction. After that, we experienced an explosion of interest in olfaction, with fast and still growing developments in the study of OBPs and other soluble proteins (Pelosi, 1994; Pelosi *et al.*, 2006; Leal, 2013). I think the most important outcome was that for the first time we showed that biochemistry could be applied to the study of olfaction. We therefore broke into the mysterious black box and stimulated an increasing number of scientists to peek inside. This fact gave confidence to scientists and prepared the ground for the discovery of olfactory receptors (ORs) about a decade later (Buck and Axel, 1991), which won the authors the Nobel Prize in 2004.

OBPs of mammals and those of insects proved completely different in structure (Bianchet *et al.*, 1996; Tegoni *et al.*, 1996; Sandler *et al.*, 2000; Tegoni

Box 1.1 (continued)

Box 1.1 (*cont.*)

et al., 2004), although similar in function (Pelosi and Maida, 1990). In both cases we are dealing with small soluble proteins, endowed with extremely stable folding. Incidentally, their exceptional resistance to heat, solvents and proteolysis has recently indicated OBPs as the most suitable biosensors for an artificial nose (Pelosi *et al.*, 2013).

Their simple structure and stability has made the study of these proteins relatively easy. Moreover, thanks to genome information and more recently to transcriptome projects (Vieira and Rozas, 2011), the number of sequences encoding OBPs has exceeded one thousand and is still growing fast. Despite such wealth of structural information, the specific function of OBPs in odour detection is still elusive. Most recent studies provided convincing evidence that OBPs are required for a correct functioning of the olfactory system and are likely involved in odour discrimination. In fact, silencing the gene encoding a specific OBP in *Drosophila* produces flies insensitive to the male pheromone, while several *Drosophila* mutants each lacking one of the 60 OBP genes have shown defects in responses to various odours (Xu *et al.*, 2005; Matsuo *et al.*, 2007; Swarup *et al.*, 2011). In aphids we have found a good correlation between avoidance behaviour of several repellents and affinity to specific OBPs (Sun *et al.*, 2012). However, how the first binding of odorant molecules to OBPs might lead to the activation of membrane-bound olfactory receptors is still an open question.

After the identification of olfactory receptors (Buck and Axel, 1991), many scientists believed that there was no room left for further major discoveries in olfaction. This might be true to some extent, because the corner bricks had already been placed. Nevertheless, a lot remains to be done. There are still several questions awaiting answers, which will require long and detailed studies, but there is also need for new brilliant ideas.

Concerning OBPs, the main open question is: how do they interact with olfactory receptors? Or, in other words: how is the chemical information encoded in the structure of odorants transferred to membrane-bound receptors? Some scientists posit that odorants interact directly with olfactory receptors, others envisage a role of OBPs as carriers of hydrophobic odorants across the aqueous nasal mucus or the sensillar lymph in the case of insects, and others think that it is the complex OBP-odorant which activates the receptor. Devising experimental protocols to prove or disprove such models is highly challenging and no convincing answer has been so far provided. Besides solving the puzzle and putting all the pieces in place, it is of fundamental and practical importance to understand whether ORs or OBPs are the recognising elements and responsible for identifying the different odours. In fact, OBPs, being small soluble proteins, are much easier to study than ORs. If discrimination of odorants occurs at the level of OBPs, we can use these proteins to measure interactions with as many odorants as we like and move quickly towards the elucidation of the olfactory code. We can also use the same proteins that can be easily prepared in bacteria, to assemble artificial devices for odour detection and recognition.

Another active field with enormous possibilities of investigation lies beyond the peripheral olfactory neurons. We know almost nothing about how the brain makes use of individual signals coming from the periphery to build 'olfactory images', so accurate and unique that they are able to promptly recall past experiences even if not supported by other sensory modalities. Then, of course, how such 'images' are stored in the memory is another fascinating topic of research. The brain certainly applies a combinatorial code to olfaction, in order to meet with thousands or even millions of odours in the environment using only a limited number of receptors. In fact, we know that each kind of odour molecule is generally able to stimulate several types of receptors at different grades of intensity, while, at the same time, each receptor can produce different levels of response to structurally related odorants. Since the number of functioning human olfactory receptors is a little more than 300, we end up with virtually an infinite number of combinations.

A third area that future research is likely to explore is that of artificial olfaction. There is a strong interest in this topic and a large number of reports have been published (Persaud and Pelosi, 1992; Pelosi, 2003; Turner and Magan, 2004; Stitzel et al., 2011; Manai et al., 2014), but we are still very far from designing an instrument able to reproduce in some way an artificial nose.

To move from a general gas detector to an instrument capable of discriminating and measuring odours in a similar way to natural organs we need to improve our knowledge and technology along two parallel lines:

(1) A better understanding of our olfactory system. In other words how our olfactory receptors read, decode and integrate the chemical information hidden in molecular structures; we are still miles away from assigning to each of the 300+ receptors its best ligands, and we know almost nothing of how their responses are mixed and integrated in the brain.
(2) Better sensing elements. So far, the instruments for the detection of odours have been based on metal oxides or conducting polymers. The former, used in commercial smoke alarms, are virtually unspecific whereas the latter presents very broad responses to odours. Both of them are orders of magnitude less sensitive than any biological system. Most recently, the use of OBPs as specific detectors in electronic devices has attracted wide attention, owing to their better specificity of response, as compared to other sensors, and to their exceptional stability as proteins. However, the issue of sensitivity still remains and represents one of the major challenges when we want to reproduce the functioning of the olfactory system with an electronic device.

1.2 How different sensory modalities concur to individual recognition

Humans can individually identify conspecifics by either using single cues (unimodal recognition) or combining them into a cross-modal recognition system (Joassin et al., 2011; see also Chapter 2). Mothers are able to recognise their babies by using single sensory modalities, such as the odour of their infant's garment (Porter et al., 1983) or the cry alone (Green and Gustafson, 1983). Humans normally rely on the visual sensory modality to identify other individuals. Face is certainly the primary identity signal (Sergent et al., 1992; Kanwisher et al., 1997; Rhodes et al., 2004), as clearly exemplified by the face photo that must necessarily accompany identity documents. Humans are also able to discriminate individuals by using other single characteristics, such as gait or clothing, especially when the face is obscured (Bruce and Young, 1986).

The concurrent use of different sensory modalities becomes important to optimise individual recognition under certain circumstances. In the forest, our primate ancestors would cope with faces partially or totally covered by the canopy; communication calls distorted or deviated by trunks, branches and twigs; and odours whose perception could be prevented or weakened by distance, or altered by the exposure to natural elements, such as heavy rain and wind. In the urban forest, the city, today's people have to deal with faces mingling in the crowd, voices covered by other voices, traffic noise, advertising announcements, music, and body odours mixed with artificial pheromones (perfumes), flue gas, and garbage smell. Hence, combining different pieces of information coming from different sensory modalities can be advantageous under less than optimal viewing or hearing conditions to determine individual identities (Blank et al., 2011). Such a cross-modal process, which combines and integrates different cues, has been favoured by natural selection because often, under natural conditions, no one cue carries complete information on individual identity.

The human brain cortex possesses both unimodal and multimodal processing areas. Unimodal regions (e.g., olfactory/piriform, auditory and visual areas) elaborate single cues and multimodal regions (mainly located in the hippocampus and adjacent areas) combine and integrate single cues leading to cross-modal recognition (Gottfried et al., 2004; Joassin et al., 2011). Sensory combination allows maximising the information delivered by single sensory modalities whereas integration allows reducing the variance in the sensory estimate to increase its reliability (Ernst and Bülthoff, 2004). The recognition process is optimised by direct structural connections existing between single-cue processing areas (Blank et al., 2011).

Cortex areas for multimodal processing are present in non-human primates (Gil-Da-Costa et al., 2004; Romanski, 2007; Ghazanfar et al., 2008) and other mammals, in which the response elicited by multisensory cues is greater than unisensory responses and sometimes greater than their arithmetic sum (Alvarado et al., 2007). In macaques, the audiovisual integration neural circuitry corresponds closely to areas in the human brain that support cross-modal representation of conspecifics (cf. Gil-Da-Costa et al., 2004; Campanella and Belin, 2007).

Neurophysiologic evidence of cross-modal recognition is supported by behavioural results. Both monkeys and apes can associate the sound of different call types with images of conspecifics and heterospecifics producing these calls (rhesus macaques, *Macaca mulatta*: Ghazanfar and Logothetis, 2003; tufted capuchin monkeys, *Cebus apella*:[1] Evans *et al.*, 2005; chimpanzees, *Pan troglodytes*: Parr, 2004). Squirrel monkeys (*Saimiri sciureus*) have been shown to spontaneously integrate auditory and visual identity cues from their one highly familiar human caretaker (Adachi and Fujita, 2007). Chimpanzees can, through intensive training, learn to associate calls from known individuals with images of those individuals (Bauer and Philip, 1983; Kojima *et al.*, 2003; Izumi and Kojima, 2004).

Olfaction is rarely considered in cross-modal recognition studies on primates. In their evolutionary history, haplorrhines have developed acute vision (trichromacy) and retained or improved their acoustic capacity (Fleagle, 2013). In contrast, olfactory abilities have been put in the corner. Social primates are mostly represented by acoustic- and vision-oriented species and, consequently, the studies on multimodal communication have focused more on these sensory modalities than on smell, considered as less relevant than other senses to monkeys' and apes' communication.

For most mammals olfaction is the dominant sense, with their behaviour being heavily influenced by the social chemosignals secreted by individual conspecifics (Wyatt, 2014).

The first evidence that animals can integrate multiple cues to build a representation of an individual came from a study on smell use in hamsters (Johnston and Bullock, 2001), which produce at least five different individually distinctive odours. The experimenters exposed a male familiar with two females, A and B, to the vaginal secretions of female A. Once habituated to the vaginal secretions, the male was tested with either A's or B's flank secretions. Males exposed to A's flank secretions investigated them less than males tested with B's flank secretions. This phenomenon, known as across-odour habituation, led the authors to conclude that 'when a male was habituated to one odour he was also becoming habituated to the integrated representation of that individual'. However, direct physical contact, not just odours, is necessary to establish in a subject the full representation of the object, whose identity will be recalled using multi-cue odour memory (Johnston and Peng, 2008).

Box 1.2 | **by Eckhard W. Heymann**

Speaking of which: odour communication in tamarins

I got in touch with olfactory communication early on during my studies at the University of Giessen. One of my teachers and then later supervisor of my PhD thesis, Heinrich Sprankel, had done research on the histology of scent and other skin glands of tree shrews and tarsiers (Sprankel, 1961, 1971), and he was treating olfactory communication both in his ethology

Box 1.2 (continued)

---

[1] *Cebus apella* has been reclassified as *Sapajus apella* (Lynch Alfaro *et al.*, 2012). In this book we use *Cebus apella* because this species is mentioned as such in the cited articles.

Box 1.2 (*cont.*)

and primatology classes. This and my behavioural observations in the course of my diploma thesis on tree shrews, where sternal scent marking is a very prominent behaviour, primed my interest in the subject.

When I started to study the behaviour of tamarins in 1982, it was obvious from the very beginning that any attempt to understand the social life of these creatures would fail if their olfactory communication was not considered. My interest was additionally fomented by Gisela Epple who came to the German Primate Center in 1983. Her work on the chemical composition of scent marks in callitrichids and on the proximate mechanisms of callitrichid olfactory communication (e.g., Epple, 1979, 1980; Epple *et al.*, 1981) had answered a number of questions. But it was also clear that there was a wide field of unanswered questions and unresolved problems. Furthermore, except for some anecdotal information (Izawa, 1978; Lindsay, 1979) nothing was known on callitrichid olfactory communication from the wild. What further increased my interest and triggered my ambitions to contribute to filling this huge gap of knowledge was the complete neglect of olfactory communication in the first detailed and systematic field studies on tamarins that emerged in the late 1970s and early 1980s (Dawson, 1977; Neyman, 1977; Terborgh, 1983; Terborgh and Goldizen, 1985). Obviously, here laid a field in front of me that was worth being ploughed.

Finally, what certainly also contributed to my interest in olfactory communication was the fact that – from my point of view – this topic requires 'square thinking', leaving trotted paths and mainstream reasoning – things that I increasingly like(d) to do. For example, the widespread hypothesis of pheromones from scent marking as a mechanism of reproductive 'suppression' of subordinate callitrichid females (e.g., Abbott, 1984) had never convinced me. This hypothesis could *not* explain why male callitrichids scent mark, why non-reproductive females sometimes show higher rates of scent marking than breeding females, etc. Maybe scent marks could work to suppress other females' reproduction in a 2 m^3 cage, but how should this work in a 40 ha or more home-range area?

The occupation with the topic led me to recognise that olfactory communication (and more generally olfaction) in primates is a hugely neglected field (Heymann, 2006), something that is now slowly changing.

There are three outcomes of my research that I should like to emphasise. The first one is the female bias in rates of scent marking and the recognition that this bias may have been shaped through sexual selection (Heymann, 1998, 2003). While I had already seen such a bias during my very first observations of tamarins, I could not actually make sense out of it until I came across Darwin's (1871, p. 131) statement that 'odoriferous glands [have] been acquired through sexual selection' and Blaustein's (1981, p. 1007) suggestion that odours 'are probably functionally equivalent to secondary sexual characteristics' and that 'sexual selection should act upon these odours just as it acts upon visually conspicuous characters'. Having read this, things became clear to me: in animals where females strongly compete for the single reproductive position in a group, and where males take the largest share in infant care apart from lactation – i.e., where there is a partial reversal of sex roles – the female bias in rates of scent marking (and the size of scent glands) could be interpreted as secondary sexual characteristics, possibly subjected to sexual selection. That sexual selection can strongly act upon females as well is now widely accepted (see Clutton-Brock, 2007), but that sexual selection shaped patterns of olfactory communication in tamarins and other callitrichids (Heymann, 2003) must remain a hypothesis for the moment, albeit a plausible one.

The second interesting outcome is the finding that scent marking in tamarins bears little relationship with territoriality (Heymann, 2000; Lledo Ferrer *et al.*, 2011, 2012), although this has created some debate and opposition (Gosling & Roberts, 2001; Roberts, 2012). Given a university training oriented towards classical ethology, I had too long been sticking to the hypothesis of a territorial function of scent marking. Only after I had found a plausible answer for the sex bias in tamarin scent marking could I also make a different 'sense out of scents' (to quote the title of a review by Epple and coworkers [1993]) with regard to the spatial patterns of scent marking.

The third interesting finding is the striking difference in patterns of scent marking between the sympatrically living *Saguinus mystax* and *Saguinus fuscicollis*, particularly the lack in the former and presence in the latter of allomarking (Heymann, 2001). Since the two species have a long separate phylogenetic history (Matauschek *et al.*, 2011; Heymann *et al.* unpublished data), it is most unlikely that these differences relate to reproductive isolation. Rather, I would speculate that the differences are linked to subtle differences in the social system, particularly in the mating system, of these sympatric species, although admittedly at the moment I do not yet have any clue how this relationship is structured.

The main future directions of this investigation domain are the following:

(1) Comparative analyses of the factors that influence the relative importance of life-history traits, social systems, activity patterns, ecology and phylogeny for shaping patterns of olfactory communication (Heymann, 2009; Delbarco-Trillo *et al.*, 2011). Such analyses will help identify specific factors and traits that can then be examined in more detail in both observational and experimental studies. However, since for many primate species, data on their olfactory communication are lacking or scanty, such analyses will remain preliminary until a broader database becomes available.

Box 1.2 (continued)

Box 1.2 (*cont.*)

(2) An understanding of the olfactory 'vocabulary'. So far, studies report rates of scent marking and other olfactory behaviours and/or the chemical composition of olfactory signals. But whether, for example, the combination of different olfactory signals (such as secretions from different scent glands and urine) conveys a message different from a single mark, or the addition of visual signals, is largely unknown, and relevant studies are only slowly emerging (Palagi and Norscia, 2009). It will also be interesting to know whether primates are capable of modulating the amount (or even the chemical content) of scent gland secretions or urine to convey different messages in relation to different behavioural contexts. Such studies will, however, be technically very difficult and require the development both of devices with which to measure olfactory signals with much precision and of clever experiments to 'playback' different olfactory signals and measure the response of potential receivers.

(3) More baseline data for a broader range of species from the different primate radiations, collected within a theoretical framework (e.g., sexual selection theory (e.g., Jannett, 1986; Heymann, 2003)). Such studies must include not only signal emission, but also reception and response to olfactory signals on the behavioural ('to whom it may concern' (Kappeler, 1998)) and the neurophysiological level (Ferris *et al.*, 2001). Furthermore, information on the chemical composition of scent marks and their longevity/degradation under natural conditions is needed.

(4) Challenge of the paradigm of scent marking as being a cheap way of communication (e.g., Krebs and Davies, 1993). This will involve metabolic studies to identify production costs, for example the question of where proteins and unsaturated fatty acids found in scent gland secretions (e.g., Epple *et al.*, 1993) are derived from. Also, detailed behavioural studies on the 'broadcasting costs' under natural conditions (e.g., how many scent marks have to be placed to reach a receiver) will be needed.

(5) Examining whether olfactory signals are related to individual quality (Endler, 1993; Zala *et al.*, 2004; Charpentier *et al.*, 2010). This will include exploring the link between intrinsic factors like the Major Histocompatibility Complex (MHC) (Setchell *et al.*, 2011) or extrinsic factors like diet (Ferkin *et al.*, 1997) and individual scent profiles, and examining variation of scent profiles over individual life trajectories. Apart from genetic and chemical analyses, dietary manipulation and its effect on the composition and attractiveness of scent marks will be necessary.

As becomes clear even from these few selected points, future studies on primate olfactory communication will have to be even more interdisciplinary

> enterprises than many have already been, linking behavioural ecologists, biochemists and chemists, geneticists, physiologists and neurobiologists, among others, and have to integrate observational and experimental approaches both in the wild and in captivity.
>
> Since the world of olfactory communication is so different from our predominating visual and acoustic channels of communication, students in this field have to be very creative and to think square. This may result in surprising new insights into olfactory communication, for example the finding of scent asymmetries (Dapporto, 2008), which in turn will lead to new ideas and research questions. Putting these into practical research is much more challenging than for other modes of communication. But the insights that will be obtained will provide refreshing perspectives on primate behavioural biology and social life.

1.3 Living in a smell-centred world: the case of *Lemur catta*

Despite the increasing importance of the visual modality over the course of primate evolution (Barton, 1998), there is evidence that smell keeps on being used as a sensory modality to convey information. Scent marking has been reported in a number of catarrhine species (Old World monkeys and apes), specialised scent glands have been discovered in both gibbons (*Hylobates syndactylus*) and mandrills (*Mandrillus sphinx*) (Geissman, 1987; Setchell *et al.*, 2010a; Freeman *et al.*, 2012). Moreover, the pig-tailed macaque (*Macaca nemestrina*) displays olfactory discriminatory abilities similar to those of squirrel monkeys (Hübener and Laska, 1998), and probably uses olfactory cues for mate assessment (Clarke *et al.*, 2009; Setchell *et al.*, 2010b).

Strepsirrhines – including lemurs, lorises and galagos – are the most scent-oriented group of the primate order. They possess, within the nasal cavity, a large number of turbinates, whose sensory membrane sends inputs to the olfactory nerves and to the olfactory bulbs, in the frontal lobe of the brain. In addition, lemurs, lorises, tarsiers and many New World monkeys have an additional sense that seems to be particularly important in sexual communication (Fleagle, 2013). The vomeronasal organ (or Jacobson's organ) is a chemical sensing organ that lies in the anterior part of the roof of the mouth in many mammals (Harrington, 1977, 1979). It is stimulated by substances found in the urine of female primates, and permits other individuals to determine chemically the reproductive status of a female. To use the vomeronasal organ is possible because of the median cleft and moist region that extends from the base of the nasal opening to the insight of the upper lip. This external nasal structure allows flehmen, observed in many mammals and reported in lemurs (Harrington, 1977, 1979; Dugmore *et al.*, 1984; Wyatt, 2014). The animal inhales with the mouth open, moving the tongue to the mouth's roof while curling their upper lip to facilitate exposure of the vomeronasal organ to scent molecules.

In strepsirrhines, the pivotal role of olfaction reflects in their specialised gland system (throat, arm, wrist, scrotum and vaginal glands), which is used to withdraw and release information in their smell-centred world (Osman-Hill, 1953; Jolly, 1966a). The large number of turbinates, the fully functional vomeronasal organ, and the odour-producing glands make strepsirrhines' olfactory system the most complex of the primate world, and social lemurs a particularly suitable model to explore smell-guided recognition in primates.

The ability to recognise individuals by their scent represents one of the fundamental mechanisms in regulating several aspects of lemur social life. Experiments on olfactory individual recognition have focused on *Lemur catta*, a diurnal strepsirrhine with the most complex social network and marking system. In fact, ring-tailed lemur males possess highly specialised brachial and antebrachial glands that are used to mark objects in the environment and to impregnate their own tail before waving it towards conspecifics (Figures 1.1, 1.2 and 1.3). Moreover, both males and females apply genital secretions during ano-genital marking (Figure 1.4: Jolly, 1966a; Tattersall, 1982; Kappeler, 1998).

In 1975, Mertl performed Habituation/Dishabituation Tests on *Lemur catta* using antebrachial secretions showing that males distinguish between the scents from different individuals. However, scent discrimination could be due to a combination of variables (kinship, rank and familiarity of the donors) rather than chemical individuality (Thom and Hurst, 2004). Thus, while the experiment demonstrated

Figure 1.1 *Lemur catta* male. The antebrachial gland on the wrist is visible. Pistoia Zoo, Italy. Photo: Elisabetta Palagi.

Figure 1.2 Wrist marking performed by a *Lemur catta* male in the Berenty forest, Madagascar. Photo: Elisabetta Palagi.

Figure 1.3 Tail anointing by a *Lemur catta* male in the Berenty forest, Madagascar. Photo: Elisabetta Palagi.

Figure 1.4 Genital marking by a *Lemur catta* female in the Berenty forest, Madagascar. Photo: Ivan Norscia.

the ability to distinguish between two scents (individual discrimination), it did not demonstrate olfactory individual recognition; that is, the discrimination of one out of many known individuals by its scent. To provide the final, empirical demonstration of olfactory individual recognition in *Lemur catta*, Palagi and Dapporto (2006) set up an experimental design to show that all the three components of a recognition system are present in lemurs: (1) a set of cues produced by the sender (expression component; Tsutsui, 2004), (2) the perception of these cues by the receiver (perception component; Mateo, 2004) and (3) a functional response by the receiver (action component; Liebert and Starks, 2004).

Expression component

To verify the presence of the expression component, Palagi and Dapporto (2006) performed gas chromatographic analyses on brachial secretions, collected by squeezing the brachial glands, because specialised gland secretions are generally 'hard-wired' in the genome (fixed information) rather than expressed by metabolic and environmentally dependent factors (Bradbury and Vehrencamp, 1998; Hurst and Beynon, 2004).

Chemical analyses showed the existence of an individual profile for the scent mark of each male (expression component), thus providing a basis for *Lemur catta* individual recognition. This uniqueness (odour fingerprint) is probably due to the differences in the relative concentration of the diverse compounds (the ten subjects shared 91.67% of the peaks) rather than in qualitative variations. Similar results were also found for common marmosets *Callithrix jacchus* (Smith *et al.*, 2001). As predicted for scents encoding individual identity, the brachial signatures did not show any similarity pattern according to ranking status, age or group provenance of the donors, thus suggesting that social and environmental situations do not affect the chemical composition of the brachial glands. The volatile components of male secretions were also individually unique throughout the seasons.

Perception component

To verify the occurrence of the perception component, Palagi and Dapporto (2006) employed Habituation/Discrimination Tests (HDT) (Johnston and Jerningan, 1994), buffering the confounding factors that may allow discriminating between different scents because they derive from different conditions (differences due to kinship, familiarity, rank, age, reproductive status and rearing conditions) but not necessarily from different subjects. Palagi and Dapporto (2006) found that animals were able to distinguish between secretions belonging to two unfamiliar males having the same age, ranking position and group of origin (Figure 1.5). This finding strongly suggests that the discrimination ability depends on the recognition of the unique scent characterising each individual (odour fingerprint, expression component). However, the presence of the expression and perception components is not sufficient to prove individual recognition. In fact, while these tests show that animals possess different odours and are able to discriminate familiar from unfamiliar scents, they do not provide any information on whether the animal matches the perceived scent with its owner. Specifically, an animal could investigate more unfamiliar scents just because they are new and not because they elicit in the animal the representation of

Figure 1.5 *Lemur catta* individual facing the choice between two male antebrachial secretions (A and B). Drawing: Carmelo Gómez González.

the odour owner's identity. Thus, a final step is necessary: assessing the occurrence of the third component of the individual recognition system, namely the action component.

Action component

To verify the presence of the action component, Palagi and Dapporto (2006) used a functional bioassay based on territorial competition (Hurst and Beynon, 2004). This trial was possible by testing individuals of two neighbouring captive groups. Territorial defence is very strong in *Lemur catta* and can result in the death or serious injury of several group members (Jolly, 1966a; Jolly and Pride, 1999). Olfactory behaviour plays a fundamental role: owners extensively mark their territory (mainly at boundaries) and spend a considerable amount of time seeking and investigating conspecific depositions (Jolly, 1966a; Mertl-Millhollen, 1986; Kappeler, 1998; Palagi *et al.*, 2005a; Palagi and Norscia, 2009). Generally, an odour belonging to a novel unfamiliar individual (a potential competitor) elicits more intense olfactory responses compared to a scent belonging to a group mate (Ramsay and Giller, 1996; Palagi *et al.*, 2005a). Instead, Palagi and Dapporto (2006) observed that in the two study groups, the odour of a well-known competitor, despite its familiarity, elicited a stronger response compared to an unfamiliar donor. This result suggested the occurrence of a higher-order processing system that categorises stimuli according to their significance and not strictly according to their sensory features (Johnston and Jerningan, 1994).

In conclusion, all the three components (expression, perception and action components) necessary to individually recognise conspecifics are present in ring-tailed lemurs. Thus, lemurs can identify other individuals using their scent as we use a picture on a passport (Figure 1.6).

Why has individual recognition been positively selected in lemurs? Due to its social and ecological features, *Lemur catta* can gain considerable advantages by individual recognition. This species shows multimale/multifemale group arrangements with a complex social structure, very high territorial competition, strict linear hierarchy, and frequent target aggression (Jolly, 1966; Gould *et al.*, 2003; Palagi and Dapporto, 2007; Norscia and Palagi, 2015; Chapter 4). Gosling (1982) proposed that 'the function of territory marking is to provide an olfactory association between the resident and the defended area which allows intruders to identify the resident when they meet and thus reduce the frequency of escalating agonistic encounters'. The ability to recognise the identity of conspecifics is particularly useful when animals can recall and use information from previous encounters to moderate future responses. Within group members, stable linear dominance hierarchies are predicted to develop when individuals can remember the outcome of prior encounters (for an extensive review see Beacham, 2003). Moreover, individual recognition may play an important role to avoid inbreeding and maintain coalitions and reciprocal alliances with specific group members (Trivers, 1971). It is therefore clear that *Lemur catta* can very much benefit from an efficient individual recognition. Indeed, females

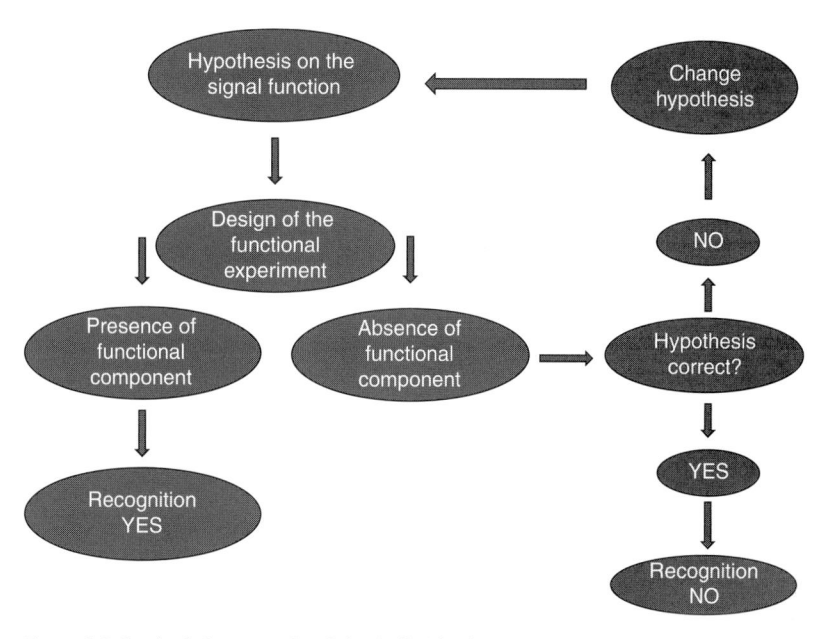

Figure 1.6 Logical framework of the individual recognition experiments.

absolutely avoid mating with their relatives (Pereira and Weiss, 1991). Moreover, now and then some group members repeatedly attack 'a single target individual'; these unprovoked aggressions can last from a few days to several months and they generally end with the forced eviction of the victim (Vick and Pereira, 1989; Palagi et al., 2005b). Furthermore, reconciliation (a form of affiliative interaction between former antagonists occurring shortly after an agonistic event; de Waal and van Roosmalen, 1979; Chapter 5) has been described in ring-tailed lemurs (Palagi et al., 2005b; Palagi and Norscia, 2015). The possibility of reconciling with a former opponent necessarily requires the ability to recognise them.

Why is olfaction the favourite sensory modality for individual recognition in ring-tailed lemurs? Owing to the complex olfactory system of *Lemur catta*, it appears more than plausible that pheromonal polymorphism, which gives each individual its unique olfactory signature, is strongly selected in this species. The capability to recognise the individual ownership, other than simply perceive the spatial and temporal pattern of scent depositions, may provide to visitor and resident lemurs continuous and fundamental information useful to make reproductive and competitive decisions (Gosling, 1982; Bradbury and Vehrencamp, 1998; Hurst and Beynon, 2004; Palagi et al., 2005a, b).

1.4 From olfactory to cross-modal recognition: a working hypothesis for lemurs

After tasting the little cakes (madeleines) he used to eat in childhood at his uncle's in Combray, Marcel Proust (in *In Search of Lost Time*) provides a clear literary example

> **Box 1.3** | **by Ipek G. Kulahci and Asif A. Ghazanfar**
> **Speaking of which: a multimodal approach to individual recognition**
>
> The ability to recognise others is the basis for complex social interactions. Just until a few years ago, the only evidence of individual recognition based on cues from multiple modalities (multisensory individual recognition) came from humans. We now know that similar to humans, other species including monkeys, horses and crows can also match the identity information found in auditory cues to those found in visual cues (Proops et al., 2009; Sliwa et al., 2011; Kondo et al., 2012).
>
> The majority of the multisensory individual recognition studies focus exclusively on visual-auditory recognition. However, given that many species are able to recognise others via scents, we decided to investigate whether olfactory cues are used in multisensory recognition. Ring-tailed lemurs (*Lemur catta*) rely on contact calling and scent marking for social interactions, group cohesion and movement (Jolly, 1966b; Oda, 1999). Lemurs also use contact calls and scents to identify different individuals (Macedonia, 1986; Palagi and Dapporto, 2006), which makes them fascinating subjects with which to address the role of olfactory cues in multisensory recognition.
>
> Multisensory individual recognition provides a gateway into the understanding of whether animals form mental representations of familiar individuals. To achieve multisensory recognition, animals need to learn signals from multiple modalities, and be able to associate cues from one modality with another even if that other cue is not encountered at the same time. One potential mechanism which would make this possible involves forming multisensory representations of familiar individuals. For example, when we hear the voice of someone we know, we can instantly visualise their face (Seyfarth and Cheney, 2009). By forming a representation of this individual, we can link information about his/her visual features, such as face and body, with information about his/her auditory features, such as voice. Animals may also form similar multisensory representations which link together identity information from multiple sensory modalities (Proops et al., 2009; Sliwa et al., 2011).
>
> We ran a series of trials in which we presented lemurs with the scent of a familiar female together with the playback of either her contact call (matched trials) or another female's contact call (mismatched trials). Lemurs spent more time attending to the scents and the contact calls when both belonged to the same female, in comparison to when the scent belonged to one female and the call belonged to a different female (Kulahci et al., 2014).
>
> Because the lemurs' responses to the matched trials differed from their responses to the mismatched trials, our study shows that lemurs can match the identity information found in the scents to those found in the contact calls. One of the most interesting outcomes was the increased attention to the matched cues. It is possible that congruent information is more salient

than information that is incongruent. For example, when lemurs detect the scent of a female soon after hearing her contact call, the identity information in scents and calls provide congruent information that reinforce each other. Such reinforcement would be especially beneficial when information in one modality is accessible for a longer period (or at a longer distance) than in the other. This is the case with scents and calls; vocalisations are transient and scents are usually available for longer periods than vocalisations. Therefore, contact calls may provide a time-stamp and a context to the scent mark. Hearing a female and detecting her scent mark in the same place and at the same time could inform the group members that she is currently close by.

Our results demonstrate that studies of multisensory recognition need not focus only on visual or acoustic cues, as olfactory cues also play a role in multisensory recognition. We do not know much about the mechanisms that allow animals to learn to associate cues from multiple sensory modalities. For example, one area that remains poorly investigated is whether animals are able to learn multiple identity cues (such as facial features, scents and vocalisations) all at once. In ring-tailed lemurs, there would be more opportunities for learning identity cues in multiple modalities at different times than learning them all at once. If lemurs indeed learn olfactory and vocal information at different times, then the only way that they can match the two would be by using multisensory mental representations which link together identity cues from these two modalities.

All multisensory individual recognition studies conducted so far have included only two modalities. The focus in particular has been on visual and auditory cues. Our study has demonstrated that olfactory cues can play an important role in multisensory recognition. This finding paves the way for an exciting potential study which incorporates three sensory modalities including vision, olfaction and audition. Lemurs would be a great study system for testing 'trimodal' individual recognition. To address whether identity cues from three modalities can be matched to each other, we can use a series of trials which include lemur pictures or videos to serve as visual cues, scent marks placed on artificial objects as olfactory cues and vocal playbacks as auditory cues. One exciting aspect of trimodal recognition involves exploring whether information in the visual modality is what animals use for linking together cues in the olfactory and auditory modalities. This possibility is of particular interest for understanding the mechanisms behind the formation of multisensory representations. For example, lemurs can learn others' scents either by observing a scent marking individual and/or by smelling an individual's scent directly on that individual, both of which require a visual input. Similarly, lemurs may see a group member vocalise, and thus be able to link visual identity with vocal identity. Having visual cues in common between

Box 1.3 (continued)

> **Box 1.3** (*cont.*)
>
> olfactory and auditory signal production might make it possible to use visual identity information to link together olfactory and auditory identity cues.
>
> A second future direction emerges from our result that the matched identity cues attract greater interest from lemurs than the mismatched identity cues. The time-stamp effect of vocalisations that we outlined earlier would be especially strong if scent marks themselves contain temporal cues (perhaps through their degradation) that the lemurs can detect. We currently do not know whether scent marking species can 'time-stamp' a scent mark and determine whether or not it has been freshly placed. Experiments which present a scent marking species with a range of fresh versus degraded scent marks can test whether this is the case. Scent marks play a critical role in multiple contexts ranging from reproduction to intragroup and intergroup interactions (Kappeler, 1998; Oda, 1999; Scordato and Drea, 2007); therefore, determining whether the cues found in a scent mark are recent or outdated could be quite beneficial.

of what scientists call cross-modal memory: 'And suddenly the memory returns... The sight of the little madeleine had recalled nothing to my mind before I tasted it; perhaps because I had so often seen such things in the interval, without tasting them, on the trays in pastry-cooks' windows, that their image had dissociated itself from those Combray days to take its place among others more recent...'. The author explains that 'the visual memory...being linked to that taste...tried to follow it into my conscious mind', thus describing – de facto – a cross-modal process that links a sensory modality (taste) to another (sight).

Even if a sensory modality can prevail in driving individual recognition, depending on each species' biology, more modalities normally concur to the identification process. What changes is the relative 'dose' of each sensory modality, the extent to which each one contributes to the whole process.

In cross-modal recognition tests, subjects are usually presented with an item in one sensory modality and are subsequently tested for the ability to recognise the same item using a second sensory modality (Ettlinger and Wilson, 1990).

Sliwa *et al.* (2011) demonstrated cross-modal individual recognition in rhesus macaques (*Macaca mulatta*), both inter- and intraspecific, showing that they were able to spontaneously match the faces of known individuals (both conspecifics and humans) to their voices. There are grounds to hypothesise that New World monkeys are also capable of cross-modal individual recognition. Capuchin monkeys (*Cebus capucinus*) exhibit auditory-visual cross-modal perception of conspecific vocalisations. In fact, Evans *et al.* (2005) found that capuchins hearing a particular vocalisation looked longer at a video of a matching facial stimulus than at a video of a non-matching facial stimulus. Cross-modal recognition, obtained by combining olfactory and acoustic cues, was demonstrated in *Lemur catta* by Kulahci *et al.* (2014; see box 1.3 for further details).

In lemurs, visual cues are probably also implied in individual recognition. For example, Marechal *et al.* (2010) demonstrated that lemurs are able to distinguish known and unknown individuals based on facial features (perception component). The authors explained that their study represented a first step to demonstrate individual recognition based on visual cues in lemurs.

References

Abbott, D. H. (1984). Behavioral and physiological suppression of fertility in subordinate marmoset monkeys. *American Journal of Primatology*, 6, 169–186.

Adachi, I. & Fujita, K. (2007). Cross-modal representation of human caretakers in squirrel monkeys. *Behavioural Processes*, 74, 27–32.

Alvarado, J. C., Stanford, T. R., Vaughan, J. W. & Stein, B. E. (2007). Cortex mediates multisensory but not unisensory integration in superior colliculus. *The Journal of Neuroscience*, 27, 12775–12786.

Barton, R. A. (1998). Visual specialization and brain evolution in primates. *Proceedings of the Royal Society B: Biological Sciences*, 265, 1933–1937.

Bauer, H. R. & Philip, M. M. (1983). Facial and vocal individual recognition in the common chimpanzee. *The Psychological Record*, 33, 161–170.

Beacham, J. L. (2003). Models of dominance hierarchy formation: effect of prior experience and intrinsic traits. *Behaviour*, 140, 1275–1303.

Bianchet, M. A., Bains, G., Pelosi, P., *et al.* (1996). The three dimensional structure of bovine odorant-binding protein and its mechanism of odor recognition. *Nature Structural Biology*, 3, 934–939.

Blank, H., Anwander, A. & von Kriegstein, K. (2011). Direct structural connections between voice- and face-recognition areas. *The Journal of Neuroscience*, 31(36), 12906–12915.

Blaustein, A. R. (1981). Sexual selection and mammalian olfaction. *American Naturalist*, 117, 1006–1010.

Bradbury, J. K. & Veherencamp, S. L. (1998). *Principles of Animal Communication*. Sinauer Associates, Sunderland.

Bruce, V. & Young, A. (1986). Understanding face recognition. *British Journal of Psychology*, 77, 305–327.

Buck, L. & Axel, R. (1991). A novel multigene family may encode odorant receptors: a molecular basis for odor recognition. *Cell*, 65, 175–187.

Campanella, S. & Belin, P. (2007). Integrating face and voice in person perception. *Trends in Cognitive Sciences*, 11, 535–543.

Charpentier, M. J. E., Crawford, J. C., Boulet, M. & Drea, C. M. (2010). Message 'scent': lemurs detect the genetic relatedness and quality of conspecifics via olfactory cues. *Animal Behaviour*, 80, 101–108.

Clarke, P. M. R., Barrett, L. & Henzi, S. P. (2009). What role do olfactory cues play in chacma baboon mating? *American Journal of Primatology*, 71, 1–10.

Clutton-Brock, T. (2007). Sexual selection in males and females. *Science*, 318, 1882–1885.

Dapporto, L. (2008). The asymmetric scent: ringtailed lemurs (*Lemur catta*) have distinct chemical signatures in left and right brachial glands. *Naturwissenschaften*, 95, 987–991.

Darwin, C. (1871). *The Descent of Man and Selection in Relation to Sex*. Murray, London.

Dawson, G. A. (1977). Composition and stability of social groups of the tamarin, *Saguinus oedipus geoffroyi*, in Panama: ecological and behavioral implications. In: D. G. Kleiman (ed.),

The Biology and Conservation of the Callitrichidae. Washington: Smithsonian Institution Press, pp. 23–37.

de Waal, F. B. & van Roosmalen, A. (1979). Reconciliation and consolation among chimpanzees. *Behavioral Ecology and Sociobiology*, 5, 55–66.

Delbarco-Trillo, J., Burkert, B. A., Goodwin, T. E. & Drea, C. M. (2011). Night and day: the comparative study of strepsirrhine primates reveals socioecological and phylogenetic patterns in olfactory signals. *Journal of Evolutionary Biology*, 24, 82–98.

Dugmore, S. J., Bailey, K. & Evans, C. S. (1984). Discrimination by male ring-tailed lemurs (*Lemur catta*) between the scent marks of male and those of female conspecifics. *International Journal of Primatology*, 5, 235–245.

Endler, J. A. (1993). Some general comments on the evolution and design of animal communication systems. *Philosophical Transactions of the Royal Society of London B*, 340, 215–225.

Epple, G. (1979). Odor communication in the tamarin *Saguinus fuscicollis* (Callitrichidae): behavioral and chemical studies. In: F. Ritter (ed.), *Chemical Ecology: Odour Communication in Animals*. Amsterdam: Elsevier/North Holland Biomedical Press, pp. 117–130.

Epple, G. (1980). Relationships between aggression, scent marking and gonadal state in a primate, the tamarin *Saguinus fuscicollis*. In: D. Müller-Schwarze & R. M. Silverstein (eds), *Chemical Signals – Vertebrates and Aquatic Invertebrates*. New York: Plenum Press, pp. 87–105.

Epple, G., Golob, N. F., Cebul M-S. & Smith, A. B., III (1981). Communication by scent in some Callitrichidae (Primates) – an interdisciplinary approach. *Chemical Senses*, 6, 377–390.

Epple, G., Belcher, A. M., Küderling I, *et al.* (1993). Making sense out of scents: species differences in scent glands, scent-marking behaviour, and scent-mark composition in the Callitrichidae. In: A. B. Rylands (ed.), *Marmosets and Tamarins: Systematics, Behaviour, and Ecology*. Oxford: Oxford University Press, pp. 123–151.

Ernst, M. O. & Bülthoff, H. H. (2004). Merging the senses into a robust percept. *Trends in Cognitive Sciences*, 8, 162–9.

Ettlinger, G. & Wilson, W. A. (1990). Cross-modal performance: Behavioural processes, phylogenetic considerations and neural mechanisms. *Behavioural Brain Research*, 40, 169–192.

Evans, T. A., Howell, S. & Westergaard, G. C. (2005). Auditory-visual cross-modal perception of communicative stimuli in tufted capuchin monkeys (*Cebus apella*). *Journal of Experimental Psychology: Animal Behavior Processes*, 31, 399–406.

Ferkin, M. H., Sorokin, E. S., Johnston, R. E. & Lee, C. J. (1997). Attractiveness of scents varies with protein content of the diet in meadow voles. *Animal Behaviour*, 53, 133–141.

Ferris, C. F., Snowdon, C. T., King, J. A., *et al.* (2001). Functional imaging of brain activity in conscious monkeys responding to sexually arousing cues. *Neuroreport*, 12, 2231–2236.

Fleagle, J. G. (2013). *Primate Adaptation and Evolution*, 3rd ed. San Diego: Academic Press. 596 p.

Freeman, N. J., Pasternak, G. M., Rubi, T. L., Barrett, L. & Henzi, S. P. (2012). Evidence for scent marking in vervet monkeys? *Primates*, 53, 311–315.

Geissman, T. (1987). A sternal gland in the siamang gibbon (*Hylobates syndactylus*). *International Journal of Primatology*, 8, 1–15.

Ghazanfar, A. A. & Logothetis, N. K. (2003). Neuroperception: facial expressions linked to monkey calls. *Nature*, 423, 937–938.

Ghazanfar, A. A., Chandrasekaran, C. & Logothetis, N. K. (2008). Interactions between the superior temporal sulcus and auditory cortex mediate dynamic face/voice integration in rhesus monkeys. *The Journal of Neuroscience*, 28, 4457–4469.

Gil-da-Costa, R., Braun, A., Lopes, M., *et al.* (2004). Toward an evolutionary perspective on conceptual representation: Species-specific calls activate visual and affective processing systems in the macaque. *PNAS*, 101(50), 17516–17521.

Gosling, L. M. (1982). A reassessment of the function of scent marking in territories. *Zeitschrift für Tierpsychologie*, 60, 89–118.

Gosling, L. M. & Roberts, S. (2001). Testing ideas about the function of scent marks in territories from spatial patterns. *Animal Behaviour*, 62, F7–F10.

Gottfried, J. A., Smith, A. P. R., Rugg, M. D. & Dolan, R. J. (2004). Remembrance of odors past: human olfactory cortex in cross-modal recognition memory. *Neuron*, 42, 687–695.

Gould, L., Sussman, R. W. & Sauther, M. L. (2003). Demographic and life-history patterns in a population of ring-tailed lemurs (*Lemur catta*) at Beza Mahafaly Reserve, Madagascar: A 15-year perspective. *American Journal of Physical Anthropology*, 120, 182–194.

Green, J. A. & Gustafson, G. E. (1983). Individual recognition of human infants on the basis of cries alone. *Developmental Psychobiology*, 16, 485–493.

Harrington, J. E. (1977). Discrimination between males and females by scent in *Lemur fulvus*. *Animal Behaviour*, 25, 147–151.

Harrington, J. E. (1979). Responses of *Lemur fulvus* to scents of different subspecies of *L. fulvus* and to scents of different species of lemuriformes. *Zeitschrift für Tierpsychologie*, 49, 1–9.

Heymann, E. W. (1998). Sex differences in olfactory communication in a wild primate, *Saguinus mystax* (Callitrichinae). *Behavioral Ecology and Sociobiology*, 43, 37–45.

Heymann, E. W. (2000). Spatial patterns of scent marking in wild moustached tamarins, *Saguinus mystax*: no evidence for a territorial function. *Animal Behaviour*, 60, 723–730.

Heymann, E. W. (2001). Interspecific variation of scent-marking behaviour in wild tamarins, *Saguinus mystax* and *Saguinus fuscicollis*. *Folia Primatologica*, 72, 253–267.

Heymann, E. W. (2003). Scent marking, paternal care, and sexual selection in callitrichines. In: C. B. Jones (ed.), *Sexual Selection and Reproductive Competition in Primates: New Perspectives and Directions*. Norman: American Society of Primatologists, pp. 305–325.

Heymann, E. W. (2006). The neglected sense – olfaction in primate behavior, ecology, and evolution. *American Journal of Primatology*, 68, 519–524.

Heymann, E. W. (2009). Primate olfactory communication: the influence of life history and sexual selection. *Folia Primatologica*, 80, 370.

Hübener, F. & Laska, M. (1998). Assessing olfactory performance in an Old World primate, *Macaca nemestrina*. *Physiology and Behavior*, 64, 521–527.

Hurst, J. L. & Beynon, R. J. (2004). Scent wars: the chemobiology of competitive signaling in mice. *BioEssays*, 26, 1288–1298.

Izawa, K. (1978). A field study of the ecology and behavior of the black-mantled tamarin (*Saguinus nigricollis*). *Primates*, 19, 241–274.

Izumi, A. & Kojima, S. (2004). Matching vocalizations to vocalizing faces in a chimpanzee (*Pan troglodytes*). *Animal Cognition*, 7(3), 179–184.

Jannett, F. J., Jr (1986). Morphometric patterns among microtine rodents. I. Sexual selection suggested by relative scent gland development in representative voles (*Microtus*). In: D. Duvall, D. Müller-Schwarze & R. Silverstein (eds), *Chemical Signals in Vertebrates 4*. New York: Plenum Press.

Joassin, F., Maurage, P. & Campanella, S. (2011). The neural network sustaining the crossmodal processing of human gender from faces and voices: an fMRI study. *Neuroimage*, 54, 1654–1661.

Johnston, R. E. & Bullock, T. A. (2001). Individual recognition by use of odors in golden hamsters: The nature of individual representations. *Animal Behaviour*, 61, 545–557.

Johnston, R. E. & Jernigan, P. (1994). Golden hamsters recognize individuals, not just individual scents. *Animal Behaviour*, 48, 129–136.

Johnston, R. E. & Peng, A. (2008). Memory for individuals: Hamsters (*Mesocricetus auratus*) require contact to develop multicomponent representations (concepts) of others. *Journal of Comparative Psychology*, 122, 121–131.

Jolly, A. (1966a). *Lemur Behaviour: a Madagascar Field Study*. Chicago: The University of Chicago Press.

Jolly, A. (1966b). Lemur social behavior and primate intelligence. *Science*, 153, 501–506.

Jolly, A. & Pride, E. (1999). Troop histories and range inertia of *Lemur catta* at Berenty, Madagascar: a 33-year perspective. *International Journal of Primatology*, 20, 359–373.

Kanwisher, N., McDermott, J. & Chun, M. M. (1997). The fusiform face area: a module in human extrastriate cortex specialized for face perception. *The Journal of Neuroscience*, 17, 4302–4311.

Kappeler, P. M. (1998). To whom it may concern: the transmission and function of chemical signals in *Lemur catta*. *Behavioral Ecology and Sociobiology*, 42, 411–412.

Kojima, S., Izumi, A. & Ceugniet, M. (2003). Identification of vocalizers by pant hoots, pant grunts and screams in a chimpanzee. *Primates*, 44(3), 225–230.

Kondo, N., Izawa, E. & Watanabe, S. (2012). Crows cross-modally recognize group members but not non-group members. *Proceedings of the Royal Society B: Biological Sciences*, 279, 1937–1942.

Krebs, J. R. & Davies, N. B. (1993). *An Introduction to Behavioural Ecology*, 3rd ed. Oxford: Blackwell Science.

Kulahci, I. G., Drea, C. M., Rubenstein, D. I. & Ghazanfar, A. A. (2014). Individual recognition through olfactory-auditory matching in lemurs. *Proceedings of the Royal Society B: Biological Sciences*, 281, 20140071.

Leal, W. S. (2013). Odorant reception in insects: roles of receptors, binding proteins, and degrading enzymes. *Annual Review of Entomology*, 58, 373–391.

Liebert, A. E. & Starks, P. T. (2004). The action component of recognition systems: a focus on the response. *Annales Zoologici Fennici*, 41, 747–764.

Lindsay, N. B. D. (1979). A report on the field study of Geoffroy's tamarin *Saguinus oedipus geoffroyi*. *Dodo Journal of the Jersey Wildlife Preservation Trust*, 16, 27–51.

Lledo Ferrer, Y., Peláez, F. & Heymann, E. W. (2011). The equivocal relationship between territoriality and scent marking in wild saddleback tamarins, *Saguinus fuscicollis*. *International Journal of Primatology*, 32, 974–991.

Lledo Ferrer, Y., Peláez, F. & Heymann, E. W. (2012). Territorial polemics: a response to Roberts. *International Journal of Primatology*, 33, 762–768.

Lynch Alfaro, J. W., De Sousa E Silva, J. & Rylands, A. B. (2012). How different are robust and gracile capuchin monkeys? An argument for the use of *Sapajus* and *Cebus*. *American Journal of Primatology*, 74, 273–286.

Macedonia, J. M. (1986). Individuality in a contact call of the ringtailed lemur (*Lemur catta*). *American Journal of Primatology*, 11, 163–179.

Manai, R., Scorsone, E., Rousseau, L., et al. (2014). Grafting odorant binding proteins on diamond bio-MEMS. *Biosensors and Bioelectronics*, 60, 311–317.

Marechal, L., Genty, E. & Roeder, J. J. (2010). Recognition of faces of known individuals in two lemur species (*Eulemur fulvus* and *E. macaco*). *Animal Behaviour*, 79, 1157–1163.

Matauschek, C., Roos, C. & Heymann, E. W. (2011). Mitochondrial phylogeny of tamarins (*Saguinus*, Hoffmannsegg 1807) with taxonomic and biogeographic implications for the *S. nigricollis* species group. *American Journal of Physical Anthropology*, 144, 564–574.

Mateo, J. M. (2004). Recognition systems and biological organization: the perception component of social recognition. *Annales Zoologici Fennici*, 41, 729–745.

Matsuo, T., Sugaya, S., Yasukawa, J., Aigaki, T. & Fuyama, Y. (2007). Odorant-binding proteins OBP57d and OBP57e affect taste perception and host-plant preference in *Drosophila sechellia*. *PLoS Biology*, 5, e118.

Mertl, A. S. (1975). Discrimination of individuals by scent in a primate. *Behavioral Biology*, 14, 505–509.

Mertl-Millhollen, A. S. (1986). Territorial scent marking by two sympatric lemur species. In: D. Duvall, D. Müller-Schwarze and R. M. Silverstein (eds), *Chemical Signals in Vertebrates 4*, New York: Plenum Press, pp. 385–395.

Neyman, P. F. (1977). Aspects of the ecology and social organization of free-ranging cotton-top tamarins (*Saguinus oedipus*) and the conservation status of the species. In: D. G. Kleiman (ed.) *The Biology and Conservation of the Callitrichidae*. Washington: Smithsonian Institution Press, pp. 39–71.

Norscia, I. & Palagi, E. (2015). The socio-matrix reloaded: from hierarchy to dominance profile in wild lemurs. *PeerJ*, 3, e729. https://dx.doi.org/10.7717/peerj.729.

Oda R. (1999). Scent marking and contact call production in ring-tailed lemurs (*Lemur catta*). *Folia Primatologica*, 70, 121–124.

Osman-Hill, W. C. (1953). *Primates (Comparative Anatomy and Taxonomy). I. Strepsirhini*. Edinburgh: University Press, pp. xxiv + 798.

Palagi, E. & Dapporto, L. (2006). Beyond odor discrimination: demonstrating individual recognition by scent in *Lemur catta*. *Chemical Senses*, 31(5), 437–443.

Palagi, E. & Dapporto, L. (2007). Females do it better. Individual recognition experiments reveal sexual dimorphism in *Lemur catta* (Linnaeus 1758) olfactory motivation and territorial defence. *Journal of Experimental Biology*, 210, 2700–2705.

Palagi, E. & Norscia, I. (2009). Multimodal signaling in wild *Lemur catta*: economic design and territorial function of urine marking. *American Journal of Physical Anthropology*, 139, 182–192.

Palagi, E. & Norscia, I. (2015). The season for peace: reconciliation in a despotic species (*Lemur catta*). *PLoS ONE*, 10, e0142150. http://dx.doi.org/10.1371/journal.pone.0142150.

Palagi, E., Dapporto, L. & Borgognini-Tarli, S. (2005a). The neglected scent: on the marking function of urine in *Lemur catta*. *Behavioral Ecology and Sociobiology*, 58, 437–445.

Palagi, E., Paoli, T. & Borgognini-Tarli, S. (2005b) Aggression and reconciliation in two captive groups of *Lemur catta*. *International Journal of Primatology*, 26, 279–294.

Parr, L. A. (2004). Perceptual biases for multimodal cues in chimpanzee (*Pan troglodytes*) affect recognition. *Animal Cognition*, 7, 171–178.

Pelosi, P. (1994). Odorant-binding proteins. *Critical Reviews in Biochemistry and Molecular Biology*, 29, 199–228.

Pelosi, P. (2003). Physiological and artificial biosensors for odour recognition systems. In: L. Barsanti, V. Evangelista, P. Gualtieri, V. Passarelli and S. Vestri (eds), *Molecular Electronics: bio-sensors and bio-computers* (NATO ASI Series). Kluwer Academic Publishers.

Pelosi, P. & Maida, R. (1990). Odorant binding proteins in vertebrates and insects: similarities and possible common function. *Chemical Senses*, 15, 205–215.

Pelosi, P., Baldaccini, N. E. & Pisanelli, A. M. (1982). Identification of a specific olfactory receptor for 2-isobutyl-3-methoxypyrazine. *Biochemical Journal*, 201, 245–248.

Pelosi, P., Zhou, J.-J., Ban, L. P. & Calvello, M. (2006). Soluble proteins in insect chemical communication. *Cell and Molecular Life Sciences*, 63, 1658–1676.

Pelosi, P., Mastrogiacomo, R., Iovinella, I., Tuccori, E. & Persaud, K. C. (2013). Structure and biotechnological applications of odorant-binding proteins. *Applied Microbiology and Biotechnology*, 98, 61–70.

Pereira, M. E. & Weiss, M. L. (1991). Female mate choice, male migration, and the threat of infanticide in ringtailed lemurs. *Behavioral Ecology and Sociobiology*, 28, 141–152.

Persaud, K. C. & Pelosi, P. (1992). Sensor arrays using conducting polymers for an artificial nose. In: J. W. Gardner & P. N. Bartlett (eds), *Sensors and Sensory Systems for an Electronic Nose*, Berlin: Springer, pp. 237–256.

Porter, R. H., Cernoch, J. M. & McLaughlin, F. J. (1983). Maternal recognition of neonates through olfactory cues. *Physiology and Behavior*, 30, 151–154.

Proops, L., McComb, K. & Reby, D. (2009). Cross-modal individual recognition in domestic horses (*Equus caballus*). *PNAS*, 106, 947–951.

Ramsay, N. F. & Giller, P. S. (1996). Scent-marking in ring-tailed lemurs: responses to the introduction of 'foreign' scent in the home range. *Primates*, 37, 13–23.

Rhodes, G., Jeffery, L., Watson, T. L., et al. (2004) Orientation-contingent face aftereffects and implications for face-coding mechanisms. *Current Biology*, 14, 2119–2123.

Roberts, S. C. (2012). On the relationship between scent-marking and territoriality in callitrichid primates. *International Journal of Primatology*, 33, 749–761.

Romanski, L. M. (2007). Representation and integration of auditory and visual stimuli in the primate ventral lateral prefrontal cortex. *Cerebral Cortex*, 17(Suppl 1), i61–i69.

Sacks, O. W. (1970). *The Man Who Mistook His Wife for a Hat and Other Clinical Tales.* New York: Touchstone.

Sakkalou, E. & Gattis, M. (2012). Infants infer intentions from prosody. *Cognitive Development*, 27, 1–16.

Sandler, B. H., Nikonova, L., Leal, W. S. & Clardy, J. (2000). Sexual attraction in the silkworm moth: structure of the pheromone-binding-protein-bombykol complex. *Chemistry & Biology*, 7, 143–151.

Scordato, E. S. & Drea, C. M. (2007). Scents and sensibility: information content of olfactory signals in the ringtailed lemur, *Lemur catta*. *Animal Behaviour*, 73, 301–314.

Sergent, J., Ohta, S. & MacDonald, B. (1992). Functional neuroanatomy of face and object processing: a positron emission tomography study. *Brain*, 115, 15–36.

Setchell, J. M., Vaglio, S., Moggi-Cecchi, J., et al. (2010a) Chemical composition of scent-gland secretions in an Old World monkey (*Mandrillus sphinx*): influence of sex, male status and individual identity. *Chemical Senses*, 35, 205–220.

Setchell, J. M., Charpentier, M. J. E., Abbott, K. M., Wickings, E. J. & Knapp, L. A. (2010b) Opposites attract: MHC-associated mate choice in a polygynous primate. *Journal of Evolutionary Biology*, 23, 136–148.

Setchell, J. M., Vaglio, S., Abbott, K. M., et al. (2011). Odour signals major histocompatibility complex genotype in an Old World monkey. *Proceedings of the Royal Society B: Biological Sciences*, 278, 274–280.

Seyfarth, R. M. & Cheney, D. L. (2009). Seeing who we hear and hearing who we see. *PNAS*, 106, 669–670.

Sliwa, J., Duhamel, J. R., Pascalis, O. & Wirth, S. (2011). Spontaneous voice-face identity matching by rhesus monkeys for familiar conspecifics and humans. *PNAS*, 108, 1735–1740.

Smith, T. E., Tomlinson, A. J., Mlotkiewicz, J. A. & Abbott, D. H. (2001). Female marmoset monkeys (*Callithrix jacchus*) can be identified from the chemical composition of their scent marks. *Chemical Senses*, 26, 449–458.

Sprankel, H. (1961). Histologie und biologische Bedeutung eines jugulo-sternalen Duftdrüsenfeldes bei Tupaia glis Diard 1820. *Verhandlungen der Deutschen Zoologischen Gesellschaft*, 25, 198–206.

Sprankel, H. (1971). Zur vergleichenden Histologie von Hautdrüsenorganen im Lippenbereich bei *Tarsius bancanus borneanus* Horsfield 1821 und *Tarsius syrichta carbonarius* Linnaeus 1758. In: H. Kummer (ed.), *Proceedings of the 3rd International Congress of Primatology*, Zurich (1970), vol 1. S. Karger, Basel, pp. 189–197.

Stitzel, S. E., Aernecke, M. J. & Walt, D. R. (2011). Artificial noses. *Annual Review of Biomedical Engineering*, 13, 1–25.

Sun, Y. F., De Biasio, F., Qiao, H. L., et al. (2012) Two odorant-binding proteins mediate the behavioural response of aphids to the alarm pheromone (E)-ß-farnesene and structural analogues. *PLoS ONE*, 7, e32759 http://dx.doi.org/10.1371/journal.pone.0032759.

Swarup, S., Williams, T. I. & Anholt, R. R. (2011). Functional dissection of Odorant binding protein genes in *Drosophila melanogaster*. *Genes, Brain and Behavior*, 10, 648–657.

Tattersall, I. (1982). *The Primates of Madagascar.* New York: Columbia University Press.

Tegoni, M., Ramoni, R., Bignetti, E., Spinelli, S. & Cambillau, C. (1996). Domain swapping creates a third putative combining site in bovine odorant binding protein dimer. *Nature Structural Biology*, 3, 863–867.

Tegoni, M., Campanacci, V. & Cambillau, C. (2004). Structural aspects of sexual attraction and chemical communication in insects. *Trends in Biochemical Sciences*, 29, 257–264.

Terborgh, J. (1983). *Five New World Primates: A Study in Comparative Ecology*. Princeton: Princeton University Press,

Terborgh, J. & Goldizen, A. W. (1985). On the mating system of the cooperatively breeding saddle-back tamarin (*Saguinus fuscicollis*). *Behavioral Ecology and Sociobiology*, 16, 293–299.

Thom, M. D. & Hurst, J. L. (2004). Individual recognition by scent. *Annales Zoologici Fennici*, 41, 765–787.

Trivers, R. L. (1971). Parental investment and sexual selection. In: B. Campbell (ed.), *Sexual Selection and the Descent of Man*. Chicago: Aldine, pp. 139–179.

Tsutsui, N. E. (2004). Scent of self: the expression component of self/non-self recognition systems. *Annales Zoologici Fennici*, 41, 713–727.

Turner, A. P. & Magan, N. (2004). Electronic noses and disease diagnostics. *Nature Reviews Microbiology*, 2, 161–166.

Vick, L. G. & Pereira, M. E. (1989). Episodic targeting aggression and the histories of *Lemur* social groups. *Behavioral Ecology and Sociobiology*, 25, 3–12.

Vieira, F. G. & Rozas, J. (2011). Comparative genomics of the odorant-binding and chemosensory protein gene families across the Arthropoda: Origin and evolutionary history of the chemosensory system. *Genome Biology and Evolution*, 3, 476–490.

Vogt, R. G. & Riddiford, L. M. (1981). Pheromone binding and inactivation by moth antennae. *Nature*, 293, 161–163.

Wyatt, T. D. (2014). *Pheromones and Animal Behavior*, 2nd Edition. Cambridge University Press, 424 pages.

Xu, P., Atkinson, R., Jones, D. N. & Smith, D. P. (2005). *Drosophila* OBP LUSH is required for activity of pheromone-sensitive neurons. *Neuron*, 45, 193–200.

Zala, S. M., Potts, W. K. & Penn, D. J. (2004). Scent-marking displays provide honest signals of health and infection. *Behavioral Ecology*, 15, 338–344.

2 What do you mean? Multimodal communication for a better signal transmission

Surrounding things transmit their images to the senses and the senses transfer them to the Sensation. Sensation sends them to the Common Sense, and by it they are stamped upon the memory and are there more or less retained according to the importance or force of the impression.

Leonardo Da Vinci, ca. 1510

2.1 What is multimodal communication?

The American journalist Sydney Harris briefly and effectively tagged the difference between information and communication. According to common definitions, information refers to facts about a situation, an individual or an event. Communication is 'the process of sharing information, especially when this increases understanding between people or groups' (Cambridge Dictionary) or 'the imparting or exchanging of information by speaking, writing, or using some other medium' (Oxford Dictionary). These definitions are centred on human communication when, in fact, communication is a widespread – and necessary – phenomenon spanning the whole animal kingdom. The light pulses of fireflies (Bradbury and Vehrencamp, 1998), the dance of honeybees (von Frisch, 1967; Seeley, 1997), the claw-waving displays of crustaceans (Dingle, 1969), the songs of birds (Vehrencamp, 2000), the alarm calls of different primates (Cheney and Seyfarth, 1990), and the articulate language of humans (Savage-Rumbaugh et al., 1998) are just some examples of information transmission between or among conspecifics.

Communication is an essential prerequisite for sociality and has evolved along with the development of social systems, from the simplest to the most complex (Freeberg et al., 2012). Communication involves the transmission of a signal, which is any action or trait produced by one animal, the sender, that provides information used by another animal, the receiver (Wilson, 1975; Endler, 1993; Hebets and Papaj, 2005). A signal is complex if it can be disassembled into different, single components, each of them able to elicit a response in the receiver. Such a response, of course, can be different from the response elicited by the complex signal. For example, when the telephone rings we stop doing what we are doing to answer. So, the ring has elicited a response in the receiver (us). Once we have picked up the phone, we can just hear a beep sequence if the signal is lost or we can hear the voice of a friend and start a conversation. In both cases, our behaviour has been modified but in a different way: getting back to do what we were doing or engaging in a conversation and maybe going out to have dinner with our friend! Our behaviour can be therefore changed in the long run.

When the traffic officer stops us while we are driving, he or she lifts up the signal disc to catch our attention. Then, the officer can simply invite us to stop aside because an ambulance is passing, using a whistle and a gesture, or approach and verbally inform us of a violation and, eventually, fine us. In both cases our behaviour changes but in the latter case the effect of the signal (visual stopping and verbal notification of the fine) persists in the long run (for example, we have to go pay the fine!).

When a female of *Lemur catta* deposits a scent signal, she may urinate, lift up the tail (in a sort of question mark shape, Figure 2.1) and leave. The visual signal (tail up) can attract a female of another troop towards the urinated spot, and the olfactory signal contained in the urine can provide information on who 'owns' the territory (the first female is clearly claiming the territory as hers!). The second female can choose the pathway to follow, depending on the message perceived. Is the other female dominant? Is the scent deposition overlapping a previous one, from the same or another female, from the same or another group? In terms of communication strategy, the lemur female does not differ much from the traffic officer!

In all of our examples (friend calling, traffic officer, lemur female), the signal is complex because it is made up of different signalling components. However,

Figure 2.1 *Lemur catta* releasing urine drops in the tail-up posture. Bronx Zoo, USA. Photo: Elisabetta Palagi.

the message contained in the first complex signal (friend calling) is conveyed by a single sensory modality: the acoustic modality (telephone ringing and friend's voice). Another example of a complex signal driven by a single sensory modality is given by the multicoloured face of mandrills (*Mandrillus sphinx*). By using reflectance data on the blue and red colours, Renoult *et al.* (2011) found that blue saturation, red saturation and the contrast between blue and red colours are all correlated with dominance, but dominance is most accurately indicated by the blue-red contrast.

In the case of the traffic officer and the lemur female, the message is based on multiple sensory modalities. The message is driven by visual and olfactory channels (tail up/urine) in the case of the lemur female and by visual and acoustic channels (traffic disc, gesture/voice, whistle) in the case of the traffic officer. Therefore, it is possible to distinguish unimodal and multimodal communication, depending on whether single or multiple sensory modalities are recruited. In any message, signals need to be successfully processed through either one or more channels to effectively convey information from senders to receivers (Grafe *et al.*, 2012).

Intuitively, it is clear that the use of multimodal communication (also to direct traffic!) spans the whole primate group, from strepsirrhines to *Homo sapiens.*

Great apes, including chimpanzees, bonobos, gorillas and orangutans, tactically deploy signals using different sensory domains (Figure 2.2). For example, they can communicate through the auditory or tactile modalities, combined or not with the visual modality depending on whether the receiver is attending to them (Tomasello

Figure 2.2 Use of different sensory cues (olfactory, visual and acoustic) to convey information in bonobos. La Vallée de Singes, France. Photo: Elisabetta Palagi.

et al., 1994; Hostetter *et al.*, 2001; Leavens *et al.*, 2004; Parr 2004; Slocombe and Zuberbuhler, 2005; Liebal *et al.*, 2006; Poss *et al.*, 2006; Leavens, 2007; Pollick and de Waal 2007). Chimpanzees make their signals more vigorous in the presence of food, by combining vocalisations and gestures more frequently (Leavens and Hopkins, 2007). Moreover, they can choose the communication channels not only to maximise the probability that the intended receivers obtain the information but also to avoid that the message is intercepted by unintended receivers. In this respect, they can opportunistically adjust down the visual and acoustic 'loudness' of their gestures depending on the attending audience (Hobaiter and Byrne, 2011; Demuru *et al.*, 2015). Gorillas increase the use of multimodal signals, using a sequence of silent, audible and tactile gestures, when playmates are not in strict body contact (Genty and Byrne, 2010).

Recently it has been observed that one of the most widespread monkey facial displays, lipsmacking, serving to communicate affiliation, can be enriched by an acoustic component. Specifically, geladas (*Theropithecus gelada*) make a derived vocalisation (called *wobble*) that they produce while lipsmacking (Bergman, 2013). Similar to in great apes, the multimodal signalling is tactically used in specific situations. Wobbles are, in fact, primarily produced by adult males during affinitive interactions with females (Gustison *et al.*, 2012). Gelada baboons also use multimodal, vocalised yawns for intragroup and intergroup communication (Palagi *et al.*, 2009; Figure 2.3). Before copulation, the subordinate males of howler monkeys can exhibit audiovisual signals at a higher rate compared to the rate of audiovisual signals released by dominants in order to increase mating opportunities (Jones and van Cantfort, 2007).

Does multimodal signalling always match with multimodal communication? Multimodal information can certainly derive from different sensory modalities employed and combined by the sender. However, multimodal information can also be inferred by the receiver based on unimodal signalling. In this case communication

Figure 2.3 Vocalised yawn in an individual of *Theropithecus gelada*. Rheine NatureZoo, Germany. Photo: Achim Johann.

is – eventually – multimodal, despite the signal being unimodal. How is this possible? Different primates can recognise visual clues conveyed by vocal signals, which include components informing the receiver on the anatomical features of the sender. For example, macaques use formants (i.e., vocal tract resonances) as acoustic cues to assess age-related body size differences among conspecifics (Ghazanfar *et al.*, 2007). They do so by cross-linking the body size information embedded in the formant spacing of vocalisations (Fitch, 1997) with the visual size of animals which are likely to produce such vocalisations (Ghazanfar *et al.*, 2007). This process is more effective when the receiver has already accumulated experience on the signal perceived and is able to relate it with other signals belonging to other sensorial classes. This process is also the foundation of individual recognition systems and it is not exclusively related to acoustic messages. As explained in Chapter 1, lemurs are able to figure out the identity of the individual emitting a specific olfactory signal by processing the chemical characteristics contained in that signal (Palagi and Dapporto, 2006a). Picturing the face of a conspecific based only on olfactory stimuli is possible following a previous matching, made by a subject in the past, between visual and olfactory signals, combined together and used in a multimodal way. The way this combination can occur is still under investigation (see Box 1.3 by Kulahci and Ghazanfar).

Multimodal communication is not limited to the primate world. On the contrary it is widespread in the animal kingdom. It clearly provides the strongest advantage in the social context in which effective communication is essential to the survival of the entire group, colony or society. Multimodal communication, via a phenomenon of convergent evolution, has been adopted any time that animals have organised themselves in structured societies. Complex societies heavily relying on complex multimodal communication can be found in invertebrates and vertebrates, spanning bees, canids and humans (de Waal and Tyack, 2003).

Box 2.1 | **by Bridget M. Waller and Katie E. Slocombe**

Speaking of which: lipsmacking as an example of multimodal signalling in monkeys

Our interest in multimodal communication sprung from the realisation that we (the authors) had somewhat different approaches to our research, despite, on the surface, appearing to be rather similar scientists. We published in the same journals, studied similar species, shared colleagues and had similar academic backgrounds. The only obvious difference being that one of us studied primate facial expression and the other primate vocalisation, and this seemed to lead to a different set of assumptions and a different approach to our research. In order to investigate this intriguing observation further, we conducted a systematic review of the primate communication literature (with gestural colleague, Katja Liebal) to determine the extent to which differences exist between research papers, depending on the modality under study (Slocombe *et al.*, 2011). There were two take-home messages from this review.

First, the dominant methods, study setting (wild or captive) and species under study differed significantly between facial expression, vocalisation and gesture papers. Our current understanding of primate communication, therefore, is based on a biased dataset that could cause erroneous conclusions (e.g., gestures are like this, but vocalisations are not). As such cross-modal comparisons are extremely influential in certain fields (e.g., language evolution) these biases could be highly problematic. Second, only 5% of the 553 papers examined primate communication as a multimodal phenomenon.

In short, therefore, our review convinced us that we needed to develop a more integrated multimodal approach to studying communication in order to (1) avoid methodological inconsistencies and (2) better understand primate communication as the holistic process it really is. Ultimately, this then led to us writing a book attempting to integrate the current state of knowledge across communicative modality boundaries and call for a more multimodal approach to primate communication (Liebal et al., 2013).

Truly multimodal research, where different communicative modalities are treated as an integrated whole, is still very much in its infancy. We have a very long tradition of studying primate communication as facial expression, gesture, vocalisation and olfaction separately, and often there are good practical reasons for studying them independently. Therefore, currently it is difficult to make strong conclusions about how and whether multimodality affords greater complexity or functionality of communication. One study we conducted recently (with Jerome Micheletta and colleagues), however, was an attempt to try and determine the communicative value of combining modalities and components during social interaction (Micheletta et al., 2013).

Micheletta and colleagues (Micheletta et al., 2013) examined lipsmacks produced by wild crested macaques (*Macaca nigra*). Lipsmacks are displays used in largely affiliative interactions, and can be composed of a range of visual and auditory components. The lips are smacked together rapidly (sometimes causing a slapping sound), and this lip movement can then be accompanied by visual components (scalp retraction, bared-teeth, head turn) and a vocal component (a soft grunt). The various components had sometimes been commented on in the previous literature, but the function of the composite characteristics had not been explored. We examined whether the composition of the lipsmack (in terms of the number and modality of components) was associated with differential social outcomes. The results showed that multimodal lipsmacks (those containing some or all of the visual components *plus* the soft grunt) were more likely to be followed by affiliative social contact (grooming) between the sender of the lipsmack and another individual. This was not simply an additive effect of extra components increasing the salience of the display, as the total number of visual components did not predict

> **Box 2.1** (*cont.*)
>
> whether social contact would occur. Interestingly, one visual component increased the likelihood of social contact more than others – the head turn. In sum, this paper shows that examining primate communication as dynamic, composite displays can reveal complexity and function that may have been previously overlooked. Importantly, the findings suggest that multimodality can increase the signal value of displays (has additive function: *sensu* Partan and Marler, 1999).
>
> When the importance of multimodal communication was first heralded by Sarah Partan and Peter Marler in their seminal paper (Partan and Marler, 1999), they classified multimodal signals in a way that highlighted how they *could* be used to increase signal complexity. Multimodal signals are classified as either redundant (equivalent to or simply enhancing the original response) or non-redundant (the combined signals modulate each other in some way, or the original responses to the components are combined). Importantly, within non-redundant signals, the authors illustrated that entirely new meanings for signals could emerge when multimodal signals are formed (emergence). Here is where the potential for complexity is at its greatest, and where we feel the most interesting discoveries are yet to be found.
>
> In essence, if a limited range of unimodal signals can be combined into different combinations of multimodal signals with new emergent meanings, the potential number of new signals is considerable. Interestingly, such emergent signals have never been found in animal communication (even in human communication multimodal signals with emergent meaning are poorly documented). It is of course possible that primates do not use multimodal signals in this way, but if they do, this could be very relevant for the evolution of complex communication. There is growing evidence that, unlike in many other species, primate communication can be underpinned by complex cognitive characteristics such as referentiality, flexibility and intentionality (Liebal *et al.*, 2013), which means that emergent multimodal signals could have been a stepping stone to the evolution of complex communication such as language (Waller *et al.*, 2013). Specifically, using emergent signals in an intentional and flexible way would increase the productivity and generativity of a communication system enormously, characteristics which are hallmarks of human language.

2.2 Why multimodal communication is used

Multimodal communication is costly, not only because it can be physiologically and behaviourally demanding but also because it can increase the risk of being detected by a competitor (e.g., for food or for a sex mate) or a predator. When a ring-tailed lemur male anoints its tail using the antebrachial organ and immediately after waves the tail towards a conspecific, the male utilises multimodal communication. This can be risky in the presence of a predator. The signal is easily detectable and costly because maintaining the specialised marking apparatus and using it in combination with a complex visual behavioural repertoire is energetically expensive and time consuming. Given its costs, multimodal communication is expected to provide the animal with some advantages which, under certain circumstances, overcome the disadvantages. If not, the behaviour would have been selected against over the course of evolution.

When, during play, the attention of a dog (*Canis familiaris*) has shifted away from the potential playmate, the other dog first tries to get the partner's attention by barking, touching or moving into the other's visual field. If the partner does not respond to such play solicitation, the inviting dog can continue with attempts to get the partner's attention, often by alternating among different behaviours, such as bumping, biting or pawing, which are also used when the partner is socially engaged with someone else (Horowitz, 2009; Palagi *et al.*, 2015).

In a series of experimental trials (Kulahci *et al.*, 2008), bumblebees (*Bombus impatiens*) were trained to discriminate between different types of artificial flowers. A group of bumblebees was trained to recognise flowers differing in either visual (shape) or olfactory (scent) modalities. Another group was trained to distinguish flowers differing in both visual and olfactory modalities. The bees trained on 'multimodal flowers' learned the rewarding flowers faster than the bees trained on 'unimodal flowers', differing only in shape. Moreover, bees trained on multimodal flowers were more accurate in deciding the speed that could increase the interception of rewarding flowers (Figure 2.4; Kulahci *et al.*, 2008).

Experimental trials performed on lemurs (*Propithecus coquereli*, *Varecia variegata* and *Lemur catta*) demonstrated that these animals rely on different sensory cues to select food, depending on their feeding ecology and food quality (Rushmore *et al.*, 2012). When animals were presented with high-quality food, folivores (*Propithecus coquereli*) required both sensory cues combined to reliably identify their preferred food items, generalists (*Lemur catta*) could identify their preferred food using either cue alone, and frugivores (*Varecia variegata*) could identify their preferred fruits using olfactory, but not visual, cues alone. However, when the lemurs – regardless of their feeding ecology – had to choose between high-quality and low-quality food items, the availability of both visual and olfactory cues for both food types made the difference. As occurs for bumblebees, the integration of different pieces of information provided by the two sensory modalities allows optimisation of dietary choice (Rushmore *et al.*, 2012).

In mandrills, *Mandrillus sphinx*, males are much larger than females, and possess large canine teeth (Setchell *et al.*, 2001; Setchell and Dixson, 2002) and a suite of sexually selected traits, including bright red, blue and violet skin colouration (Setchell and Dixson, 2001a, b; Setchell *et al.*, 2001) and conspicuous vocalisations. This multimodal signalling increases the detection probability in large and fluid groups of mandrills living in the rain forest (Setchell and Kappeler, 2003; Setchell *et al.*, 2009), full of confounding stimuli and sensorial obstacles, and where not all the individuals are always in visual contact.

Figure 2.4 Bumblebees dealing with natural and artificial flowers. Photo: Ipek Kuhlaci.

The use of multimodal communication, as indicated by the biological examples presented above, is mirrored in cultural traditions of humans, who use songs, musical instruments and visual art as 'artificial' cues to increase the impact of the messages the rituals are meant to convey. In many tribal societies from all over the world, ritual ceremonies, such as weddings and initiation and crop fertility rites, are multimedia activities, used to communicate. They usually combine visual stimuli (e.g., body and face paintings and body, hand and feet movements) with sounds (singing, hand clapping, feet tapping, drumming, etc.), sometimes also involving olfactory stimulation (e.g., via drug inhalation or balsam and incense breathing). The combination of these signals is used to increase the effectiveness of the message to be conveyed, by recruiting the attention of the receiver through the concurrent recruitment of different sensory channels, which facilitates an emotional and behavioural response. Indeed, cultural anthropology studies of 'multimodal rituals' typically emphasise their importance for passing from one generation to another information, group tradition and emotional dispositions which the society depends on (Radcliffe-Brown, 1948; Dissanayake, 2006). Analogously, in industrialised societies the clapping and laughing accompanying sitcom jokes or an evocative music played during a documentary or a thriller are used to make the message stronger using the emotional component as an additional vector to impact on the audience.

In all these examples, the multimodal signal is used as an overstimulation to increase the probability that the receiver responds to the stimuli. Indeed, multimodal communication can increase signal detectability, or how easily a signal can be distinguished from its background (Guildford and Dawkins, 1991). Detectability can be increased by reducing the amount of time needed by an observer to respond to a given stimulus (reaction time) and/or by increasing the chance of a signal to be 'discovered' (detection probability) (Markl, 1983; Bradbury and Vehrencamp, 1998; Rowe, 1999; Gosling and Roberts, 2001a). The reaction time can be reduced when different modalities are combined since they produce an inter-sensory facilitation, which, in turn, increases detection speed (Gielen et al., 1983). When Verreaux's sifaka are scared by the presence of a predator or an intruder (e.g., a non-familiar observer), they can emit an acoustic alarm call (the 'shi-fak' vocalisation which their name derives from) accompanied by head-bobbing, a visual pattern. This multimodal 'package' is more vigorous than the simple and single acoustic signal and probably reduces the latency in the response (run away!) of the potential receivers.

Detection probability can be increased via multimodal communication, for example by adding an alerting stimulus (e.g., vigorous movements) to the main 'messenger' stimulus in order to elicit the receiver's selective attention (Bradbury and Vehrencamp, 1998; Kappeler, 1998; Wyatt, 2003). Multimodal signalling can help the sender to override environmental 'signal pollution', provoking additional sensory stimulation and possibly masking the information borne by the communicative signal. For example, if the traffic officer whistles and lifts up the signal disc to catch our attention it is more probable that we detect the 'stopping signal', even when we are listening to the radio in the car or the traffic noise is overwhelming. In the forest, ring-tailed lemurs are very active in depositing scent messages on the ground, trunks and branches within, and sometimes outside, their territories. If a lemur emits a visual signal (raising the tail up) when depositing an olfactory message (via urine drops), in the presence of other conspecifics, it increases the probability that other lemurs (possible sex mates, competitors, etc.) will find and investigate that very message in a forest area 'polluted' by many other odour messages. On the other hand, the lemur normally uses the tail-up urine deposition in open spaces and not in the canopy because branches and leaves are visual obstacles preventing the signal from being spotted. In fact, signal detectability depends not only on the signal design (e.g., unimodal or multimodal, simple or complex, etc.) but also on the conditions of the environment (Grafe et al., 2012).

Box 2.2 | **by James P. Higham**

Speaking of which: the complex scenarios of multimodal communication

Many animals, from birds and mammals to insects and spiders, communicate using multiple signals, often expressed in multiple modalities. Where these are very separated in time, this may simply be the expression of unimodal signals at different periods. However, sometimes these signals are expressed together. This can occur in two main ways. The first of these is in an obligate 'fixed' (Smith, 1977; Partan and Marler, 2005) fashion, such as when a bird or a frog gives a croak that involves a concomitant inflation of its throat

Box 2.2 (continued)

Box 2.2 (*cont.*)

sac (a visual cue; Taylor *et al.*, 2008), or when humans speak words that require concomitant movement of the lips (McGurk and MacDonald, 1976). The second is in a 'free' (Marler, 1961; Partan and Marler, 2005) non-fixed fashion, where multiple signals occur at the same time within a multimodal display, such as when a male bird of paradise gives a colourful dance to a female while making vocalisations.

Many primates, from lemurs to apes, communicate using such multimodal signals or multimodal displays. When stink fights occur between two ring-tailed lemurs, one might consider this a fixed multimodal display. Individuals rub their tails, which are marked with highly contrasting achromatic bands of white and black, on to scent glands, and use vigorous tail motions to waft odours on to each other (Sauther *et al.*, 1999). The scents cannot be delivered without the wafting movements, which constitute a highly arresting visual display, and the visual and olfactory elements of the multimodal signal seem therefore fixed. Many signals that have been traditionally considered unimodal in the primate literature are in fact really part of wider multimodal displays. For example, the evolution of female primate copulation calls in anthropoids has often been addressed as a unimodal signal (see Maestripieri and Roney, 2005 for a review), and rarely considered fully within the context of the numerous female sexual signals available to males during copulation. These can include the presence of sexual swellings (visual), head movements that involve the turning of the head backwards towards males with accompanying facial expressions such as grimaces and eyelid flashes (visual), and the secretion of scents from the vagina (olfactory). In my own research on baboon sexual swellings it became clear that considering the swelling in isolation from these other available signals was inappropriate (Higham *et al.*, 2008, 2009). In particular, the inspection of ano-genital areas appeared potentially crucial, not least because signals such as vaginal fatty acids that are potentially available to males (for rhesus macaques, see Michael and Keverne, 1968) are theoretically likely to be more honest than other signals (Hasson, 1994), because they are so tightly linked to the female's underlying physiology. Our recent work has sought to look in more detail at male inspections of swellings during the oestrus cycle in olive baboons (Rigaill *et al.*, 2013), a topic that has also been addressed in chacma baboons (Clarke *et al.*, 2009). We found that males inspected female ano-genital regions much more around the likely fertile phase (Rigaill *et al.*, 2013). As only males with close access to females (and not all males) can do this, it seems likely to have important effects on the potential information available to different male receivers (Higham *et al.*, 2009). In our recent work on rhesus macaques, we have started to move away from considering only visual signals such as facial colouration (Dubuc *et al.*, 2009; Higham *et al.*, 2010), and towards studies where both visual and vocal signals are assessed (Higham *et al.*, 2013).

In part, this change of emphasis in our own work has been led not only by an increased understanding of how difficult it might be to understand the role of one signal fully in the absence of any analysis of other available signals, but also because multimodal communication seems in many ways to be special (Higham and Hebets, 2013). There are numerous issues that make multimodal communication differ from multiple and multicomponent unimodal signals (Higham and Hebets, 2013; Semple and Higham, 2013). These may include the related efficacy issues of the physical properties of different signal channels, the environmental mediation and potential interference with these properties, and the sensory systems by which con- and heterospecifics detect these signals (Higham and Hebets, 2013; Semple and Higham, 2013). Together, these issues mean that signals in different modalities have different properties of permanence and transmission. A lemur olfactory marking on a tree trunk may not be directed at any one individual, but may persist in the local environment for many days. In contrast, lemur vocal and visual signals are much more directed, but much more transient. Forest cover may affect the transmission of visual signals, and wind and forest noise from sources such as cicadas may interfere with vocal signals, while olfactory signal transmission may be hindered by rain, which is likely to wash the scent mark from a branch or stem. Hence the different modalities are affected differently by different environmental noise types. When a second signal is being added to a first, this seems likely to lead to the second signal evolving in a different, rather than the same, modality. Here, receiver sensory systems are also critical. Visual and vocal signals may be detected from afar by well-developed visual and aural systems, but this is unlikely to be the case with olfactory apparatus. The different detectability of signals given in different modalities give the potential for 'public' and 'private' information to be available to different receivers with different levels of access to the signaller (as in the case of baboon sexual swellings described above). In contrast, multicomponent signals where all signals are detected by the same sensory system offer far less scope for private and public messages to be given simultaneously (Wilson *et al.*, 2013).

A further difference between multimodal and multiple unimodal communication comes in the form of potential cognitive and learning differences, with the former often seeming to be more memorable (e.g., humans, Lovelace *et al.*, 2003). Additionally, there are also natural links between signals and the underlying qualities that they communicate and these may lead to multimodal communication. For example, bird fitness may be signalled by carotenoid-collecting ability in some species, with the carotenoids themselves being used to generate colour signals. Such physiological links between signal information and signal form are common, and when multiple qualities are being communicated that are by their very nature very different, these

Box 2.2 (continued)

Box 2.2 (*cont.*)

links are likely to lead to signals that have very different forms. This may also favour the evolution of multimodal over multiple unimodal communication (Higham and Hebets, 2013; see also Wilson *et al.*, 2013). Further discussion of all issues that may make multimodal communication different to unimodal multiple and multicomponent signals can be found in Higham and Hebets (2013).

One of our most recent and interesting research directions in the area of multimodal communication has been to investigate game theoretic models of multiple and multimodal signals, especially those already developed in economics, and look at how they might be usefully applied to animals (Wilson *et al.*, 2013). In such models it is only possible to model multiple signals, rather than multimodal signals per se. It is up to us as biologists to determine the biological likelihood that such models are likely to lead to multimodal communication specifically (Higham and Hebets, 2013; Wilson *et al.*, 2013). One interesting early conclusion of these models is that, in the absence of further constraints, multisignal models look very much like their unisignal counterparts (Wilson *et al.*, 2013). Without further constraints, there are no fitness or efficiency advantages to be obtained from giving multiple signals. Following this conclusion, we have gone on to investigate a range of constraints that do make multiple signals more likely to evolve. These include: (1) the presence of constraints on the cost functions that are associated with signals, and their physical bandwidth limits; for example, if the number of different available signal types is not sufficient to distinguish all the variation seen in quality types; (2) orthogonal noise structures across modalities (either in sensory processing or in environmental noise, see above in this chapter); (3) the use of signalling modes that have strategic differences; (4) the communication of multiple qualities (see above in this chapter); (5) the presence of multiple signallers; and (6) the presence of multiple receivers. Such circumstances all provide biologically plausible scenarios that theoretically favour multiple signalling generally, and often multimodal signalling specifically. A commentary discussing the Wilson *et al.* (2013) results has been published, and usefully adds additional discussion of which models are more or less likely to lead specifically to multimodal communication, while also developing helpful suggestions for empirical testing (Ruxton and Schaefer, 2013).

Multimodal communication has a great number of challenges and exciting areas ripe for future investigation, and many of these have been outlined in recent publications to which I direct the reader (Higham and Hebets, 2013; Partan, 2013). I finish by re-emphasising a few of these. One key point is that the existing frameworks for the classification of multimodal signals (Partan and Marler, 1999; Partan and Marler, 2005; Smith and Evans, 2013) deal only with bimodal signals. Many multimodal signals and signalling displays are at least trimodal, and some feature even more modalities than this. We currently lack good theoretical and empirical testing frameworks for such complex communicative scenarios (Higham and Hebets, 2013). Such frameworks also tend to focus on signal function and 'information content', and lack an integration of issues related to efficacy and the environment (Higham and Hebets, 2013). We also lack an understanding of how multimodal signalling develops and whether developmental trajectories are common or very different in different species and taxa (Partan, 2013). There is also a need for more work on the sensory integration of signals in different modalities (Higham and Hebets, 2013; Partan, 2013). Another area where there has been little work is the role of multimodal signals in population processes and in creating and maintaining isolation in populations and species (see Higham and Hebets, 2013; Partan, 2013; Uy and Safran, 2013). Moreover, in a world in which anthropogenic environmental change is ubiquitous, we as yet have little idea how multimodal communication affects species' responses to environmental change, and the ability of species to adapt and survive (Partan, 2013). We have a long way to go before we have answers to these and many other questions.

2.3 Beyond olfactory signalling: multimodal communication in lemurs

As should be clear by the examples provided above, evolution did not exempt lemurs from the necessity of communicating multimodally.

Malagasy strepsirrhines use multimodal signals in both reproductive and non-reproductive contexts (Palagi *et al.*, 2005; Lewis and van Schaik, 2007; Drea and Scordato, 2008). A study showed, for example, that *Microcebus murinus* females use multimodal oestrus advertisement by associating different sensory cues. During the oestrus period, mouse lemur females have been found to increase their scent marking activity and emit complex acoustic vocalisations (oestrus trill calls), as a form of multimodal oestrus advertisement eliciting male arousal and male–male competition (Buesching *et al.*, 1998). Vaginal pink swelling (visual cue), associated with an increased scent marking activity during the mating season, has been observed in *Propithecus edwardsi* (Pochron and Wright, 2003) and *Lemur catta* (Jolly, 1966; Sclafani *et al.*, 2012). As extensively reported in Chapter 3 of this book, Lewis and van Schaik (2007) have underlined the importance of multimodal signalling in *Propithecus verreauxi*, in which the additional visual cue provided by the stained chests of the males enhances the information transmitted via the olfactory signal produced by the scent glands (see also Chapter 3). Switching from unimodal (one cue) to multimodal signalling (more integrated cues) may increase the probability of sifaka males to be promptly detected by females.

Lemur catta has furnished relevant data concerning the potential functions and patterns of multimodal communication. Drea and Scordato (2008) pointed out that the motor pattern associated with scent marking in the ring-tailed lemurs can produce composite olfactory (odour deposition) and visual (body/tail posture used for marking) signals that contain a complex combination of ephemeral and long lasting cues and serve differentiated functions.

The famous stink fights, reported by Jolly in 1966 and involving the release of odorants through vigorous tail movements, are the most elucidating example of multimodal communication in strepsirrhines (fixed multimodal pattern, as noted by Highman in Box 2.2). Unlike females, males possess two different organs for specialised scent deposition: an antebrachial organ on the wrist and the brachial organ near the axilla. During stink fights, males apply the odour onto the tail by rubbing it against their wrist spur and release it by arching their tail forward over the body and waving it towards a conspecific. This type of communication is used both in intra- and intersexual competition. During the mating season there is a general increase of the olfactory activity, both in scent detection and release. Females are overall more tolerant of males and allow them to sniff their genital area, which undergoes sexual swelling, thus generating a bimodal signal (as in the case of baboons, described in Box 2.2). This direct body investigation enhances an arousal in males, which increases their propensity to compete with other males. This is when stink fights commence. A male approaches another male while anointing and waving its tail (the typical stink-fight pattern) and the same motor pattern is mirrored by the opponent. Generally, these ritualised fights do not escalate into real aggression, but probably are useful for females to test the ability and resistance of males in order to choose their mate. It is worth noting that the mating period usually overlaps with the driest period in the wild. Hence, maintaining this type of communication via 'fake' combats is energetically demanding, especially when food and water intake is limited (and to produce secretion, water is needed!). The multimodal signal is probably honest, because it is costly and animals do not use 'cheating' signals unless they are also 'cheap' (Wilson *et al.*, 2013). Therefore, a male which is able to keep up a high level of fights can be considered as a male of good quality. Consequently, its offspring are supposed to be of good quality as well. It is well known that the frequency of male scent marking in this species strongly correlates with the ranking position of marking males (Kappeler, 1990), which are granted more copulations because they are allowed to stick around females (Palagi *et al.*, 2003, 2004).

Females have other means to test males. As a form of courtship (or at least they think so) males direct their 'stink' also towards females, which resist this approach, rebel against it, and repel the males by beating and chasing them, often causing dangerous injuries to them (Jolly, 1966; Sclafani *et al.*, 2012). The result is that only the males which are able to go on with their multimodal, stinky approach and physically resist the females' repeated chasing are the most likely fathers of the following season. In this case, multimodal communication can be a key factor for reproductive success.

Multimodal communication in *Lemur catta* is not restricted to the mating context. Through the same motor patterns of the stink fights, lemurs can fine-tune their playful interactions by adjusting the position of some body parts, such as ears and head. This multimodal behaviour, more clearly explained in Chapter 8, acquires the function of a meta-communicative signal in that it informs the playmate that what is going to occur is 'just a game' and not a serious fight.

The use of the tail as a visual signal is combined not only with the olfactory cues provided by specialised gland secretions but also with unspecialised excretions, such as urine.

The study of urine marking in *Lemur catta* provides a good example of the theoretical approach that is needed to demonstrate the multimodal nature of a signal. In fact, the complex signal must be first disassembled into its simple components. Then, such components must be tested separately to verify whether they are effective and 'self-standing' as communicative signals. Finally, the possible functions of the multimodal signal must be assessed, also highlighting the added value and information gained by combining difference sensory cues. To fulfil these investigation requirements, urine marking in ring-tailed lemurs was studied first in captivity (Palagi *et al.*, 2005; Palagi and Dapporto, 2006b) and then in the wild (Palagi and Norscia, 2009). Captivity allowed the adoption of an experimental approach in controlled conditions, which was necessary to clearly separate and test the single components. Indeed, when an olfactory cue is part of a multimodal signal, it is difficult to isolate the information encoded within the chemical matrix of the odour deposition from the information conveyed through other sensory modalities. Assessing the function of odour cues is further confounded when the olfactory behaviour varies with social or ecological context (Scordato and Drea, 2007).

In captivity it was possible to match the experimental approach to behavioural observations on the patterns of use of the signal throughout the year. Last but not least, the privileged observation quality (with reduced observational bias) provided by captive conditions allowed the marking function of urine – based on the fine distinction of the posture adopted by the lemur when urinating – to be discovered. Further investigation was performed in the wild (Palagi and Norscia, 2009), in the presence of different environmental variables and of different groups with overlapping home ranges and at the same time competing to maintain the 'ownership' of their core areas. Thanks to this approach the functions previously hypothesised – but not fully tested – for the multimodal signal in the ring-tailed lemurs were confirmed.

The aftermath of such stepwise and multifaceted investigation is the demonstration that urine marking in *Lemur catta* is a multimodal signal composed of an olfactory cue (urine) and a visual cue (tail up, increasing the detection probability) and that it is used, especially by females, to regulate intergroup relationships. Therefore, the combination of the tail as a visual signal and olfactory cues is not exclusive of males, as in the case of glandular secretions, and it is not necessarily used for intragroup communication only.

Lemur catta can urinate using two different tail positions: tail up (UT-up) and tail down (UT-down). When the first tail configuration (UT-up) is observed, only a few drops of urine are deposited, whereas when the tail is held down (UT-down) urine is released in streams. UT-up was half of the time followed with another olfactory deposition pattern (genital marking). The two marking patterns allows the concurrent use of different chemosignals (urine and glandular secretions) which, melded together, can empower the signal or change the meaning of its message, or both (Figure 2.5).

Figure 2.5 Urine release in the tail-up posture (UT-up) followed by genital marking, in *Lemur catta*. Drawing: Carmelo Gómez González.

This difference led the researchers to hypothesise that UT-up, in which the tail is held so highly visible to others, could have a signalling function and that both urine and the tail could play a role in the communication. This hypothesis was supported by the observation that lemurs sniffed more frequently UT-up than UT-down. Thus, the olfactory investigation by lemurs was elicited by the tail 'flagship' (visual cue) because, when the receivers approached the urinated spot, no olfactory hint had been obtained yet. Data also showed that UT-up underwent seasonal fluctuations, on a monthly basis, which were not observed for UT-down and that the frequency of UT-up was location dependent, being preferentially placed on hotspots, along the boundaries with a neighbouring group.

The demonstration of the multimodal nature of urine marking in *Lemur catta* started with the isolation of the olfactory component for both the deposition modalities (UT-up and UT-down). Avoiding contamination, urine was collected on filter paper and then presented to the individuals of four captive groups. To reduce the bias related to familiarity and seasonality, the individuals were presented with the urine of the same donor belonging to a foreign group, collected in the same month. Animals preferentially sniffed UT-up urine, investigating it more frequently and for a longer time than UT-down urine. This finding indicated that the urine deposited by UT-up and UT-down differs in chemical composition because the visual cue was not involved. The authors explained such difference by suggesting the UT-up posture could favour the mix of urine and genital secretions. Indeed both in captivity and in the wild a stronger temporal association of UT-up with genital marking, of both males and females, was observed.

The receiver preference for investigating UT-up, both in the presence (observational data) and in the absence (scent trials) of visual cues, indicates that UT-up is a complex signal, which is both a multimodal signal (based on two different sensory modalities: visual and olfactory cues) and a multiple signal (composed of two signals able to elicit a behavioural response on their own).

Once the multimodal nature of UT-up was assessed, the authors examined what kind of information was possibly delivered by the signal. To this purpose, a second round of scent trials was set and performed. For both modalities (UT-up and UT-down), the individuals of each group were presented with the urine coming from a donor belonging to the same group as the receiver and from a donor of a foreign group. Urine samples were collected in the same month, from animals of the same sex and age. The recognition experiments clearly showed that individuals could discriminate between urine of their own group and urine from a foreign group, a necessary prerequisite for the use of urine in intergroup communication.

The possible function and benefits of using multimodal signalling were investigated in the wild, due to some limitations to captive research, such as the absence of neighbouring groups, interspecific competition and predation, all of which can affect investigation, detection and deposition patterns of the olfactory behaviour. Moreover, in free-living populations, individuals can largely differ in their responses to chemosignals due to environmental variability, which is particularly high in Madagascar (Wright, 1999).

The role of UT-up in regulating intergroup relationships was confirmed in the wild by verifying who deposited and investigated urine, where and when. Females were more active in both depositing and investigating urine. Moreover, female UT-ups were more investigated than male UT-ups, especially by extra-group females. Immediately after intergroup fights one to three UT-up depositions were always observed and most of them (>70%) was performed in the presence of other groups. Specifically, UT-up depositions were exclusively performed by females and when at least one extra-group individual was in sight, which once again underlines the importance of incorporating a visual cue (tail up) in the signal.

These results fit with the intergroup regulation function of UT-up because, in *Lemur catta*, females are dominant and highly active in territorial defence, chasing and attacking intruders (Sussman and Richard, 1974; Budnitz and Dainis, 1975; Jolly and Pride, 1999; Nunn and Deaner, 2004). Ring-tailed lemur groups are rather permeable to extra-group males (visiting and migrating males: Jolly, 1966; Sauther, 1991) but impassable for alien females, blocked by resident females defending 'their own' resources (Nakamichi and Koyama, 1997).

Depending on the ecological correlates, signallers can deposit scents according to a 'geographical' strategy involving the delineation of the territory perimeter and the demarcation of either food resources or places most frequented by extra-group competitors (Kruuk, 1972; Peters and Mech, 1975; Gosling, 1987; Ono *et al.*, 1988; Roper *et al.*, 1993; Sun *et al.*, 1994).

In the wild (Berenty forest), the frequency of UT-up performed on the locations mapped via global positioning system correlated with the number of sightings of extra-group individuals in such locations but did not correlate with the time spent by intragroup members at the same locations for feeding and resting. However, the fact that the animals performed UT-up in the places most frequented by extra-group competitors does not indicate per se that this strategy of signal deposition is effective. To fill this gap, we verified that the majority of the UT-ups performed were actually sniffed and/or licked at least once (83.21%) within 10 min after deposition, and that animals investigated (sniffed/licked) UT-up at higher rates in the places most frequented by other groups.

Lemur catta economise on UT-up depositions, both maximising the detection probability by placing depositions on the pathways most patrolled and investigated by potential receivers and minimising the reaction time, by depositing urine in the presence of extra-group individuals. This strategy, adopted by different mammals (Brashares and Arcese, 1999; Gosling et al., 2000), can be particularly effective in habitats characterised by high temperatures and humidity (like Berenty), which strongly affect the persistence of chemosignals (Bradbury and Vehrencamp, 1998). This strategy is also consistent with the fact that resources in the study site (the secondary forest of Ankoba at Berenty) are heterogeneous and that lemur home ranges are widely overlapping due to high population density (Jolly et al., 2006; Norscia and Palagi, 2008). Habitat heterogeneity can channel animal movements leading to the tendency for scent marks to be placed along the pathways that are most likely to be frequented by extra-group individuals (Peters and Mech, 1975; Gosling and Roberts, 2001a). Under these ecological conditions, preventing groups from monopolising core areas of exclusive use, the strategy of perimeter delineation via scent marking would be ineffective. An interesting investigation direction would be assessing if and how different ecological conditions affect multimodal signalling (distribution topography and timing), which is the more expensive form of communication. Indeed, in terms of signal economy, the energy and time costs of producing, establishing and renewing multimodal signals, including olfactory communication, introduce constraints to the range of spatial patterns in territorial signalling (Begg et al., 2003). Consequently, the animals need to be parsimonious and make economic 'decisions' by choosing the optimal strategy of signal transmission. The optimal strategy, in economical terms, varies depending on the signal function and the ecological correlates.

References

Begg, C. M., Begg, K. S., Du Toit, J. T. & Mills, M. G. L. (2003). Scent-marking behaviour of the honey badger, *Mellivora capensis* (Mustelidae), in the southern Kalahari. *Animal Behaviour*, 66, 917–929.

Bergman, T. (2013). Speech-like vocalized lip-smacking in geladas. *Current Biology*, 23, R268–R269.

Bradbury, J. K. & Veherencamp, S. L. (1998). *Principles of Animal Communication*. Sinauer Associates, Sunderland.

Brashares, J. S. & Arcese, P. (1999). Scent marking in a territorial African antelope. II. The economics of marking with faeces. *Animal Behaviour*, 57, 11–17.

Budnitz, N. & Dainis, K. (1975). *Lemur catta*: ecology and behavior. In: I. Tattersall & R. W. Sussman (eds), *Lemur Biology*. New York: Plenum, pp. 219–235.

Buesching, C. D., Heistermann, M., Hodges, J. K. & Zimmermann, E. (1998). Multimodal oestrus advertisement in a small nocturnal prosimian, *Microcebus murinus. Folia Primatologica*, 69(suppl 1), 295–308.

Cheney, D. L. & Seyfarth, R. M. (1990). *How Monkeys See the World*. Chicago: University of Chicago Press.

Clarke, P. M. R., Barrett, L. & Henzi, S. P. (2009). What role do olfactory cues play in chacma baboon mating? *American Journal of Primatology*, 71, 1–10.

de Waal, F. B. M. & Tyack, P. L. (2003). *Animal Social Complexity: Intelligence, Culture, and Individualized Societies*. Cambridge, Massachusetts: Harvard University Press.

Demuru, E., Ferrari, P. F. & Palagi, E. (2015). Emotionality and intentionality in bonobo playful communication. *Animal Cognition*, 18(1), 333–344.

Dingle, H. (1969). A statistical and informational analysis of aggressive communication in the mantis shrimp, *Gonodactylus bredini. Animal Behaviour*, 17, 561–575.

Dissanayake, E. (2006). Chapter I. In: S. Brown and U. Voglsten (eds), *Music and Manipulation: on the social uses and social control of music*. Oxford and New York: Berghahn Books, pp. 31–56.

Drea, C. M. & Scordato, E. S. (2008). Olfactory communication in the ringtailed lemur (*Lemur catta*): Form and function of multimodal signals. In: L. Jane, J. L. Hurst, R. J. Beynon, C. S. Roberts & T. D. Wyatt (eds), *Chemical Signals in Vertebrates*, 11. Springer., pp. 91–102.

Dubuc, C., Brent, L. J. N., Accamando, A. K., *et al.* (2009). Sexual skin color contains information about the timing of the fertile phase in free-ranging rhesus macaques. *International Journal of Primatology*, 30, 777–789.

Endler, J. A. (1993). Some general comments on the evolution and design of animal communication systems. *Philosophical Transactions of the Royal Society B: Biological Sciences*, 340, 215–225.

Fitch, W. T. (1997). Vocal tract length and formant frequency dispersion correlate with body size in rhesus macaques. *Journal of the Acoustical Society of America*, 102, 1213–1222.

Freeberg, T. M., Dunbar, R. I. M. & Ord, T. J. (2012). Social complexity as a proximate and ultimate factor in communicative complexity. *Philosophical Transactions of the Royal Society B: Biological Sciences*, 367, 1785–1801.

Genty, E. & Byrne, R. W. (2010). Why do gorillas make sequences of gestures? *Animal Cognition*, 13, 287–301.

Ghazanfar, A. A., Turesson, H. J., Maier, J. X., *et al.* (2007). Vocal tract resonances as indexical cues in rhesus monkeys. *Current Biology*, 17, 425–430.

Gielen, S. C. A. M., Schmidt, R. A. & van der Heuval, P. J. M. (1983). On the intersensory facilitation of reaction time. *Perception and Psychophysics*, 34, 161–168.

Gosling, L. M. (1987). Scent marking in an antelope lek territory. *Animal Behaviour*, 35, 620–622.

Gosling, L. M. & Roberts, S. C. (2001a). Scent-marking by male mammals: cheat-proof signals to competitors and mates. *Advances in the Study of Behaviour*, 30, 169–217.

Gosling, L. M., Roberts, S. C., Thornton, E. A. & Andrew, M. J. (2000). Life history costs of olfactory status signalling in mice. *Behavioral Ecology and Sociobiology*, 48, 328–332.

Grafe, T. U., Preininger, D., Sztatecsny, M., Kasah, R., Dehling, J. M., *et al.* (2012). Multimodal communication in a noisy environment: a case study of the bornean rock frog *Staurois parvus. PLoS ONE*, 7(5), e37965. http://dx.doi.org/10.1371/journal.pone.0037965.

Guildford, T. & Dawkins, M. S. (1991). Receiver psychology and the evolution of animal signals. *Animal Behaviour*, 42, 1–14.

Gustison, M. L., le Roux, A. & Bergman, T. J. (2012). Derived vocalizations of geladas (*Theropithecus gelada*) and the evolution of vocal complexity in primates. *Philosophical Transactions of the Royal Society B*, 367, 1847–1859.

Hasson, O. (1994). Cheating signals. *Journal of Theoretical Biology*, 167, 223–238.

Hebets, E. A. & Papaj, D. R. (2005). Complex signal function: developing a framework of testable hypotheses. *Behavioral Ecology and Sociobiology*, 57, 197–214.

Higham, J. P. & Hebets, E. A. (2013). An introduction to multimodal communication. *Behavioral Ecology and Sociobiology*, 67, 1381–1388.

Higham, J. P., MacLarnon, A. M., Ross, C., Heistermann, M. & Semple, S. (2008). Baboon sexual swellings: information content of size and color. *Hormones and Behaviour*, 53, 452–462.

Higham, J. P., Semple, S., MacLarnon, A., Heistermann, M. & Ross, C. (2009). Female reproductive signals, and male mating behavior, in the olive baboon. *Hormones and Behavior*, 55, 60–67.

Higham, J. P., Brent, L. J. N., Dubuc, C., *et al.* (2010). Color signal information content and the eye of the beholder: a case study in the rhesus macaque. *Behavioral Ecology*, 21, 739–746.

Higham, J. P., Pfefferle, D., Heistermann, M., Maestripieri, D. & Stevens, M. (2013). Signaling in multiple modalities in male rhesus macaques: barks and sex skin coloration in relation to androgen levels, social status and mating behavior. *Behavioral Ecology and Sociobiology*, 67, 1457–1469.

Hobaiter, C. & Byrne, R. W. (2011). The gestural repertoire of the wild chimpanzee. *Animal Cognition*, 14, 745–767.

Horowitz, A. C. (2009). Attention to attention in domestic dog (*Canis familiaris*) dyadic play. *Animal Cognition*, 12, 107–118.

Hostetter, A. B., Cantero, M. & Hopkins, W. D. (2001). Differential use of vocal and gestural communication by chimpanzees (*Pan troglodytes*) in response to the attentional status of a human (*Homo sapiens*). *Journal of Comparative Psychology*, 115, 337–343.

Jolly, A. (1966). *Lemur Behaviour: a Madagascar Field Study*. Chicago: The University of Chicago Press.

Jolly, A. & Pride, E. (1999). Troop histories and range inertia of *Lemur catta* at Berenty, Madagascar: a 33-year perspective. *International Journal of Primatology*, 20, 359–373.

Jolly, A., Koyama, N., Rasamimanana, H., Crowley, H. & Williams, G. (2006). Berenty Reserve: a research site in southern Madagascar. In: A. Jolly, R. W. Sussman, N. Koyama & H. Rasamimanana (eds), *Ringtailed Lemur Biology: Lemur catta in Madagascar*. New York: Springer-Verlag Press, pp. 32–42.

Jones, C. B. & Van Cantfort, T. E. (2007). Multimodal communication by male mantled howler monkeys (*Alouatta palliata*) in sexual contexts: a descriptive analysis. *Folia Primatologica*, 78, 166–185.

Kappeler, P. M. (1990). Social status and scent-marking behaviour in *Lemur catta*. *Animal Behaviour*, 40, 774–776.

Kappeler, P. M. (1998). To whom it may concern: the transmission and function of chemical signals in *Lemur catta*. *Behavioral Ecology and Sociobiology*, 42, 411–421.

Kruuk, H. (1972). *The Spotted Hyena: a study of predation and social behaviour*. Illinois: University of Chicago Press.

Kulahci, I. G., Dornhaus, A. & Papaj, D. R. (2008). Multimodal signals enhance decision making in foraging bumble-bees. *Proceedings of the Royal Society B: Biological Sciences*, 275, 797–802. http://dx.doi.org/10.1098/rspb.2007.1176.

Leavens, D. A. (2007). Animal cognition: multimodal tactics of orangutan communication. *Current Biology*, 17, R762–R764.

Leavens, D. A. & Hopkins, W. D. (2007). Multimodal concomitants of manual gestures by chimpanzees (*Pan troglodytes*). In: K. Liebal, C. Müller and S. Pika (eds), *Gestural Communication in Nonhuman and Human Primates*. Amsterdam, Philadelphia: John Benjamins Publishing Company.

Leavens, D. A., Hostetter, A. B., Wesley, M. J. & Hopkins, W. D. (2004). Tactical use of unimodal and bimodal communication by chimpanzees, *Pan troglodytes*. *Animal Behaviour*. 67, 467–476.

Lewis, R. J. & van Schaik, C. P. (2007). Bimorphism in male Verreaux' sifaka in the Kirindy forest of Madagascar. *International Journal of Primatology*, 28, 159–182.

Liebal, K., Pika, S. & Tomasello, M. (2006). Gestural communication of orangutans (*Pongo pygmaeus*). *Gesture*, 6, 1–38.

Liebal, K., Waller, B. M., Burrows, A. M. & Slocombe, K. E. (2013). *Primate Communicaton: A Multimodal Approach*. Cambridge University Press.

Lovelace, C. T., Stein, B. E. & Wallace, M. T. (2003). An irrelevant light enhances auditory detection in humans: a psychophysical analysis of multisensory integration in stimulus detection. *Cognitive Brain Research*, 17, 447–453.

Maestripieri, D. & Roney, J. R. (2005). Primate copulation calls and post-copulatory female choice. *Behavioral Ecology*, 16, 106–113.

Markl, H. (1983). Vibrational communication. In: R. Huber & H. Markl (eds), *Neurobiology and Behavioral Physiology*. Berlin, Heidelberg, New York: Springer, pp. 332–353.

Marler, P. (1961). The logical analysis of animal communication. *Journal of Theoretical Biology*, 1, 295–317.

McGurk, H. & MacDonald, J. (1976). Hearing lips and seeing voices. *Nature*, 264, 746–748.

Michael, R. P. & Keverne, E. (1968). Pheromones in the communication of sexual status in primates. *Nature*, 218, 746–749.

Micheletta, J., Engelhardt, A., Matthews, L., Agil, M. & Waller, B. M. (2013). Multicomponent and multimodal lipsmacking in crested macaques (*Macaca nigra*). *American Journal of Primatology*, 75(7), 763–773.

Nakamichi, M. & Koyama, M. (1997). Social relationships among ring-tailed lemurs (*Lemur catta*) in two free-ranging troops at Berenty Reserve, Madagascar. *International Journal of Primatology*, 18, 73–93.

Norscia, I. & Palagi, E. (2008). Berenty 2006: census of *Propithecus verreauxi* and possible evidence of population stress. *International Journal of Primatology*, 29, 1099–1115.

Nunn, C. L. & Deaner, R. O. (2004). Patterns of participation and free riding in territorial conflicts among ringtailed lemurs (*Lemur catta*). *Behavioral Ecology and Sociobiology*, 57, 50–61.

Ono, Y., Ikeda, T., Baba, H., *et al.* (1988). Territoriality of Guenther's dikdik in the Omo National Park, Ethiopia. *African Journal of Ecology*, 26, 33–49.

Palagi, E. & Dapporto, L. (2006a). Beyond odour discrimination: demonstrating individual recognition in *Lemur catta*. *Chemical Senses*, 31, 437–443.

Palagi, E. & Dapporto, L. (2006b). Urine marking and urination in Lemur catta: a comparison of design features. *Annales Zoologici Fennici*, 43, 280–284.

Palagi, E. & Norscia, I. (2009). Multimodal signaling in wild Lemur catta: economic design and territorial function of urine marking. *American Journal of Physical Anthropology*, 139, 182–192.

Palagi, E., Telara, S. & Borgognini-Tarli, S. M. (2003). Sniffing behavior in *Lemur catta*: seasonality, sex, and rank. *International Journal of Primatology*, 24, 335–350.

Palagi, E., Telara, S. & Borgognini-Tarli, S. M. (2004). Reproductive strategies in *Lemur catta*: balance among sending, receiving, and counter-marking scent signals. *International Journal of Primatology*, 25, 1019–1031.

Palagi, E., Dapporto, L. & Borgognini-Tarli, S. (2005). The neglected scent: on the marking function of urine in *Lemur catta*. *Behavioral Ecology and Sociobiology*, 58, 437–445.

Palagi, E., Leone, A., Mancini, G. & Ferrari, P. F. (2009). Contagious yawning in gelada baboons as a possible expression of empathy. *PNAS*, 106, 19262–19267.

Palagi, E., Burghardt, G. M., Smuts, B., *et al.* (2015). Rough-and-tumble play as a window on animal communication. *Biological Reviews*.

Parr, L. A. (2004). Perceptual biases for multimodal cues in chimpanzee (*Pan troglodytes*) affect recognition. *Animal Cognition*, 7, 171–178.

Partan, S. (2013). Ten unanswered questions in multimodal communication. *Behavioral Ecology and Sociobiology*, 67, 1523–1539.

Partan, S. & Marler, P. (1999). Communication goes multimodal. *Science*, 283(5406), 1272–1273.

Partan, S. R. & Marler, P. (2005). Issues in the classification of multisensory communication signals. *American Naturalist*, 166, 231–245.

Peters, R. P. & Mech, L. D. (1975). Scent-marking in wolves. *American Scientist*, 63, 628–637.

Pochron, S. T. & Wright, P. C. (2003). Variability in adult group compositions of a prosimian primate. *Behavioral Ecology and Sociobiology*, 54, 285–293.

Pollick, A. S. & de Waal, F. B. M. (2007). Ape gestures and language evolution. *PNAS*, 104, 8184–8189.

Poss, S. R., Kuhar, C., Stoinski, T. S. & Hopkins, W. D. (2006). Differential use of attentional and visual communicative signaling by orangutans (*Pongo pygmaeus*) and gorillas (*Gorilla gorilla*) in response to the attentional status of a human. *American Journal of Primatology*, 68, 978–992.

Radcliffe-Brown, A. R. (1948). *The Andaman Islanders*. Glencoe, IL: The Free Press.

Renoult, J. P. Schaefer, H. M., Sallé, B. & Charpentier, M. J. E. (2011). The evolution of the multicoloured face of mandrills: insights from the perceptual space of colour vision. *PLoS ONE*, 6(12), e29117. http://dx.doi.org/10.1371/journal.pone.0029117.

Rigaill, L., Higham, J. P., Lee, P. C., Blin, A. & Garcia, C. (2013). Multimodal sexual signaling and mating behavior in olive baboons (*Papio anubis*). *American Journal of Primatology*, 75, 774–787.

Roper, T. J., Contradt, L., Butler, J., et al. (1993). Territorial marking with faeces in badgers (*Meles meles*): a comparison of boundary and hinterland latrine use. *Behaviour*, 127, 289–307.

Rowe, C. (1999). Receiver psychology and the evolution of multicomponent signals. *Animal Behaviour*, 58, 921–931.

Rushmore, J., Leonhardt, S. D. & Drea, C. M. (2012). Sight or scent: lemur sensory reliance in detecting food quality varies with feeding ecology. *PLoS ONE*, 7(8), e41558. http://dx.doi.org/10.1371/journal.pone.0041558.

Ruxton, G. D. & Schaefer, H. M. (2013). Game theory, multi-modal signalling and the evolution of communication. *Behavioral Ecology and Sociobiology*, 67, 1417–1423.

Sauther, M. L. (1991). Reproductive behavior of free-ranging *Lemur catta* at Beza Mahafaly Special Reserve, Madagascar. *American Journal of Physical Anthropology*, 84, 463–477.

Sauther, M. L., Sussman, R. W. & Gould, L. (1999). The socioecology of the ringtailed lemur: Thirty-five years of research. *Evolutionary Anthropology*, 8, 120–132.

Savage-Rumbaugh, S., Stuart, S. G. & Talbot, T. J. (1998). *Apes, Language, and the Human Mind*. New York, NY, US: Oxford University Press.

Sclafani, V., Norscia, I., Antonacci, D. & Palagi, E. (2012). Scratching around mating: factors affecting anxiety in wild *Lemur catta*. *Primates*, 53, 247–254.

Scordato, E. S. & Drea, C. M. (2007). Scents and sensibility: information content of olfactory signals in the ringtailed lemur, *Lemur catta*. *Animal Behaviour*, 73, 301–314.

Seeley, T. D. (1997). Honey bee colonies are group-level adaptive units. *The American Naturalist*, 150(supplement), 522–541.

Semple, S. & Higham, J. P. (2013). Primate signals: Current issues and perspectives. *American Journal of Primatology*, 75, 613–620.

Setchell, J. M. & Dixson, A. F. (2001a). Arrested development of secondary sexual adornments in subordinate adult male mandrills (*Mandrillus sphinx*). *American Journal of Physical Anthropology*, 115, 245–252.

Setchell, J. M. & Dixson, A. F. (2001b). Changes in the secondary sexual adornments of male mandrills (*Mandrillus sphinx*) are associated with gain and loss of alpha status. *Hormones and Behaviour*, 39, 177–184.

Setchell, J. M. & Dixson, A. F. (2002). Developmental variables and dominance rank in male mandrills (*Mandrillus sphinx*). *American Journal of Primatology*, 56, 9–25.

Setchell, J. M. & Kappeler, P. M. (2003). Selection in relation to sex in primates. *Advances in the Study of Behavior*, 33, 87–173.

Setchell, J. M., Lee, P. C., Wickings, E. J. & Dixson, A. F. (2001). Growth and ontogeny of sexual size dimorphism in the mandrill (*Mandrillus sphinx*). *American Journal of Physical Anthropology*, 115, 349–360.

Setchell, J. M., Charpentier, M., Abbott, K. A., Wickings, E. J. & Knapp, L. A. (2009). Is brightest best? Testing the Hamilton-Zuk hypothesis in mandrills. *International Journal of Primatology*, 30, 825–844.

Slocombe, K. E. & Zuberbuhler, K. (2005). Agonistic screams in wild chimpanzees (*Pan troglodytes schweinfurthii*) vary as a function of social role. *Journal of Comparative Psychology*, 119, 67–77.

Slocombe, K. E., Waller, B. M. & Liebal, K. (2011). The language void: the need for multimodality in primate communication research. *Animal Behaviour*, 81(5), 919–924.

Smith, C. L. & Evans, C. S. (2013). A new heuristic for capturing the complexity of multimodal signals. *Behavioral Ecology and Sociobiology*, 67, 1389–1398.

Smith, W. J. (1977). *The Behavior of Communicating: an ethological approach*. Cambridge, MA: Harvard University Press.

Sun, L., Xiao, B. & Dai, N. (1994). Scent marking behaviour in the male Chinese water deer. *Acta Theriologica*, 39, 177–184.

Sussman, R. W. & Richard, A. (1974). The role of aggression among diurnal prosimians. In: L. R. Hollowey (ed.), *Primate Aggression, Territoriality, and Xenophobia: a comparative perspective*. New York: Academic Press, pp. 49–76.

Taylor, R. C., Klein, B. A., Stein, J. & Ryan, M. J. (2008). Faux frogs: multi-modal signalling and the value of robotics in animal behavior. *Animal Behaviour*, 76, 1089–1097.

Tomasello, M., Call, J., Nagell, K., Olguin, K. & Carpenter, M. (1994). The learning and use of gestural signals by young chimpanzees: A trans-generational study. *Primates*, 35, 137–154.

Uy, J. A. C. & Safran, R. J. (2013). Variation in the temporal and spatial use of signals and its implications for multimodal communication. *Behavioral Ecology and Sociobiology*, 67, 1499–1511.

Vehrencamp, S. (2000). Handicap, index, and conventional signal elements of bird song. In: Y. Espmark, T. T. Amundsen & G. Rosenqvist (eds), *Animal Signals: Signalling and Signal Design in Animal Communication*. Trondheim: Tapir Academic Press, pp. 277–300.

von Frisch, K. (1967). *The Dance Language and Orientation of Bees*. Cambridge, MA: Harvard University Press.

Waller, B. M., Liebal, K., Burrows, A. M. & Slocombe, K. E. (2013). How can a multimodal approach to primate communication help us understand the evolution of communication? *Evolutionary Psychology*, 11, 538–549.

Wilson, A., Dean, M. & Higham, J. P. (2013). A game theoretic approach to multimodal communication. *Behavioral Ecology and Sociobiology*, 67, 1399–1415.

Wilson, E. O. (1975). *Sociobiology: The New Synthesis*. Cambridge, MA: Belknap Press of Harvard University Press.

Wright, P. C. (1999). Lemur traits and Madagascar ecology: coping with an island environment. *American Journal of Physical Anthropology*, 110(Suppl 29), 31–72.

Wyatt, T. D. (2003). *Pheromones and Animal Behaviour: Communication by Smell and Taste*. Cambridge University Press.

3 A vertical living: sexual selection strategies and upright locomotion

Could our vertical posture have influenced our sexual signals?...
Virtually all the sexual signals...are on the front of the body – the facial expressions, the lips, the beard, the nipples, the areolar signals, the breasts of the female, the pubic hair, the genitals themselves...The frontal approach means that the in-coming sexual signals and rewards are kept tightly linked with the identity signals from the partner.

Desmond Morris, 1967

3.1 Upright habits and their impact on sexual signalling

In a number of animal species, it is possible to observe that males and females strongly differ in specific morphological and/or behavioural traits. This phenomenon, called dimorphism, can be largely observed in primates showing male–female differences in parental investment, growth and developmental pattern, body and canine size, scent glands and scent marking behaviour, vocalisation and visual ornaments (e.g., skin colours, flanges and fur). Such sexually dimorphic features are defined as 'secondary' because they are under the hormonal control of primary sexual characters (reproductive organs). Therefore, secondary traits are not primarily involved in the production of offspring. Not only do they not appear to promote the survival of individuals but sometimes they appear to be detrimental to the animal which possesses them; for example, colourful ornaments (such as the notorious peacock tail) can increase animal visibility and predation probability and, consequently, reduce the chances of survival.

Since secondary sexual traits do not confer advantages or produce disadvantages, their evolutionary permanence is in apparent contrast with the classic natural selection theory proposed by Darwin in the *Origin of Species* (1859). Why have secondary sexual traits been selected, then? To explain this paradox, Darwin (1871) developed the theory of selection in relation to sex (sexual selection) which pivots around the fact that secondary sexual traits confer an advantage with respect of acquiring a mate by intimidating rivals and/or by attracting mates. In this respect, the race to maximise offspring production – promoted by sexual selection – is as important as the race to maximise survival chances – promoted by natural selection. According to a convincing recent theoretical synthesis, sexual selection is part of natural selection (Clutton-Brock, 2004) intended as a process aimed at increasing an animal's chance of surviving and producing the highest number of offspring.

3.1 Upright habits and their impact on sexual signalling

Sexual selection is based on the variability in the expression of secondary sexual features (e.g., colour brightness, tail length, etc.) across individuals of the same sex. Bimorphism comes into place when it is possible to distinguish between two categorical classes of individuals of the same sex according to the presence/absence of a given secondary sexual feature. A typical example of bimorphism provided by textbooks is the case of *Philomachus pugnax*, the ruff, whose male population is characterised by two genetically determined morphs: dark-collared males (dominants) and white-collared males (subordinates). White-collared males are tolerated because they allow maintaining variability (they can sneak copulations every now and then!) and increase the attractiveness of the lek (the reproductive arena where males are concentrated) to females (van Rhijn, 1973). To be successful, a secondary sexual trait needs to be advertised by an animal as a clearly perceivable signal (visual, acoustic or chemical), sent out to potential partners to inform on the sender's quality (Andersson, 1994).

Different evolutionary pathways have led in several primate species to the development of vertical locomotion, which can be more or less occasional (Figure 3.1). It is particularly interesting to focus on the convergent evolution of frontal signals as a necessary adaptation to optimise the communication potential between 'vertical' subjects. We particularly refer to the primates that are obliged to a vertical locomotion by anatomical constraints. Frontal visual signals can be favoured by sexual selection when three conditions are met: (1) a mating system based on strong mate choice by either sex and high levels of intrasexual competition (Kappeler and van Schaik, 2004); (2) a diurnal lifestyle, which makes visual signals detectable; (3) upright locomotion, which makes face and/or chest signals visible.

Diurnal, bipedal vertebrates often rely, for mate choice, on visual sexual signals that are placed frontally to the observer. This situation occurs quite frequently

Figure 3.1 Different primates (a, *Propithecus verreauxi*; b, *Gorilla gorilla*; c, *Pongo pygmaeus*) showing vertical locomotion. Photo: Elisabetta Palagi.

Figure 3.2 Luzon bleeding-heart (*Gallicolumba luzonica*) showing a red patch on its breast. Bronx Zoo, USA. Photo: Elisabetta Palagi.

in birds. Among the visual sexual signals used by bird females to choose mating partners some are worth mentioning, such as peacock (*Pavo cristatus*) tail (Zahavi, 2007), the level of symmetry in chest plumage of male zebra finches (*Taeniopygia guttata*) (Swaddle and Cuthill, 1994), and the size of the black feather bib on the throat of male house sparrows (*Passer domesticus*) (González et al., 2001). In particular, sparrows' melanin-based black throat patch – whose expression is related to the food quality eaten by sparrows (Veiga and Puerta, 1996) – is often referred to as a 'badge of status' and is involved in female mate choice. Females preferentially choose males with large badges because they are more able than small-badged male sparrows to acquire territories of a better quality which include safe nest sites. Moreover, large-badged males appear to be more efficient mate guarders than small-badged males (Møller, 1989). The characteristic feature of Luzon bleeding-hearts (*Gallicolumba luzonica*) is the deep red spot on its breast, which resembles a bleeding wound. This species is monogamous and courtship begins with the male chasing the female on the ground. When the female stops, the male begins his courtship display by inflating his breast to emphasise the red spot (Figure 3.2; Burley, 1981).

In primates only a few species meet the three conditions mentioned above. Orangutans, belonging to the family of pongids, are one of the most sexually dimorphic apes with dimorphism in size, adornments and vocal signals (Fleagle, 2013). Fully mature orangutan males are characterised by an irreversible

Figure 3.3 Flanged (left) and unflanged (right) orangutan males. Drawing: Carmelo Gómez González.

bimorphism in their frontal sexual adornments, which consist in cheek flanges and a throat pouch, a sort of chest 'badge' (Figure 3.3). Males without such secondary sexual features are generally named as 'unflanged' and, under particular social circumstances (e.g., the absence of a flanged male), can acquire in a few months the adornments typical of flanged males (Utami Atmoko and van Hooff, 2004). Also in orangutans, as in birds, the existence of two male morphs has been related to female choice. When females can escape and cannot be accessed by force, they get to choose their mate during the ovulatory phase (Nadler, 1995; Gangestad and Thornhill, 2004). Paternity analyses have revealed that both unflanged and flanged males are equally successful in siring offspring. Moreover, sexually active unflanged males are often unrelated to flanged males, this last result excluding that sexual selection can be at the basis of bimaturism in orangutans. The most endorsed hypothesis is that there are two, alternative, coexisting tactics. While flanged males use their physical attributes to attract females (somehow waiting to be chosen), unflanged males actively scan for oestrus females and try to seduce them into mating. The existence of both strategies – equally successful (Utami *et al.*, 2002) – appears to be density dependent: when orangutan population density is low, the number of flanged males – whose anatomical features make them able to perform long calls to establish long distance contact – increases (Maggioncalda *et al.*, 1999; Utami Atmoko and van Hooff, 2004). This phenomenon may be explained by the fact that actively searching for females and trying to seduce them is more difficult in a 'scarcely populated forest'.

Other than in pongids, obligate vertical locomotion combined with diurnal habits has also evolved in hominids and indriids, located at the opposite extremes of the primate evolutionary bush. All extant indriids are specialised leapers with long

Figure 3.4 Vertical postures in *Propithecus verreauxi* and *Homo sapiens*. Drawing: Carmelo Gómez González.

hindlimbs and a short, dorsally oriented ischium. They are vertical clingers and leapers and travel by leaping between vertical supports. On the ground, they move via bipedal hopping. In humans, the skeleton is characterised by an extremely short and broad pelvic bone, and extremely long femur, tibia and fibula: long lower extremities are associated with upright, bipedal locomotion (Fleagle, 2013).

Besides the different anatomical formula, set to bipedal walking in hominids and to bipedal leaping in indriids, a vertical living always implies face-to-face communication. Here, we consider *Homo sapiens* (a hominid) and *Propithecus verreauxi* (an indriid) as model species to investigate what happens to communication frontal signals when moving on two feet is mandatory (Figures 3.4 and 3.5).

3.2 Sexual competition and frontal visual signals in humans

Frontal secondary sexual signals have been very important in shaping the mating behaviour in *Homo sapiens*. This aspect has been extensively explained in *The Naked Ape* by Desmond Morris (1967) who addressed how sexual communication in humans has been also affected by the acquisition of an obligate upright posture. Although some frontal features such as beards, highbrow hair, enlarged and hairy chests and robust jaws can be used by human females (women!) to select their mates (e.g., women's selection appears to be influenced by the amount of chest hairs in males; Dixson *et al.*, 2010), such features appear to be mostly related to intrasexual competition (Sell *et al.*, 2009). Indeed, all these characteristics strongly correlate with testosterone levels which, in turn, enhance aggression propensity (Guthrie, 1970; Puts, 2010). For example, male faces with beards are rated as the most dominant

3.2 Sexual competition and frontal visual signals in humans

Figure 3.5 Two primate species showing vertical locomotion in the Berenty forest: *Homo sapiens* (on the right) and *Propithecus verreauxi* (on the left).

(Muscarella and Cunningham, 1996; Neave and Shields, 2008) and robust mandibles and greater upper bodies can be associated with fighting abilities (Sell *et al.*, 2009). Although contest competition may have predominated in shaping men's sexual signals, male mate choice has been probably more important in shaping women's (Puts, 2010). Indeed, male contest and dominance were pivotal in early hominid societies, and males most likely were the choosy sex. Indeed, male contest can have increased mating opportunities over human evolution in at least three different ways.

First, coalitional aggression could have facilitated acquisition and defence of mates against males belonging to other groups. A clear example in humans is the capture of women as a primary objective of early warfare (Darwin, 1871; Lerner, 1986; Hrdy, 1997). Coalitionary killing is also present in chimpanzees, as documented by several episodes of group violence described in different study sites from eastern to western Africa (Goodall, 1986; Nishida and Hiraiwa-Hasegawa, 1987). Male coalitionary attacks can be directed at exterminating males of another community to increase the number of females they have access to (Nishida *et al.*, 1985). On the other hand, chimpanzee males can also form coalitions to defend females from males of other communities (Ghiglieri, 1989).

Second, males could use force within a group to acquire mates and exceptionally increase mating opportunities as occurs in one-male societies (harems). In support of this second scenario comes a palaeo-anthropological study by Lockwood and coworkers (2007), who examined facial remains and dental wear stages in young adult and old adult males of *Paranthropus robustus*, detecting a difference in size and robusticity between these two male classes. This result combined with the estimates of sexual dimorphism typical of this hominin species led the authors to hypothesise that the reproductive strategy of single males of *P. robustus* was focused on monopolising groups of females (possibly forming a harem), in a manner similar to that of silverback gorillas (Lockwood *et al.*, 2007).

Third, male contest could have contributed indirectly to mating success if dominant males could acquire more resources than non-dominant ones. *Homo ergaster*, for example, has been assumed to live in groups with a formalised male dominance system possibly based on the mastery of bifacial tools, allowing extremely more efficient hunting and carcass exploitation by males. The ability to make bifaces could influence power structure, being associated with political resource and control (Porr, 2005). Therefore, also in early human species, males have probably exerted mate choice.

In spite of the original male mate choice, the ability of early humans to provide benefits (e.g., resources and protection) to females varied across individuals. Moreover, early humans probably lived in groups with many females and 'best males' could not access all of them, due to the constraints of mate monopolisation potential (cf. Kappeler and van Schaik, 2002). As a result, women needed to compete for the highest-quality men and for the benefits they could provide (Puts, 2010; Rosvall, 2011).

What features have been selected for in females to favour male selection? Frontal, juvenile features appear to be the most attractive because women's maximum reproductive potential is expressed at young age (<25 years old). In particular, gracial facial features, reduced body hair, and body fat distribution appear to be designed to attract mates. No other primate is sexually dimorphic in body fat distribution (Pond and Mattacks, 1987). When approaching sexual maturity, women deposit fat – needed for ovulation, long pregnancy and lactation – on their breast and hips. Such features, advertising fecundity, have also become secondary sexual signals, being used by men for mate selection. Indeed, although with geographic variation, the male's choice appears to be affected by the size and symmetry of women's breasts (Morris, 1967; Møller *et al.*, 1995; Singh and Young, 1995) and by the shape of women who have wider hips (Marlowe *et al.*, 2005).

Box 3.1 | **by David A. Puts**

Speaking of which: vertical locomotion and its impact on the evolution of human sexual behaviour

Traits do not evolve in isolation and can exert reciprocal selection pressures on one another. So not only might vertical locomotion influence social behaviour, but selection pressures related to social behaviour also likely influence locomotor patterns.

In our own lineage, the transition to obligate bipedal locomotion and a terrestrial lifestyle likely had ramifying social consequences. One outcome was a new set of predator threats and potential food sources, which may have selected for more intense sociality as a means of predator defence and/

or cooperative foraging (Sterelny, 2007). The liberation of the forelimbs from locomotion also freed the hands for gesturing, which some have suggested led to spoken language (Gordon et al., 1973), perhaps the most momentous evolutionary change in social behaviour in the history of any species. And of course a vertical posture emphasised different body parts as potential sexual signals (Szalay and Costello, 1991).

At the same time, social selection pressures may have played a role in the evolution of our vertical locomotion. For example, some have suggested that bipedal locomotion evolved from upright displays of body size used as threats (Jablonski and Chaplin, 1993). Given that our closest relatives, the chimpanzees, are proficient tool-users (Goodall, 1968), it is a reasonable inference that our earliest bipedal ancestors were, as well. Thus, it is possible that greater reliance on tool use with the upper limbs favoured increased locomotion with the lower limbs (Darwin, 1871). Various social behaviours likely lay at the root of this transition, including the capacity for social learning. And, again, these changes may have been associated with further changes, such as a reduction in the size of our canine teeth, so that hand-held weapons rather than teeth came to predominate in intrasexual competition (Puts, 2010).

So one can easily see from a consideration of the evolution of vertical locomotion and social behaviour how suites of traits evolve together. It is probably impossible to fully understand the evolution of one without understanding how it is related to the evolution of the other.

My research focuses on the hormonal and evolutionary bases of human sexuality and sex differences, with special focus on behaviour and psychology. One of the most interesting outcomes of my research has been the conclusion that aggression and threats of aggression ('contest competition') have been a more important aspect of men's competition for mates than prior research would suggest (Puts, 2010; Puts et al., 2012; Hill et al., 2013). Most of the past research on sexual selection in men has focused on female mate choice and suggests that men's secondary sex traits were primarily shaped by ancestral women's preferences. However, recent theoretical and empirical work in my lab and other labs has started to challenge this perspective. In general, terrestrial animals tend to engage in higher levels of contest competition (Stirling, 1975; Berry and Shine, 1980; Puts, 2010). We also tend to see substantial male contest competition in our closest relatives, chimpanzees, gorillas and orangutans, which, because of their large size, tend to be more terrestrial than many smaller primates (Leutenegger and Kelly, 1977). So based on cross-species and phylogenetic data, we might expect male contests to have been influential in shaping men's traits.

When we look at men's traits, they indeed appear better designed for winning fights or intimidating other males than for attracting females. For example, in

Box 3.1 (continued)

Box 3.1 (*cont.*)

psychological studies where voice quality or facial hair was manipulated using computer software, deep voices and beards were far more effective at making men appear dominant to other men than they were at making men attractive to women (Puts *et al.*, 2007; Wolff and Puts, 2010; Dixson and Vasey, 2012). And men engage in much higher levels of physical aggression than women do (Archer, 2004; Ellis *et al.*, 2008), and have 60% greater muscle mass and 75% greater upper body muscle mass (Lassek and Gaulin, 2009) – all traits characteristic of species with male contest competition. In species with male contests, males also tend to evolve weapons, such as large claws, horns, antlers, or, in primates, large canine teeth (Andersson, 1994). Humans do not possess any of these traits, and in fact we are the only ape species lacking large canines (Wood *et al.*, 1991). Yet even technologically unsophisticated societies possess weapons capable of dispatching the largest and most powerful animals on the planet. We seem to have replaced large canines with the capacity to manufacture even more deadly weapons, such as clubs, spears, bows and arrows, and the like. In sum, although female choice was no doubt an important selective pressure shaping men's phenotypes over human evolution, male contest competition has been underappreciated and may have been as least as influential.

Another exciting outcome of research in my lab has been the finding that ovarian hormones play an important role in women's mating behaviour and psychology. Females in many non-human primate species go through oestrus, the phase of the ovulatory cycle immediately preceding ovulation that is accompanied by changes in appearance, odour and behaviour. But because women do not exhibit genital swelling, conspicuous behavioural changes or other obvious signs, it was widely believed that women do not undergo oestrus (Alexander and Noonan, 1979; Symons, 1979). However, the work of many researchers around the world, including members of my lab, has shown that women's psychology, behaviour and appearance change in predictable ways across the ovulatory cycle (Gangestad and Thornhill, 2008; Gildersleeve *et al.*, 2014). For example, when evaluating men's attractiveness, women prefer more masculine male faces (Penton-Voak *et al.*, 1999), voices (Puts, 2005; Feinberg *et al.*, 2006) and bodies (Little *et al.*, 2007) during the fertile phase of the cycle than outside the fertile phase. These preference shifts seem to be driven by changes in progesterone, oestrogen and perhaps testosterone (Jones *et al.*, 2005; Puts, 2006; Welling *et al.*, 2007; Roney *et al.*, 2011; Pisanski *et al.*, 2014). Women's appearance also changes over the cycle so that their faces and voices are more attractive when their oestrogen levels are high and their progesterone levels are low – a hormonal state corresponding to peak fertility in the cycle (Puts *et al.*, 2013). Although these cyclic changes are not necessarily large (we sometimes need many participants and carefully controlled studies to see the effects), they are pervasive across many aspects of women's mating psychology, behaviour and appearance.

There are several unresolved questions regarding changes in women's phenotypes over the ovulatory cycle. For example, it will be important to discover which traits change over the cycle, the patterns of these changes, and what hormones are driving them. This information will clarify the phenotypes that we need to explain and help us understand why these changes evolved in the first place. Are they holdovers from more pronounced oestrous changes in ancestral primate species? Do they represent an adaptive suite of traits designed by selection in our own lineage for playing a mixed reproductive strategy, obtaining investment from a long-term mate but recruiting high-quality genes from another male during the fertile part of the cycle?

A related unanswered question regards why cues to ovulation have apparently been suppressed in women. This is one of the most fundamental and long-standing questions on the evolution of human sexuality. Perhaps a lack of conspicuous ovulatory cues functions to facilitate a mixed reproductive strategy by making women harder for men to mate guard during the fertile part of the cycle (Benshoof and Thornhill, 1979; Symons, 1979)? Perhaps suppressing ovulatory cues functions to force males to invest in their mates by making men unable to copulate only during the fertile part of the cycle and then leave without investing (Alexander and Noonan, 1979)? Or perhaps it functions to allow some males to invest in their own offspring by preventing more dominant males from monopolising copulations during the fertile part of the cycle (Strassman, 1981)? Cross-species comparison will be especially helpful in determining how we resemble other primates, which selection pressures we share, and which selection pressures and aspects of the phenotype are derived in our lineage.

More cross-species and cross-cultural work is also needed to explore the relative importance of different mechanisms of sexual selection in humans and in non-human primates. Which ecological and species characteristics are best at explaining cross-species variation in the forms of sexual selection, and what do these variables predict about our own species? Across cultures, especially across traditional societies, are men's mating and reproductive success more strongly tied to their dominance and success in male contests, or to their attractiveness to women? With so many key questions left unanswered, and with so much work left to be done, the future of research on sexual selection in human and non-human primates promises to be an exciting one.

3.3 Frontal visual signals in lemurs: the case of sifaka

Upright locomotion has also evolved in the diurnal, arboreal strepsirrhine species *Propithecus verreauxi* (Verreaux's sifaka), which moves via bipedal hopping (when on the ground) and vertical leaping (Figure 3.6; Jolly, 1966; Brockman, 1999).

Figure 3.6 *Propithecus verreauxi* individuals displacing through vertical leaping (Berenty forest). Photo: Ivan Norscia.

The sifaka society shows an interesting twist with respect to humans': whereas the latter is likely to come from an originally male-dominated society where males were the most choosy sex (Baer and McEachron, 1982; Lee et al., 2011), the sifaka are characterised by exclusive female dominance (all adult females being dominant over males; see Chapter 4) and a slight male–female dimorphism favouring females, in which body mass can be higher. As a result, contrary to humans, in the sifaka sexual coercion by males over females is not possible and females are the choosing sex (Richard, 1992; Lewis, 2002; Norscia et al., 2009). Interestingly, in humans the most evident secondary sexual markers are carried by females whereas, in the sifaka, males are 'in charge' of carrying them, as better detailed later on (Lewis, 2005; Lewis and van Schaik, 2007; Dall'Olio et al., 2012). Hence, in both species, relevant secondary sexual markers, advertising the quality of the carrier, are born by the 'chosen' sex.

In order to better understand what follows, it is worth noting that in the sifaka, females usually experience a single, seasonal oestrus period (2–3 days) per year, which leads to mating synchronisation within sifaka populations (Richard, 1992; Lewis and van Schaik, 2007). Both sexes can mate with multiple partners in their own group and even in the neighbouring groups when the group does not offer enough mating opportunities (Pochron and Wright, 2003). Around a month before females become sexually receptive, males start roaming and visiting other groups in search of mates (Brockman, 1999) and a biological market comes into play (see Chapter 8). Both new incomers and resident males form the 'pool' which the females can draw on to select mates (Norscia et al., 2009).

The sociobiological features of the sifaka – and specifically the short and synchronised oestrus and the female-based dominance and mate choice, leading to the impossibility for males to access females by force – exacerbate male intrasexual competition in this species.

Lewis (2005) reported bimorphism in male sifakas, which can show either a stained or unmarked chest. The bimorphism is based on a frontal visual signal produced by an intense olfactory activity. Males (but not females) possess specialised throat scent glands and perform throat scent marking by rubbing throat and chest up against a substrate, often multiple times within a single marking event (Lewis and van Schaik, 2007). About half of the scent marks by sifaka males are overmarks, in which a scent mark is placed on or near a female scent mark and thus, in cases of intense activity, the staining of the chest is probably a combination of a male's own glandular secretions, female ano-genital secretions, female urine and dirt (Figure 3.7; Lewis, 2005; Lewis and van Schaik, 2007). The frontal visual signal arises from an olfactory investment. In this respect, the scent marking activity not only provides olfactory information but also produces and maintains an additional, visual signal. The recruitment of different sensorial modalities empowers the effectiveness of the signal used by stain-chested males to optimise their reproductive opportunities. It is not surprising that, in the lemur world, a visual cue is paired with specialised olfactory activity. In fact, in strepsirrhines, glandular scent marking has a variety of social functions such as advertisement and territorial defence (*Propithecus verreauxi*, Lewis, 2005; *Propithecus edwardsi*, Pochron et al., 2005; *Lemur catta*, Mertl-Millhollen, 2006), intergroup communication (*Propithecus verreauxi*, Lewis, 2005), advertisement of social dominance (*Lemur catta*, Kappeler, 1990), signalling of reproductive condition (*Lemur catta*, Palagi et al., 2003, 2004), and mate selection (*Nycticebus pygmaeus*, Fisher et al., 2003; *Propithecus verreauxi*, Norscia et al., 2009).

Lewis and van Schaik (2007) described phenotypic variation between sifaka males (stained- versus clean-chested males) as a form of reversible bimorphism in that it can be gained or lost depending on males' fluctuating scent-releasing glandular potential. The authors did not find any clear evidence that the two morphs of males differ in their intrinsic physical characteristics, such as body size and maxillary canine length.

Dall'Olio et al. (2012) investigated whether bimorphism can be linked to the possibility to maximise mating opportunities, as expected when sexual selection applies. The study was conducted in the riverine forest of Berenty, in south Madagascar, where sifaka groups range from one to ten individuals, according to a complete survey conducted in November–December 2006 (Norscia and Palagi, 2008). Dall'Olio et al. (2012) found that stained-chested males had a higher

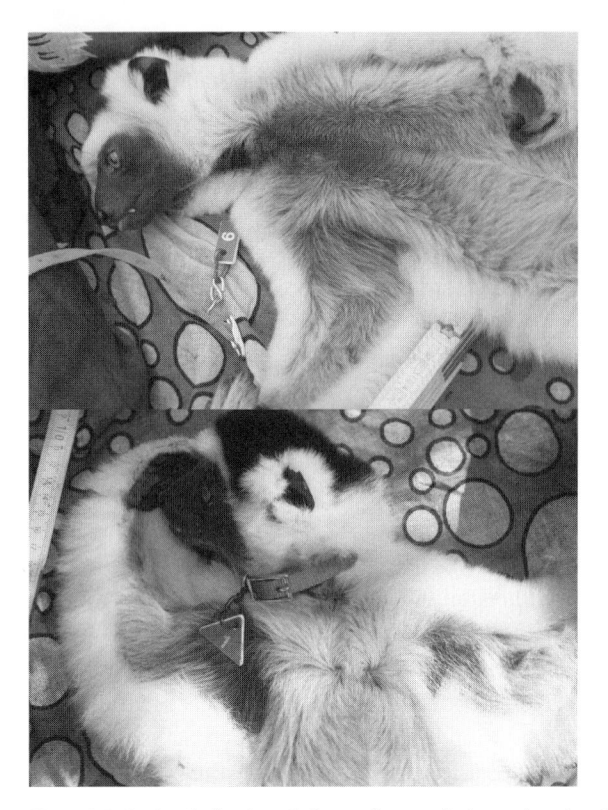

Figure 3.7 Stained- (top) and clean- (bottom) chested males. Kirindy Mitea forest, Madagascar. Photo: Rebecca J. Lewis.

throat- and genital-marking activity than clean-chested males during the mating season but not during the birth season. Moreover, the authors found that females copulated more frequently with stained-chested than with clean-chested males (including both group resident and non-resident). Specifically, in the mating season a scent marking dichotomy between the two different morphs of sifaka males existed (stained-chested males scent marked more frequently than clean-chested ones). This dichotomy disappeared during the birth season, when males were not sexually stimulated and males' intrasexual competition decreased due to the lack of its main driving force, the presence of receptive females (see Chapter 8). It has been observed that one of the proximate causes of the scent marking dichotomy in the mating season is the difference in the concentration of testosterone levels between stained- and clean-chested males, which also differ in their testes mass. This difference vanishes in the birth season, thus indicating that stained- and clean-chested males do not differ in their testosterone levels, as well (Lewis, 2009). This is consistent with the lack of difference in the frequency of scent marking rates between the two morphs of males in the birth season (Dall'Olio et al., 2012).

Additionally, the disappearance of the scent marking dichotomy outside the mating season (with stained-chested males showing a higher throat-marking frequency than clean-chested males during the mating season but not during the birth season) confirms that male chest-badge signal can be linked to male intrasexual competition (ultimate cause), in agreement with previous literature (Lewis, 2005).

The following question is: do females use males' bimorphism to choose mates? Yes, they do. Dall'Olio *et al.* (2012) found that indeed stained-chested males gain a higher number of copulations than clean-chested males, also supporting the hypothesis that stained-chested males are generally dominant in their social groups, thus being 'good quality' males (Lewis and van Schaik, 2007). Moreover, copulations involved both in-group and out-group stained-chested males, thus suggesting that the chest badge can also be functional to females to gather information on the new incomers, which – as specified above – form part of females' mating 'pool'. Chest-badge information can be actually used by females and be successful to males. At Beza Mahafaly, for example, it has been observed that most (29 of 52) sifaka males get to sire at least one offspring outside their resident group (Lawler, 2007).

Of course, the choice pattern by females can result from the information derived from different cues expressed by males which, united, build up a complete and accurate portrait of a male's potential quality. In this respect, the badge could represent the complete information puzzle, which grants females an immediate access to cues that cannot be timely accessed otherwise. In other words, since the badge depends on testosterone, scent marking and dominance, it can provide an 'overview' of a male's physical state. To demonstrate the function of a potential communicative signal, however, an experimental approach is required, but such approach is not feasible with this species and observational data remain, by far, the only available source of information (Figure 3.8).

3.4 Frontal signals and…business cards

In this chapter, we presented three cases of primates possessing vertical locomotion: a strepsirrhine species, *Propithecus verreauxi*, and two apes, *Pongo pygmaeus* and *Homo sapiens*. Orangutans and humans are phylogenically close but occupy completely different ecological niches, with the former being arboreal and the latter being terrestrial. Instead, sifaka and orangutans are phylogenetically distant but are both arboreal.

Humans are characterised by original male dominance, sexual dimorphism (with men generally larger than women), male monomorphism (it is not possible to biologically distinguish different clear-cut classes of males), the most relevant secondary sexual signals carried by females, and males being the original choosing sex. Orangutans show male dominance, strong sexual dimorphism (with males much larger than females), permanent bimorphism of males (with males being either flanged or unflanged) carrying the secondary sexual signal, and females probably being the prominent choosing sex. Sifaka show female dominance, slight sexual dimorphism in favour of females (showing a large body mass under certain conditions), reversible bimorphism of males (which can be clean- or stained-chested) carrying the secondary sexual signal, and females being the choosing sex.

Figure 3.8 A sifaka female (on the left) faces a clean-chested and a stained-chested male. Drawing: Carmelo Gómez González.

Despite the interspecific differences in phylogeny, ecology and sociobiology, the way the reproductive quality is advertised to potential mates is similar in the three primate species. The orientation of the body in space appears to be the main variable influencing secondary sexual signalling, in both visually and olfactory-oriented species (apes versus lemurs). Since genitals are mostly hidden, all vertical 'movers' replicate sexual signals in other body parts. In particular, the prominent secondary sexual signals are located frontally and in the upper part of the body (breast and face). Interestingly, in lemurs the olfactory message released via throat marking is 'translated' into a visual one, the chest badge, possibly because upright locomotion, in a diurnal species, makes the use of visual cues particularly effective in reaching the target (potential, vertically oriented, mates). Sifaka females may assess the quality of the males of their group at a first glance by using the chest badge, which summarises the multifaceted message released by a complex anatomical and behavioural olfactory system (specialised and unspecialised secretions, marking and over-marking frequency, top scent releasing, etc.). Indeed, quality assessment via chest badge is energetically convenient and much less time consuming. Moreover, the information delivered by the chest badge is probably only available to females when they have to assess the quality of unknown extra-group males. In this case, for example, it is not possible for females to check the scent marking activity of each male regularly, over time. In this respect, the chest badge could be seen as a sort of informative business card, which the females promptly read and use to 'hire' the best mates. Interestingly, in the Old World vervet monkeys (*Chlorocebus aethiops*) it has been observed that chest-rubbing is mostly performed by high-ranking males during the breeding season, probably as a form of scent marking (Freeman *et al.*, 2012). However, in this case males do not show any chest badge, which would be not visible and therefore useless given the quadrupedal locomotion typical of the species.

References

Alexander, R. & Noonan, K. M. (1979). Concealment of ovulation, parental care, and human social evolution. In: N. Chagnon and W. Irons (eds), *Evolutionary Biology and Human Social Behavior: An Anthropological Perspective*. North Scituate, MA: Duxbury Press, pp. 436–453.

Andersson, M. (1994). *Sexual Selection*. Princeton University Press, Princeton. 624 p.

Archer, J. (2004). Sex differences in aggression in real-world settings: a meta-analytic review. *Review of General Psychology*, 4, 291–322.

Baer, D. & McEachron, D. L. (1982). A review of selected sociobiological principles: application to hominid evolution: I. The development of group social structure. *Journal of Social and Biological Structures*, 5, 69–90.

Benshoof, L. & Thornhill, R. (1979). The evolution of monogamy and concealed ovulation in humans. *Journal of Social and Biological Structures*, 2, 95–106.

Berry, J. F. & Shine, R. (1980). Sexual size dimorphism and sexual selection in turtles (order Testudines). *Oecologia (Berlin)*, 44, 185–191.

Brockman, D. K. (1999). Reproductive behavior of female *Propithecus verreauxi* at Beza Mahafaly, Madagascar. *International Journal of Primatology*, 20, 375–398.

Burley, N. (1981). Mate choice by multiple criteria in a monogamous species. *American Naturalist*, 117, 515–528.

Clutton-Brock, T. (2004). What is sexual selection? In: P. Kappeler and C. van Schaik (eds), *Sexual Selection in Primates*. pp. 24–36.

Dall'Olio, S., Norscia, I., Antonacci, D. & Palagi, E. (2012). Sexual signalling in *Propithecus verreauxi*: male 'chest badge' and female mate choice. *PLoS ONE*, 7(5), e37332. http://dx.doi.org/10.1371/journal.pone.0037332.

Darwin, C. (1859). *On the Origin of Species by Means of Natural Selection*. London: John Murray.

Darwin, C. (1871). *The Descent of Man and Selection in Relation to Sex*. London: John Murray.

Dixson, B. J. & Vasey, P. L. (2012). Beards augment perceptions of men's age, social status, and aggressiveness, but not attractiveness. *Behavioral Ecology*, 23, 481–490.

Dixson, B. J., Dixson, A. F., Bishop, P. & Parish, A. (2010). Human physique and sexual attractiveness in men and women: a New Zealand–U.S. comparative study. *Archives of Sexual Behaviour*, 39, 798–806.

Ellis, L., Hershberger, S., Field, E., et al. (2008). *Sex Differences: Summarizing More than a Century of Scientific Research*. New York: Taylor and Francis.

Feinberg, D. R., Jones, B. C., Law Smith, M. J., et al. (2006). Menstrual cycle, trait estrogen level, and masculinity preferences in the human voice. *Hormones and Behavior*, 49, 215–222.

Fisher, H. S., Swaisgood, R. R. & Fitch-Snyder, H. (2003). Countermarking by male pygmy lorises (*Nycticebus pygmaeus*): do females use odour cues to select mates with high competitive ability? *Behavioral Ecology and Sociobiology*, 53, 123–130.

Fleagle, J. G. (2013). *Primate Adaptation and Evolution*, 3rd ed. San Diego: Academic Press. 596 p.

Freeman, N. J., Pasternak, G. M., Rubi, T. L., Barrett, L. & Henzi, S. P. (2012). Evidence for scent marking in vervet monkeys? *Primates*, 53, 311–315.

Gangestad, S. W. & Thornhill, R. (2004). Females multiple mating and genetic benefits in humans: investigations of design. In: P. Kappeler & K. van Schaik (eds), *Sexual Selection in Primates*. pp. 90–113.

Gangestad, S. W. & Thornhill, R. (2008). Human oestrus. *Proceedings of the Royal Society B: Biological Sciences*, 275, 991–1000.

Gildersleeve, K., Haselton, M. G. & Fales, M. R. (2014). Do women's mate preferences change across the ovulatory cycle? A meta-analytic review. *Psychological Bulletin*, 140, 1205–1259.

Ghiglieri, M. P. (1989). Hominid socio-biology and hominid social evolution. In: P. G. Heltne & L. A. Marquardt (eds), *Understanding Chimpanzees*. Cambridge, MA: Harvard University Press, pp. 370–379.

González, G., Sorci, G., Smith, L. C. & de Lope, F. (2001). Testosterone and sexual signalling in male house sparrows (*Passer domesticus*). *Behavioral Ecology and Sociobiology*, 50, 557–562.

Goodall, J. (1968). The behavior of free-living chimpanzees in the Gombe Stream Reserve. *Animal Behavior Monographs*, 1, 165–311.

Goodall, J. (1986). *The Chimpanzees of Gombe: Patterns of Behavior*. Cambridge, MA: Harvard University Press.

Gordon, W. H., Andrew, R. J., Carini, L., et al. (1973). Primate communication and the gestural origin of language [and comments and reply]. *Current Anthropology*, 14, 5–24.

Guthrie, R. D. (1970). Evolution of human threat display organs. In: T. Dobzansky, M. K. Hecht & W. C. Steers (eds), *Evolutionary Biology*, New York: Appleton-Century-Crofts, pp. 257–302.

Hill, A. K., Hunt, J., Welling, L. L. M., et al. (2013). Quantifying the strength and form of sexual selection on men's traits. *Evolution and Human Behavior*, 34, 334–341.

Hrdy, S. B. (1997). Raising Darwin's consciousness: female sexuality and the prehominid origins of patriarchy. *Human Nature*, 8, 1–49.

Jablonski, N. G. & Chaplin, G. (1993). Origin of habitual terrestrial bipedalism in the ancestor of the Hominidae. *Journal of Human Evolution*, 24, 259–280.

Jolly, A. (1966). *Lemur Behavior: A Madagascar Field Study*. Chicago: University of Chicago Press.

Jones, B. C., Perrett, D. I., Little, A. C., et al. (2005). Menstrual cycle, pregnancy and oral contraceptive use alter attraction to apparent health in faces. *Proceedings of the Royal Society B: Biological Sciences*, 272, 347–354.

Kappeler, P. M. (1990). Social status and scent-marking behaviour in *Lemur catta*. *Animal Behaviour*, 40, 774–775.

Kappeler, P. M. & van Schaik, C. P. (2002). Evolution of primate social system. *International Journal of Primatology*, 23, 707–740.

Kappeler, P. M. & van Schaik, C. P. (2004). *Sexual Selection in Primates*. Cambridge University Press, 284 p.

Lassek, W. D. & Gaulin, S. J. C. (2009). Costs and benefits of fat-free muscle mass in men: relationship to mating success, dietary requirements, and natural immunity. *Evolution and Human Behavior*, 30, 322–328.

Lawler, R. R. (2007). Fitness and extra-group reproduction in male Verreaux's sifaka: an analysis of reproductive success from 1989–1999. *American Journal of Physical Anthropology*, 132, 267–277.

Lee, I. C., Pratto, F., & Johnson, B. T. (2011). Intergroup consensus/disagreement in support of group-based hierarchy: An examination of socio-structural and psycho-cultural factors. *Psychological Bulletin*, 137, 1029–1064.

Lerner, G. (1986). *The Creation of Patriarchy*. Oxford: Oxford University Press.

Leutenegger, W. & Kelly, J. T. (1977). Relationship of sexual dimorphism in canine size and body size to social, behavioral, and ecological correlates in anthropoid primates. *Primates*, 18, 117–136.

Lewis, R. J. (2002). Beyond dominance: the importance of leverage. *Quarterly Review of Biology*, 77, 149–164.

Lewis, R. J. (2005). Sex differences in scent-marking in sifaka: mating conflict or male services? *American Journal of Physical Anthropology*, 128, 389–398.

Lewis, R. J. (2009). Chest staining variation as a signal of testosterone levels in male Verreaux's sifaka. *Physiology and Behavior*, 96, 586–592.

Lewis, R. J. & Kappeler, P. M. (2005). Seasonality, body condition, and timing of reproduction in Propithecus verreauxi verreauxi in the Kirindy Forest. *American Journal of Primatology*, 67, 347–364.

Lewis, R. J. & van Schaik, C. P. (2007). Bimorphism in male Verreaux' sifaka in the Kirindy forest of Madagascar. *International Journal of Primatology*, 28, 159–182.

Little, A. C., Jones, B. C. & Burriss, R. P. (2007). Preferences for masculinity in male bodies change across the menstrual cycle. *Hormones and Behavior*, 51, 633–639.

Lockwood, C. A., Menter, C. G., Moggi-Cecchi, J. & Keyser, A. W. (2007). Extended male growth in a fossil hominin specis. *Science*, 318, 1443–1446.

Maggioncalda, A. N., Sapolsky, R. M. & Czekala, N. M. (1999). Reproductive hormone profiles in captive male orangutans: implications for understanding developmental arrest. *American Journal of Physical Anthropology*, 109, 19–32.

Marlowe, F., Apicella, C. & Reed, D. (2005). Men's preference for women's profile waist-to-hip ratio in two societies. *Evolution and Human Behavior*, 26, 458–468.

Mertl-Millhollen, A. S. (2006). Scent marking as resource defense by female *Lemur catta*. *American Journal of Primatology*, 68, 605–621.

Møller, A. P. (1989). Natural and sexual selection on a plumage signal of status and on morphology in house sparrows, *Passer domesticus*. *Journal of Evolutionary Biology*, 2, 125–140.

Møller, A. P., Soler, M. & Thornhill, R. (1995). Breast asymmetry, sexual selection, and human reproductive success. *Ethology and Sociobiology*, 16, 207–219.

Morris, M. (1967). *The Naked Ape*. London: Cape.

Muscarella, F. & Cunningham, M. R. (1996). The evolutionary significance and social perception of male pattern baldness and facial hair. *Ethology and Sociobiology*, 17, 99–117.

Nadler, R. D. (1995). Sexual behavior of orangutans (*Pongo pygmaeus*). In: R. D. Nadler, B. M. F. Galdikas, L. K. Sheeran and N. Rosen (eds), *The Neglected Ape*. New York, NY: Plenum Press, pp. 2223–237.

Neave, N. & Shields, K. (2008). The effects of facial hair manipulation on female perceptions of attractiveness, masculinity, and dominance in male faces. *Personality and Individual Differences*, 45, 373–377.

Nishida, T. & Hiraiwa-Hasegawa, M. (1987). Chimpanzees and bonobos: Cooperative relationships among males. In: B. B. Smuts, D. L. Cheney, R. M. Seyfarth, R. W. Wrangham & T. T. Struhsaker (eds), *Primate Societies*. Chicago: University of Chicago Press, pp. 165–177.

Nishida, T. Hiraiwa-Hasegawa, M., Hasegawa, T. & Takahata, Y. (1985). Group extinction and female transfer in wild chimpanzees in the Mahale National Park, Tanzania. *Zeitschrift für Tierpsychologie*, 67, 284–301.

Norscia, I. & Palagi, E. (2008). Berenty 2006: census of *Propithecus verreauxi* and possible evidence of population stress. *International Journal of Primatology*, 29, 1099–1115.

Norscia, I., Antonacci, D. & Palagi, E. (2009). Mating first, mating more: biological market fluctuation in a wild prosimian. *PLoS ONE*, 4(3), e4679. http://dx.doi.org/10.1371/journal.pone.0004679.

Palagi, E., Telara, S. & Borgognini-Tarli, S. (2003). Sniffing behaviour in *Lemur catta*: seasonality, sex, and rank. *International Journal of Primatology*, 24, 335–350.

Palagi, E., Telara, S. & Borgognini-Tarli, S. (2004). Reproductive strategies in *Lemur catta*: the balance among sending, receiving, and countermarking scent signals. *International Journal of Primatology*, 25, 1019–1031.

Penton-Voak, I. S., Perrett, D. I., Castles, D. L., et al. (1999). Menstrual cycle alters face preference. *Nature*, 399, 741–742.

Pisanski, K., Hahn, A. C., Fisher, C. I., et al. (2014). Changes in salivary estradiol predict changes in women's preferences for vocal masculinity. *Hormones and Behavior*, 66, 493–497.

Pochron, S. T. & Wright, P. C. (2003). Variability in adult group compositions of a prosimian primate. *Behavioral Ecology and Sociobiology*, 54, 285–293.

Pochron, S. T., Morelli, T. L., Terranova, P. et al. (2005). Patterns of male scent marking in *Propithecus edwardsi* of Ranomafana National Park, Madagascar. *American Journal of Primatology*, 65, 103–115.

Pond, C. M. & Mattacks, C. A. (1987). The anatomy of adipose tissue in captive *Macaca* monkeys and its implications for human biology. *Folia Primatologica*, 48, 164–185.

Porr, M. (2005). The making of the biface and the making of the individual. In: C. S. Gamble & M. Porr (eds), *The Hominid Individual in Context: Archaeological investigations of Lower and Middle Palaeolithic landscapes, locales and artefacts*. London/New York: Routledge, pp. 68–80.

Puts, D. A. (2005). Mating context and menstrual phase affect women's preferences for male voice pitch. *Evolution and Human Behavior*, 26, 388–397.

Puts, D. A. (2006). Cyclic variation in women's preferences for masculine traits: potential hormonal causes. *Human Nature*, 17, 114–127.

Puts, D. A. (2010). Beauty and the beast: Mechanisms of sexual selection in humans. *Evolution and Human Behavior*, 31, 157–175.

Puts, D. A., Hodges, C. Cárdenas, R. A. & Gaulin, S. J. C. (2007). Men's voices as dominance signals: vocal fundamental and formant frequencies influence dominance attributions among men. *Evolution and Human Behavior*, 28, 340–344.

Puts, D. A., Jones, B. C. & DeBruine, L. M. (2012). Sexual selection on human faces and voices. *Journal of Sex Research*, 49, 227–243.

Puts, D. A., Bailey, D. H., Cárdenas, R. A., et al. (2013). Women's attractiveness changes with estradiol and progesterone across the ovulatory cycle. *Hormones and Behavior*, 63, 13–19.

Richard, A. F. (1992). Aggressive competition between males, female-controlled polygyny and sexual monomorphism in a Malagasy primate, *Propithecus verreauxi*. *Journal of Human Evolution*, 22, 395–406.

Roney, J. R., Simmons, Z. L. & Gray, P. B. (2011). Changes in estradiol predict within-women shifts in attraction to facial cues of men's testosterone. *Psychoneuroendocrinology*, 36, 742–749.

Rosvall, K. A. (2011). Intrasexual competition in females: evidence for sexual selection? *Behavioral Ecology*, 22, 1131–1140.

Sell, A., Cosmides, L., Tooby, J., et al. (2009). Human adaptations for the visual assessment of strength and fighting ability from the body and face. *Proceedings of the Royal Society B: Biological Sciences*, 276, 575–584.

Singh, D. & Young, R. K. (1995). Body weight, waist-to-hip ratio, breasts, and hips: role in judgments of female attractiveness and desirability for relationships. *Ethology and Sociobiology*, 16, 483–507.

Sterelny, K. (2007). Social intelligence, human intelligence and niche construction. *Philosophical Transactions of the Royal Society B: Biological Sciences*, 362, 719–730.

Stirling, I. (1975). Factors affecting the evolution of social behavior in the Pinnipedia. *Rapports et Procès-Verbaux des Rèunions du Conseil Permanent International pour l'Exploration de la Mer*, 169, 205–212.

Strassman, B. I. (1981). Sexual selection, paternal care, and concealed ovulation in humans. *Ethology and Sociobiology*, 2, 31–40.

Swaddle, J. P. & Cuthill, I. C. (1994). Female zebra finches prefer males with symmetric chest plumage. *Proceedings of the Royal Society B: Biological Sciences*, 258, 267–271.

Symons, D. (1979). *The Evolution of Human Sexuality*. New York: Oxford University Press.

Szalay, F. S. & Costello, R. K. (1991). Evolution of permanent estrus displays in hominids. *Journal of Human Evolution*, 20, 439–464.

Utami Atmoko, S. & van Hooff, J. A. R. A. M. (2004). Alternative male reproductive strategies: male bimaturism in orangutans. In: P. M. Kappeler & C. P. van Schaik (eds), *Sexual Selection in Primates: New and Comparative Perspectives*. Cambridge University Press, pp. 196–207.

Utami, S. S., Goossens B, Bruford, M. W., de Ruiter, J. R. & van Hooff, J. A. R. A. M. (2002). Male bimaturism and reproductive success in Sumatran orang-utans. *Behavioral Ecology*, 13, 643–652.

van Rhijn, J. G. (1973). Behavioural dimorphism in male ruffs, *Philomacus pugnax* (L.). *Behaviour*, 47, 153–229.

Veiga, J. P. & Puerta, P. (1996). Nutritional constraints determine the expression of a sexual trait in the house sparrow, *Passer domesticus*. *Proceedings of the Royal Society of London: Biological Sciences*, 263, 229–234.

Welling, L. L., Jones, B. C., DeBruine, L. M., *et al.* (2007). Raised salivary testosterone in women is associated with increased attraction to masculine faces. *Hormones and Behavior*, 52, 156–161.

Wolff, S. E. & Puts, D. A. (2010). Vocal masculinity is a robust dominance signal in men. *Behavioral Ecology and Sociobiology*, 64, 1673–1683.

Wood, B. A., Li, Y. & Willoughby, C. (1991). Intraspecific variation and sexual dimorphism in cranial and dental variables among higher primates and their bearing on the hominid fossil record. *Journal of Anatomy*, 174, 185–205.

Zahavi, A. (2007). Sexual selection, signal selection and the handicap principle. In: B. G. M. Jamieson (ed.), *Reproductive Biology and Phylogeny of Birds*. Enfield, New Hampshire: Science Publishers, pp. 143–160.

Part II

How conflicts shape societies

4 Bossing around the forest: power asymmetry and hierarchy

Power must be understood in the first instance as the multiplicity of force relations immanent in the sphere in which they operate and which constitute their own organization: as the process which, through ceaseless struggle and confrontations, transforms, strengthens, or even reverses them; as the support which these force relations find in one another, thus forming a chain or a system, or on the contrary, the disjunctions and contradictions which isolate them from one another; and lastly, as the strategies in which they take effect...Power...is the name that one attributes to a complex strategic situation in a particular society.

Michel Foucault, *History of Sexuality*, 1976

4.1 I've got the power!

A society is not the simple sum of single individuals but an emergent property of their multiple and intertwined relationships, which are able to generate complex relational networks. Dominance relationships within social groups can be based upon the ability of using force, individually or by forming coalitions (intrinsic and derived dominance, respectively; Hand, 1986) or upon leverage (when involving inalienable resources that cannot be taken by force; Lewis, 2002), as extensively explained in Chapter 8.

Over the centuries, power has been one of the main subjects of discussion for philosophers from Niccolò Machiavelli in the sixteenth century (e.g., *The Prince*) to Michel Foucault in the twentieth century (e.g., *History of Sexuality and The Subject and Power*), men of letters such as Vittorio Alfieri (e.g., *Della Tirannide,* nineteenth century), contemporary linguists such as Noam Chomsky (e.g., *Understanding Power: The Indispensable*, 2002) and scientists such as Frans de Waal (e.g., *Chimpanzee Politics: Power and Sex Among Apes,* 1982), to cite a few by heart. Social hierarchies, deriving from the different distribution of power among individuals, drive the behaviour in many species, including humans. History is full of examples of the impact that extreme power imbalance can have on societies. The holocaust, the abduction of Sabine women, and the ancient Egyptian pyramidal hierarchy – from farmers and labourers to the pharaoh – are just some of them. In today's society, a corporation is structured as an organisational chart with the Chief Executive Officer (CEO) at the root of the hierarchy.

In humans, social position can affect parental investment, and mental and physical health, for example contributing to hypertension, depression and the increase of stress levels (Adler *et al.*, 2000, 2008). Interestingly, there is evidence that the human brain is 'shaped' to detect and process hierarchy. Zink *et al.* (2008), using functional magnetic resonance imaging (fMRI) in an interactive, simulated social context, observed that in both stable and unstable social hierarchies, viewing a superior individual differentially engaged perceptual-attentional saliency, and cognitive systems, and particularly the dorsolateral prefrontal cortex. In an unstable hierarchy setting, additional regions related to emotional processing (amygdala), social cognition (medial prefrontal cortex) and behavioural readiness were recruited. Moreover, it has been shown that individuals lower in social status are more likely to engage neural circuitry often involved in 'mentalising' or thinking about others' thoughts and feelings (Muscatell *et al.*, 2012). In short, social status affects the degree of emotional involvement of individuals and their interest in what others may feel or think.

The studies on non-human primates clearly indicate that the connection between hierarchy and the 'social brain' has a long evolutionary history. Like humans, non-human primates are affected by social status, in that rank can dramatically influence global health, social stress level and reproductive potential (Preuschoft and van Schaik, 2000; Sapolsky *et al.*, 2000; Sapolsky, 2005). In a group perspective, dominance rank scaffolds the quality of animal relationships and permeates all behavioural spheres, including aggression, affiliation, parental care and sexual activity (Clutton-Brock *et al.*, 1984; Ogola Onyango *et al.*, 2008; Palagi *et al.*, 2008a; Norscia *et al.*, 2009). From an ecological perspective, the structure of dominance relationships can influence reproductive success (Pusey *et al.* 1997; von Holst *et al.*, 2002), resource access (Clutton-Brock, 1982; Krebs and Davies, 1987), territory quality (Fox *et al.*, 1981), predation risk (Hall and Fedigan, 1997) and energy budgets (Isbell and Young, 1993; Koenig, 2000).

According to how power is distributed among individuals, animal societies have been traditionally classified from despotic to egalitarian – or democratic to totalitarian/authoritarian in humans – mainly depending on the level of benefit asymmetry associated with rank and on how such asymmetry is expressed (Vehrencamp, 1983; Hemelrijk, 1999; Flack and de Waal, 2004; Sueur *et al.*, 2011).

Dominance relationships were first studied in 1922 when Schjelderup-Ebbe observed that in every pair of hens there was one (the dominant) that could peck the other (the subordinate) but could not be pecked in return. As a result, in any group of hens individuals could be ranked so that those of high rank dominated most of their companions. Not only were dominance relationships not random but sometimes individuals could be ranked according to a 'perfect order', defined as 'linear hierarchy'.

In the 1930s and 1940s an intense investigation on the subject was conducted by different authors (e.g., see Allee, 1931; Guhl *et al.*, 1945) and was especially focused on understanding what variables, such as physical and context- and/or hormonal-related factors, can underpin the formation of animal (and specifically hens') linear hierarchies. Overall, it was clear that although physical features may explain dominance to a certain extent, with heavier animals and animals with larger combs winning many conflicts, success and physical characteristics were not always strictly correlated. For this reason, Landau (1951) concluded that social factors – and not only individual inherent characteristics – must somehow act to promote hierarchy formation. Landau developed an index to evaluate the 'strength' of a hierarchy, based on the number of individuals dominated by each member of the group and on the group size.

Drews (1993) provided the first structural definition of social dominance, pointing out that it emerges and stabilises from repeated dyadic agonistic interactions in which winners and losers are consistently found.

Although dominance can be based on the confrontation of individuals in agonistic interactions (Bernstein, 1981; Drews, 1993), other correlates have proved useful to assess dominance relationships (Lewis, 2002). Back in the 1970s, researchers studied hierarchical relationships in human groups, mostly considering preschool children (McGrew, 1972; Missakian, 1976; Strayer and Strayer, 1976), but also adolescent groups (Savin-Williams, 1977, 1979, 1980). Besides non-verbal behavioural measures of dominance (e.g., aggression), some investigations have also adopted verbal indices of dominance. For example, if A gives an order to B and B obeys, then A is rated as dominant to B. Such studies suggested that human hierarchies are frequently linear or near-linear (Missakian, 1976; Savin-Williams, 1977, 1979, 1980). Later on, other indicators, such as occupational status, income, education, and the ability to acquire and control objects have been used to assess social dominance, delivering more complex results (Hawley, 1999; Singh-Manoux *et al.*, 2005).

In chimpanzees, approach frequencies or typical vocalisations such as pant grunts can inform on dominance relationships (chimpanzee: Murray, 2007; Wroblewski *et al.*, 2009). In New and Old World monkeys different behaviours have served the same purpose, such as formalised signals (in rhesus macaques; de Waal and Luttrell, 1985), the direction of approach-retreats (in chacma baboons, *Papio ursinus*: Cheney, 1977; Kitchen *et al.*, 2005; tufted capuchin, *Cebus apella*: Parr *et al.*, 1997) and genital display and inspection (in squirrel monkeys: Alvarez, 1975). Clearly, due to the differences characterising the biology of each species, no consensus has been reached – and it is unlikely that consensus will be reached – on the measure of dominance relationships in animal societies.

4.2 Hierarchy: easy to say, not easy to define

Cambridge Dictionaries Online defines hierarchy as 'a system in which people or things are arranged according to their importance'. Yet, ordering items can be anything but easy.

Based on human social groups, Maiya and Berger-Wolf (2011) defined hierarchy as a rooted, directed tree represented by a graph, $G_H = (V_H; E_H)$. An edge (v; w) $\in E_H$ denotes that v is dominant over w with v being referred to as the parent and w being referred to as the child. Two nodes with the same parent are referred to as siblings. If a path exists from some node v to some node w, then v is an ancestor of w and w is a descendant of v (Figure 4.1).

Starting from hierarchy, considered as a generative model for social networks, the authors drew several interaction models (direct model, distance model,

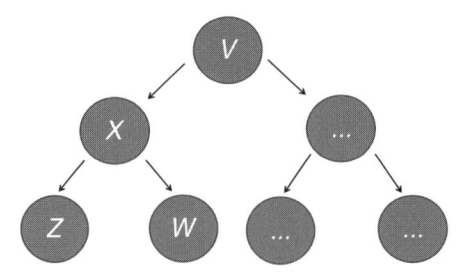

Figure 4.1 Graph showing an example of hierarchical relationships; v is dominant over x with v being the parent and x the child. The two nodes, z and w, with the same parent are siblings. v is defined as the ancestor of w and w is a descendant of v.

team-led and manager-led models, etc.) to derive networks among people, with a non-linear order, based on the assumption that the probability of interaction decays as the distance between individuals within a hierarchy grows. Other authors have proposed different methods to deduct hierarchy relationships, such as graph-theoretic centrality measures (Wasserman *et al.*, 1994) or Bayesian inference (Kemp and Tenenbaum, 2008). Going into the details of each one of them would be beyond the scope and the domain of this chapter but it is worth pointing out that the debate is open on how to implement correct procedures to mathematically determine hierarchies using models based on interindividual interactions or their representation.

Animal societies can show a nested or multilevel hierarchy composed of an ensemble of sets ordered by inclusion and with a unique maximal set. For example, if sets are A, B, C, then A is a subset of B and A is a subset of C, and either B is included in C or C is included in B. More commonly, societies show a one-level hierarchy, only involving a single and separated set of individuals. For example, geladas (*Theropithecus gelada*), hamadryas baboons (*Papio hamadryas*), black-and-white snub-nosed monkeys (*Rhinopithecus bieti*) and golden snub-nosed monkeys (*Rhinopithecus roxellana*) live in nested fission–fusion societies normally consisting, in increasing order of size, of one-male groups as basal social entities (modules), which join together to constitute a second level of social organisation, the band. More bands can form higher grouping levels (e.g., herds/troops and/or communities) (Kawai *et al.*, 1983; Grüter and Zinner, 2004).

Various methods have been elaborated to rank individuals and determine the dominance hierarchy. As explained by Hemelrijk *et al.* (2005), individuals can be ranked by the total frequency of their attacks or victories (AttFr), their average individual dominance index (ADI), the David's score (DS, explained below), the number of individuals against whom they win more often than lose (Netto) and by minimising the number and strength of hierarchical inconsistencies, following the so-called I&SI method (de Vries and Appleby, 2000).

Within social groups, hierarchy can be either linear (A>B>C>D) – as found in some of Schjelderup-Ebbe's groups of hens – or non-linear (e.g., triangular: A>B and B>C but C>A, pyramidal: A>[B=C=D], or class system based: [A+B]>[C=D+E+F]). Such

A	B	C	D	E	F	G	H	I	L	
0	1	2	6	5	5	6	2	12	8	A
0	0	3	10	5	14	16	6	12	6	B
1	1	0	4	2	7	31	6	24	32	C
0	3	0	0	7	13	6	1	5	5	D
0	0	0	0	0	4	0	4	15	3	E
0	0	0	0	0	0	6	10	17	4	F
0	0	2	1	0	0	0	6	13	16	G
0	0	0	0	0	0	3	0	10	10	H
0	0	0	0	0	0	12	11	0	6	I
0	0	0	0	0	0	0	5	4	0	L

Figure 4.2 An example of a sociomatrix. The individuals of a group are reported in lines and columns. Lines report the number of conflicts won by each individual and columns the number of losses.

features derive from the relational properties of networks of dyads rather than from properties of individuals or single dyads (Preuschoft and van Schaik, 2000).

To quantitatively measure the hierarchy of a social group, sociomatrices are often used. They report the number of conflicts between group members, with the winners listed in rows and losers in columns. For each pair of individuals (dyad) the number of aggressive events that subject i has won over subject j does not correspond – if not by chance – to the number of aggressive events that subject j has won over subject i. Therefore, the matrix is not symmetrical. This matrix is different from the so-called 'dominance matrix', which can be prepared once the hierarchy has been assessed. A dominance matrix has either a '0' or a '1' in all off-diagonal cells: a '1' in the i, j cell indicates that individual i dominates individual j and a '0' indicates that j dominates i (Chase, 1980).

In the sociomatrix (Figure 4.2), it is important to introduce only the 'decided' interactions; that is, conflicts with a clear winner and loser, typically indicated by the fact that the loser eventually give up and leaves the competition arena (e.g., by fleeing).

The construction of a sociomatrix is a starting point to derive a linear hierarchy from a set of binary dominance relationships. The linearity of a hierarchy depends on the number of established relationships and on the degree to which they are transitive (Landau, 1951; Kendall, 1962; Appleby, 1983; de Vries, 1995). A transitive triad, for example, is a set of three individuals that are all connected to each other, in which the asymmetrical relationships are transitive (if A>B and B>C, then A>C) (Shizuka and McDonald, 2012). The coefficient of linearity, expressed by the corrected Landau's index (h'; Landau, 1951), has been used over time to determine the structure of dominance relationships in social groups and make comparisons (Palagi *et al.*, 2008b; Paoli and Palagi, 2008; Hewitt *et al.*, 2009). If the hierarchy is linear, the matrix can then be reorganised to fit animal rank order (de Vries *et al.*, 1993) and obtain a dominance matrix if needed. However, the linearity coefficient has been found not sufficient to describe how much a society is despotic because hierarchies sharing similar levels of linearity (h') can differ in the extent of power asymmetry between individuals (Flack and de Waal, 2004).

For this reason another property of dominance hierarchy has been recognised, and the concept of steepness has been introduced (de Vries *et al.*, 2006). Like the slope of a mountain or a hill, hierarchy can be more or less steep, depending on the entity of power difference between individuals. If we picture a linear hierarchy as a staircase, its steepness will depend on the height of each of the steps that separate one level from another and which can have, in our case, different heights. In operational terms, steepness derives from the size of the absolute differences between adjacently ranked individuals in their overall success in winning dominance encounters. When these differences are large the hierarchy is steep; when they are small the hierarchy is shallow. Whereas linearity is based on the binary dyadic dominance relationships (BDR), steepness requires a cardinal rank measure (Flack and de Waal, 2004; de Vries *et al.*, 2006). The steepness measure elaborated by de Vries *et al.* (2006) uses the David's score to rank individuals. This score, different from other scoring systems (e.g., Clutton-Brock's index) is based on the unweighted and weighted sum of the individual's dyadic proportions of wins combined with the unweighted and weighted sum of its dyadic proportions of losses (Gammel *et al.*, 2003). The advantage of the David's score is that the overall success of an individual is determined by weighting each dyadic success measure by the unweighted estimate of the interactant's overall success, so that relative strengths of the other individuals are taken into account. Thus, defeating a high-ranking animal is weighted heavier than defeating a low-ranking one (de Vries *et al.*, 2006).

As pointed out by de Vries *et al.* (2006), the comparison of the hierarchical structure of different groups using the steepness values has a limitation related to the observational zeros and can lead to incorrect interpretations if the observational effort differs between different groups. This situation may, in fact, lead to different proportions of zero dyads in the study groups due to observational bias and not to a real lack of interaction between individuals. As has been shown by Klass and Cords (2011) using both simulated and empirical data from four wild monkey groups, the steepness measure is negatively influenced by the proportion of zero dyads in the matrix and the rank order becomes inconsistent at 26–38% unknown relationships. Specifically, the steepness is lower when matrices contain more zero dyads. If the zero dyads accurately reflect the absence of clear dominance–subordination relationships among individuals, interpreting this lower steepness as an indication of less despotic hierarchy is correct. On the contrary, when these zero dyads are (at least partly) due to less observational effort, this interpretation is questionable. Therefore, the comparison of dominance matrices via steepness requires similar and adequate observational effort (de Vries *et al.*, 2006).

Starting from the analyses of 101 dominance matrices published in 55 studies (spanning invertebrates to mammals), Shizuka and McDonald (2012, 2014) presented a new measure for determining the level of hierarchy transitivity, not sensitive to observational zeros. This measure, called triangle transitivity (t_{tri}), is based on the transitivity of dominance relations among sets of three individuals that all interact with each other (A dominates over B, B over C, and therefore A dominates over C). Triangle transitivity and linearity are essentially equivalent when dominance relations of all dyads are known but – as discussed above – such a condition is not always met in observational studies. As the authors pointed out, larger groups are normally characterised by a higher number of null dyads so linearity in large groups may decrease as a side effect of these dyads (Shizuka and McDonald, 2012). It should be also considered, however, that the probability that individuals interact with each and all the other individuals of the group actually decreases as the group size increases and, consequently, in very large agglomerates not all individuals necessarily interact (so null dyads depends on the actual absence of interactions and not on missing data).

The method by Shizuka and McDonald (2012) is based on the analyses of social networks, with a non-linear order, used for humans and follows a logic similar to that of de Vries (1995). However, the procedure is conducted without filling in null dyads (missing data) with randomised dominance relations. In fact, filling in null dyads artificially creates cyclic triads (and not transitive dyads), a particular form of triad in which directional relations form a cycle, e.g., A dominates B, B dominates C, and C dominates A (A>B>C>A). Cycle triads lead to systematic underestimates of the degree of transitivity of relationships, thus decreasing the level of linearity. In conclusion, the authors state that triangle transitivity has two major advantages: it does not require unobserved relations (null dyads) to be randomly filled in, and its expected value is constant across group sizes. Since the triangle transitivity is an index referring to the whole group, it is suitable for inter-group comparisons but not for calculating inter-individual dominance differences within groups. Clearly, each hierarchy measure has advantages and drawbacks and, similar to the discussion over hierarchy determination in humans reported above, the debate is open on how to reliably assess animal hierarchies using, comparing and improving different existing ranking criteria, such as the I&SI method and the David's score (Balasubramaniam *et al.*, 2013; de Vries *et al.*, 2006), Elo-rating (Neumann *et al.*, 2011) and the Adams' Bayesian approach (Adams, 2005), and adopting new ones, such as the triangle transitivity (Shizuka and McDonald, 2012).

Box 4.1 | **by Takeshi Furuichi**

Speaking of which: female social status of wild bonobos

When I was an undergraduate student, I was interested in the 'snow monkeys' (Japanese macaques, *Macaca fuscata*) living in the Shimokita Peninsula, Japan, which is in the most northern part of the range of non-human primates (Furuichi *et al.*, 1982). Every year I would spend a few months there to follow the monkeys in the forest, mainly during the winter when it was easier to track their footprints in the snow. I was most interested in the behaviour of the alpha male because its status was rather peripheral when ranging in natural forest. This situation contrasted with the studies carried out on artificially

Box 4.1 (continued)

Box 4.1 (*cont.*)

provisioned groups (Chance, 1967) which described the alpha male as the 'central' individual receiving special attention by group members. In the wild, group members did not pay much attention to the alpha male and he sometimes followed the group in the rear part. My observations made me wonder on what the alpha status meant in the natural condition.

Even though the life history and the social status of female Japanese macaques had been intensively studied by that time, not much was known about the life history of males because it was very difficult to follow specific individuals moving from group to group throughout their life. Therefore, I began to study the social behaviour and life history of males in wild groups on Yakushima Island as part of my PhD investigation. It was extremely difficult, as I had expected. Although I could reveal, within a group, some tendencies of changes in social behavioural patterns of males at different ages (Furuichi 1984, 1985), I had no valid means to follow males once they left the group. I captured more than 30 males of different groups using an anaesthetic blow pipe and released them after tattooing, but they never settled down in adjacent groups after emigrating from those groups, meaning that they tended to move very far beyond the adjacent groups.

When I shifted my target to wild bonobos at Wamba in the Democratic Republic of the Congo, I became interested again in the life history of the migrating (and therefore most difficult) sex: the dispersing females in male-philopatric groups. Despite having obtained many interesting results by comparing the social and sexual behaviour of females of different age living in the same group (Furuichi, 1989), after more than 30 years of investigation there are still a lot of unknown issues: why do females leave their natal group whereas males remain with their mother (Kano, 1992; Hashimoto *et al.*, 1996; Gerloff *et al.*, 1999), why do females start visiting other groups or ranging alone when they are very young (the youngest female recorded started temporarily ranging apart from the natal group when 4 years old, and left the group when 5 years old, in an early juvenile stage!), where do they travel and what do they do in the period in between their permanent exit from their natal group at 5 to 10 years old and the moment they settle down in a new group at 9 to 14 years old... (Hashimoto *et al.* 2008, Sakamaki *et al.*, in press).

Even though further research is needed to find the answers to these questions, during my investigation I found that the social life of females has got two faces: one is the affiliative face representing the peaceful nature of bonobo social relations; the other is the competitive face of mothers supporting their male offspring (Furuichi, 2011).

During my investigation I observed that all the females born in the study group left before adulthood, which means that all the adult females of the group had immigrated from other groups and were unrelated. However, unlike

Figure 4.3 Aggregation of bonobo females (*Pan paniscus*) in the central part of the party. Wamba, Congo. Photo: Takeshi Furuichi.

chimpanzee females, bonobo females form very close associations with one another (White, 1988; Furuichi 1989, 2009). They tend to aggregate in the central part of the mixed-sex party while ranging, and form a large aggregation during grooming sessions (Figure 4.3). Females tend to avoid males when they perform agonistic displays but sometimes females counterattack males using the support of other females (Parish, 1994; Furuichi, 1997; Stevens *et al.*, 2007; Surbeck *et al.*, 2010). In the feeding context, females tend to be dominant over males: females more often occupy the best positions at the centre of the fruiting trees, and males frequently leave those positions to females when females arrive. However, when considering – in a period of stable dominance – all the antagonistic interactions including male–female replacement, the total wins and losses were almost equal between sexes.

In female bonobos, sexual swelling (allegedly signalling oestrus) is prolonged into the non-conceivable nursing period and during pregnancy. This prolonged swelling largely lowers the proportion of males competing for oestrous females, making the agonistic interactions among males much less frequent and severe compared to chimpanzees. Probably, the most valuable thing for the male's reproductive success is being preferred or accepted by females as mating partners rather than being able to win agonistic interactions with other males, which may contribute to the high social status of females. Males seldom attack females physically, though females sometimes do so against

Box 4.1 (continued)

Box 4.1 (*cont.*)

males, and there have been no observations of infanticide by males (Furuichi and Hashimoto, 2002; Paoli *et al.*, 2006; Furuichi, 2011). During intergroup encounters, the males of a group frequently show agonistic display against the other group whereas females join the opposite group, feed together on the same tree, exchange grooming with extra-group females, or copulate with the males of the other group. Thus intergroup encounters in bonobos are more peaceful than in chimpanzees (Idani, 1990; Kano, 1992). Thus the prolonged period of the alleged oestrus and the ecological conditions allowing the aggregation of females in large mixed-sex parties are the key factors contributing to the peaceful nature of this species.

Contrary to such contributions to their peaceful nature, female bonobos sometimes show agonistic behaviours as competitive mothers. In the study groups of bonobos, young adult or even adolescent males with high-ranking mothers tended to acquire high-ranking status among adult males. While following E1 group at Wamba I observed that an adolescent male, Ten, started to challenge the alpha male, Ibo, who won most of the time. Ten's mother, Sen, the second-ranking female and Ibo's mother, Kame, the alpha female frequently intervened to support their sons during the agonistic interactions. Once Sen eventually had defeated Kame in a dyadic confrontation, Ten was able to gain the alpha male position. Interestingly, males unrelated to alpha females occupied the alpha position only when the alpha females did not have adolescent or adult sons. The current alpha male, Nobita, is son of the alpha female, Kiku (Figure 4.4). Unlike in chimpanzees, we observed that many of the prime adult bonobo males were in low-ranking positions, possibly because they did not receive support from females after the death of their mothers due to old age (Furuichi, 1997, 2011; Furuichi *et al.*, 2012).

When Ten started challenging other males in 1983, his mother Sen started showing very aggressive behaviours towards others. An adult female, Halu, with an adolescent male, Haluo, in 1987 began to show aggressive behaviours more frequently than before but her aggressiveness decreased in 1989 when she lost Haluo at the age of 11 years. In 2013, a late adolescent male, Jiro, was repeatedly challenging the alpha male, Nobita, and was defeated most of the time. However, when the alpha female Kiku, mother of Nobita, became weakened due to injuries caused by the attacks of other females and probably due to pregnancy in old age, the mother of Jiro, Jacky, became very aggressive towards other males in support of Jiro. Even after Kiku had recovered her power, Jacky continued to frequently show aggressive behaviours. During a field study in 2014, I observed very frequent agonistic interactions in this group, and females such as Kiku and Jacky were playing crucial roles in such interactions. If I had carried out a study of the type and frequency of agonistic interactions during this period, I might have concluded that females

Figure 4.4 Individuals of *Pan paniscus*. The old adult alpha female Kiku (centre) with her alpha male adult son Nobita (right) and newborn baby. Wamba, Congo. Photo: Takeshi Furuichi.

were dominant over males in bonobos (as some previous studies reported; Parish, 1994), unlike my previous conclusion of equal status between males and females (Furuichi, 1997).

Although based on three observed cases only, my investigation suggests that females can become very aggressive to favour their sons when they start challenging the dominant male to gain the alpha position. In the natural condition where dominance rank does not seem to influence the amount of food obtained, females may not be able to increase the number of their offspring even if they acquire a high rank among females (Furuichi, 1997). However, females may still be able to increase the number of grandoffspring by helping their sons to access high-ranking positions. This hypothesis of female competition for the number of grandoffspring was also supported, in another study site (LuiKotale), by the investigation carried out by Martin Surbeck about the mothers' support to their sons during intermale competition over oestrous females (Surbeck *et al.*, 2010).

Female bonobos leave their natal group and visit several groups before settling down. They show different faces according to their status, and age and sex of their offspring. The females who entered my study group in 1983 are

Box 4.1 (continued)

> Box 4.1 (*cont.*)
>
> still alive and live in the same group after 30 years. Therefore, we may need observations by several generations of researchers before we come to feel that we know the life history of female bonobos fairly well.

4.3 Dominance profile in lemurs: beyond linear hierarchy

During our studies in Madagascar we had the opportunity to investigate different lemur groups, of different species. All of them were characterised by absolute female dominance and linear hierarchy (Palagi *et al.*, 2008b; Norscia and Palagi, 2011; Sclafani *et al.*, 2012). Are these two features sufficient to draw the comprehensive dominance profile of a social group? To answer this question we came up with the idea of making the concurrent use of different analytical approaches to evaluate dominance relationships and determine the level of tolerance of different social groups (Norscia and Palagi, 2015). We observed for more than 1200 hours, in the period a round mating, several sympatric groups of the Berenty forest (Madagascar), and in particular two groups of *Lemur catta* (ring-tailed lemurs; Figure 4.5), two groups of *Propithecus verreauxi* (Verreaux's sifaka; Figure 4.6), and a single group of introduced *Eulemur fulvus rufus* × *collaris* (brown lemurs; Figure 4.7). They were an excellent study model as they shared the same habitat and an apparently similar hierarchical structure.

Figure 4.5 A group of *Lemur catta* in the Berenty forest, Madagascar. Photo: Ivan Norscia.

4.3 Dominance profile in lemurs: beyond linear hierarchy

Figure 4.6 A group of *Propithecus verreauxi* in the Berenty forest, Madagascar. Photo: Elisabetta Palagi.

Figure 4.7 A group of *Eulemur rufus* × *collaris* in the Berenty forest, Madagascar. Photo: Ivan Norscia.

As a first step, based on the aggressive events between dyads, we calculated the Landau's corrected index, which did not help in differentiating the three species, because a *P. verreauxi* group showed a linearity coefficient higher than a *L. catta* group and three groups of different species showed comparable linearity coefficient values. If we ended our investigation here, we would conclude that the study groups all had a similar level of despotism.

As a second step we used spontaneous and unsolicited submissive acts (avoidance) to verify whether the three study species differed in the degree of formalisation of dominance relationships. Indeed, as reported in the first section, hierarchies can be, and have been, measured according to different parameters, other than the direction of aggression (Flack and de Waal, 2004; Kitchen *et al.*, 2005). The two *L. catta* groups stand out because they maintained linearity whereas other groups did not. This outcome, deriving from a higher frequency of unidirectional dyadic avoidance patterns in *L. catta* groups, can indicate a greater acceptance of the inferior social rank to dominants by subordinates (deference), a greater intolerance by dominants to subordinates, or both. We defined hierarchy as aggression-based if it is exclusively unveiled by overt aggression and submission-based if its detection does not necessarily depend on an arena of aggressive encounters. According to this definition, linear hierarchy is both aggression- and submission-based in *L. catta* groups and aggression-based in *P. verreauxi* and *E. rufus* × *collaris* groups (Norscia and Palagi, 2015). While avoidance allows comparison because it is present in all species (it is not a species-specific pattern), for *P. verreauxi* it would be also interesting to perform an analysis based on submissive chatters because they are used as signals of subordination in this species (Lewis and van Schaik, 2007).

Of the two troops of ring-tailed lemurs maintaining linear hierarchy with both aggression and avoidance, one showed a behaviour-dependent hierarchy. In fact, the ranking order obtained by fitting avoidance sociomatrices did not match fully with that generated by aggression sociomatrices. Even though the same leader and exclusive female dominance were maintained, many individuals possessed a different position in the submission- and aggression-based hierarchies. Such findings indicated that the power asymmetry perceived by individuals does not necessarily go in tandem with the asymmetry appearing from aggressive interactions (actual asymmetry), even in social groups that are characterised by crystallised dominance relationships, such as ring-tailed lemur groups. We believe that this twofold approach, which considers both submissive and aggressive interactions, can actually reveal divergences between perceived and actual power asymmetry in species that are classically labelled as despotic (e.g., baboons, Rowell, 1967; mandrills, Wickings and Dixson, 1992).

By applying different methods, we also observed that the top-ranking female remained the same only in the two ring-tailed lemur groups. The matching between the ranking order obtained via binary dyadic dominance relationships and that obtained by normalised David's scores was almost total for *L. catta* groups, intermediate for *P. verreauxi* and weaker for *E. rufus* × *collaris*. A lower level of matching corresponded with a higher number of subjects occupying a different ranking position in the hierarchies obtained by the two different methods. It is worth remarking that the differences in the hierarchical arrangements are not context- or behaviour-dependent (e.g., feeding priority, grooming direction, outcome of aggressive encounters) as we observed in the case of *Lemur catta* (aggression- versus submissive-based hierarchy) and of other species (e.g., Alvarez, 1975; de Waal and Luttrell, 1985). In this case, different analytical methods were applied to the same exact dataset (aggression sociomatrices).

With a similar approach, Balasubramaniam *et al.* (2012) applied different analytical procedures (David's score, binary dyadic dominance relationships, and Bayesian methods) to six datasets for female macaques (three despotic and three tolerant groups) and obtained different ranking orders. Via correlation, the authors found that the largest inconsistencies and variability in rank orders were associated with tolerant groups, probably reflecting their behavioural characteristics (e.g., dyads having more frequent reversals in directions of aggression) rather than computational inconsistencies between methods.

As a third step, we separated the social groups as a function of their hierarchy steepness (Norscia and Palagi, 2015). We compared the differences in the David's scores between adjacently ranked individuals of each group, across the five study groups. This method allowed segregating the *L. catta* groups from the group of *E. rufus* × *collaris*, with ring-tailed lemur groups showing the steepest hierarchy gradient. Conversely, the *P. verreauxi* groups and the *E. rufus* × *collaris* group showed similar steepness levels. These results are in line with the observations of Balasubramaniam *et al.* (2012) who noted that steepness measures on different macaque species were more continuous than other measures (e.g., counter-aggression) and did not fully match the species' separation into different tolerance grades (from 1, extremely despotic to 4, extremely tolerant).

As a further step we calculated the triangle transitivity (Shizuka and McDonald, 2012) for each group and related it with group cohesion levels (Norscia and Palagi, 2015). Indeed, the distribution of group fission events can reflect variation between the strength of social relationships that connect the leader and other group members (King *et al.*, 2008). During our field observation, we randomly checked for the cohesion of our study groups by recording – three to four times a day – the interindividual spatial distance (more or less than 20 m) between group members. *A posteriori* (after determining animals' rank), for each cohesion bout we calculated the proportion of individuals within the fixed distance from the dominant female over the total animal number.

We observed that the lower transitivity values characterising the group of *E. rufus* × *collaris* and a group of *P. verreauxi* corresponded to lower group cohesion around the dominant. The top triangle transitivity was recorded for the other group of *P. verreauxi*, due to the low numerosity of the group (only six individuals) and absence of non-interacting dyads. High levels of triangle transitivity were found in *L. catta*, which reflect the rigid ranking order and the strong group cohesion around the dominant.

The most interesting result is that *E. rufus* × *collaris* does not show the lowest level of triangle transitivity as the other measures would predict (weak linearity,

Table 4.1 Summary of the features of dominance profile found in *L. catta* (two groups), *P. verreauxi* (two groups) and *E. rufus × collaris* (one group) in the Berenty forest, Madagascar.

	Lemur catta	Propithecus verreauxi	Eulemur rufus × collaris
Dominance relationships within groups			
Dominant sex	Female	Female	Female
Linear	+	+	+
Formalised	+	−*	−
Steep	+	0	−
Consistent	+	0	−
Transitive	+	0	0
Cohesion around the dominant	+	0	0

+ = high; 0 = moderate or variable; − = low. * This feature was evaluated using avoidance to allow interspecific comparisons (avoidance can be found in all species). The analysis should be repeated in *P. verreauxi* also using submissive chatters. See Chapter 4 for the definition of each feature.

steepness and group cohesion). In our study (Norscia and Palagi, 2015) we could have interpreted this finding as a result biased by the presence of non-interacting dyads. In fact, if some dyads do not interact, the resulting transitivity is lowered because the formation of transitive versus cyclic triangles is skewed (Shizuka and McDonald, 2012). We posited that the comparison between triangle transitivity and linearity could provide the hierarchy assessment with an added value because it suggested that in the core group of *Eulemur rufus × collaris* (composed of individuals that actually interact with each other) relationships were more transitive than it appeared by considering linearity alone. The observation of the different cohesion levels helps in explaining this difference by reinforcing the idea that hierarchy is less cohesive in the brown lemur group because the presence of non-interacting dyads (informed by the weak group cohesion around the dominant) does not affect transitivity (non-interacting dyads excluded) as much as it affects linearity (non-interacting dyads included). From a behavioural ecology perspective, the highest group dispersion of brown lemurs is consonant with their habitat use pattern. At Berenty, they tend to extend resource exploitation in terms of diet variety (Jolly *et al.*, 2000; Pinkus *et al.*, 2006), amount of food intake (Simmen *et al.*, 2003), temporal activity (Donati *et al.*, 2009) and ranging patterns (Tanaka, 2007).

In conclusion, the use of (1) the same method applied to different behavioural databases, (2) different methods applied to the same behavioural databases and (3) different correlates of aggressive encounters resulted in different hierarchical profiles and group differentiation arrangements relative to the same study subjects (Table 4.1). Hence, a multiple analytical approach can lead to a more in-depth description of dominance style, which is a multilevel concept combining many aspects of social dominance (strength of social relationships, direction and extent

4.4 Schizophrenic hierarchy in lemurs: the dominance profile is not just plain and simple

Besides detecting the differences in the dominance style among species using different methods, as discussed in the previous part, another point deserving attention is the detection of different ranking orders in the same social group, according to the context.

Previous reports, spanning human and non-human primates, have provided clear indications that the hierarchical arrangement of social groups is not necessarily plain and simple. Human groups are a complex mix of embodied categories and observable attributes whose interconnections are contingent and lead to crosscutting social cleavages; as a consequence, multiple hierarchies, resulting from ethnic, religious, political or occupational arrangement, can emerge (Schryer, 2001). In chimpanzees (*Pan troglodytes*) males can be ranked into separate hierarchies according to either formal subordination (via pant-grunting) or agonistic dominance (Newton-Fisher, 2004). Rhesus macaques (*Macaca mulatta*) can show real and formal dominance relationships (de Waal and Luttrell, 1985), with the latter only depending on unidirectional and context-independent signals (de Waal, 1982, 1986). In the New World squirrel monkeys (*Saimiri sciureus*), Alvarez (1975) observed that hierarchy varied from quasi-linear to circular, depending on the behavioural patterns considered for rank assessment (approaching, following, withdrawing and genital inspection).

In social strepsirrhines the existence of a context or behaviour-dependent hierarchy has been reported but not quantitatively demonstrated. Richard (1974) in *Propithecus verreauxi* detected no consistent correlation between the rank of individuals ordered according to the criterion of the priority of access to food (feeding hierarchy) and their rank in hierarchies established according to the frequency of aggression, the direction and frequency of grooming, or preferential access to females during the mating season.

These hints indicate that more than one single hierarchy can coexist in the same social group at the same time. We tried to confirm this assumption by using a quantitative analysis. To this purpose we used the data gathered on two troops of ring-tailed lemurs (of 10 and 13 individuals, respectively, for a total of 389 observation hours) and two groups of Verreaux's sifaka (of 10 and 6 individuals, for a total of 640 observation hours) at Berenty (Madagascar) in November 2006 to February 2007 and March to July 2008. We collected aggressive events via focal and all occurrences sampling (see Altmann, 1974 for details on these observational methods). We assessed hierarchy linearity via binary dyadic relationships using the aggression sociomatrices, by running the test via Matman 1.0. All groups showed linearity (or a trend)[1].

---

[1] $(0.60 < h' < 0,99; 0.00 < p < 0.07)$.

Interestingly, ring-tailed lemur hierarchy stayed linear in both feeding and non-feeding contexts[2] whereas sifaka hierarchy was linear in the non-feeding context only[3]. Females were dominant in all contexts, as also confirmed by normalised David's scores (calculated via the software Steepness 1.0) for non-linear hierarchies. Although we cannot completely exclude an effect of observational bias in dyad interactions – mainly due to the fact that conflicts while groups are feeding in the

---

Box 4.2 | **by Bernard Thierry**

Speaking of which: cross-species comparisons in social styles

I entered the field of primatology in the eighties by studying the behaviour of rhesus macaques (*Macaca mulatta*). As interesting as these animals may be, one cannot help but notice that their relationships are characterised by deep social inequalities associated with rough aggression and submission, making their social life appear quite stressful by human standards. This was consistent with the literature emphasising the role of competition and dominance in primate societies, although we should bear in mind that knowledge at that time was mainly based on the study of rhesus macaques. I then had the opportunity to study Tonkean macaques (*M. tonkeana*), and was surprised to see that they commonly protest against higher-ranking adversaries, and use a facial expression similar to the human smile to initiate affiliative exchanges. In complete contrast to rhesus macaques, who make threats each time you approach them, Tonkean macaques direct appeasement signals at you from the very first meeting. Such huge contrasts between closely related species prompted me to start a comparative study of macaque social styles (Thierry, 1986).

The sociobiological theory of this period was of little help regarding cross-species variations as it mainly looked for adaptive functions applicable to all societies. What then resurfaced in my mind were my previous readings of French structuralism. Whereas this school of thought was already out of fashion, its main lesson had remained accurate: 'styles form systems'; that is to say, a society is a system made up of multiple components, none of which can undergo a change without inducing changes in other components (Lévi-Strauss, 1953). The system may therefore exist under several states, meaning that each society belongs to a series of transformations corresponding to these different states. Although a similar perspective was advocated in evolutionary biology about the processes responsible for the form of organisms (Gould and Lewontin, 1979; Webster and Goodwin, 1982), an exclusive emphasis on the individual as the unit of selection in the study of animal

---

[2] Feeding context – Group A – $N_{conflicts}$: 112, $h' = 0.79$, $p < 0.001$; Group B – $N_{conflicts} = 59$, $h' = 0.46$, $p = 0.03$. Non-feeding context – Group A – $N_{conflicts}$: 101, $h' = 0.91$, $p < 0.001$; Group B – $N_{conflicts} = 76$, $h' = 0.46$, $p = 0.045$.

[3] Feeding context – Group A – $N_{conflicts}$: 37, $h' = 0.29$, $p = $ ns; Group B – $N_{conflicts} = 34$, $h' = 0.80$, $p = $ ns. Non-feeding context – Group A – $N_{conflicts}$: 44, $h' = 0.56$, $p < 0.04$; Group B – $N_{conflicts} = 42$, $h' = 0.43$, $p = 0.049$.

4.4 Schizophrenic hierarchy in lemurs: the dominance profile is not just plain and simple

behaviour maintained, for decades, an atomistic view of social systems which considered the adaptive function of each component independently of others. The comparison of macaque societies made it necessary to simultaneously address several of their dimensions while focusing on relationships and transactions; that is, social traits requiring both the performance of a behaviour by one individual and a subsequent response by another. Such variables grasp the form of social interchanges better than mere behaviour frequencies and durations which are quite sensitive to the influence of demographic and environmental factors. It soon appeared that even though all macaques display stable linear dominance rank orders, they differ by a number of social patterns (Thierry, 1986, 2000). In species like rhesus and Japanese macaques (*M. fuscata*), conflicts are highly unidirectional: the target of aggression generally flees or submits, as the risk of biting is relatively high. Reconciliations are infrequent, scoring between 4 and 12% among unrelated individuals. Dominance and kinship relationships strictly determine who may interact with whom. This leads to the formation of closed matrilines composed of related females and their offspring. Accordingly, mothers behave protectively towards their infants, restricting their contacts to close relatives. This picture differs from that observed in Tonkean, moor (*M. maurus*) and crested macaques (*M. nigra*), all species from Sulawesi island, where the majority of aggressive acts induce protests or counterattacks from the targeted individual. The intensity of aggression is low and measures of conciliatory tendencies yield high values: reconciliations tend to occur as much as 50% of the time among unrelated partners. The dominance gradient between group members is weak and kinship networks are quite open, meaning that the individual retains a fair degree of freedom in its choice of partners. Mothers display high levels of permissiveness, allowing other females to handle and carry their infant from an early age. Last but not least, it is noteworthy that other species of macaques show social styles that rank intermediately between the other two categories. The whole genus may therefore be set along a social tolerance scale that ranges from societies exhibiting strong power asymmetry to others characterised by more relaxed dominance hierarchies (Figures 4.8 and 4.9; Thierry, 2000).

To date, no ecological model is capable of explaining the contrasts found between macaques: we do not know of any combination of environmental factors that could account for different levels of social tolerance. Quite to the contrary, social styles appear to vary in a consistent way with phylogeny, with greater similarities being found in more closely related species. Such evidence reveals a fair degree of stasis in the evolution of macaque societies. This is a likely outcome of the coupling of social traits which act as constraints within societies, limiting adaptive responses to external pressures (Thierry, 2013). As an example, the occurrence of connections between harshness of aggression,

Box 4.2 (continued)

Box 4.2 (*cont.*)

hierarchical steepness and maternal protectiveness explains why such characters are transmitted as a cluster of covarying traits through the evolutionary process. Although such a transformational view of evolution would have been criticised as group selectionism in the recent past, it is now recognised that all individuals in a group are interdependent; individual fitness is linked to the fitness of others as each can benefit from collective performances on a long-term basis (Roberts, 2005).

A promising direction for future research is the comparative study of personality. As individuals are by nature the building blocks of societies, we should expect that personality traits covary with social traits. There are hints that more tolerant macaques are more explorative and less easily aroused by psychological stressors than their less tolerant counterparts, but reliable data based on direct cross-species comparisons are still lacking on this issue. Interestingly, recent investigations of species-specific personality structures revealed that macaques located on the tolerant end of the scale present a particular dimension – sociability or connectedness – that is unseen in others and may reflect their higher propensity to affiliate with mates (Neumann *et al.*, 2013; Adams *et al.*, 2015).

The covariation of behavioural traits has so far only been quantitatively documented in macaques, and it may be asked whether there is something particular about the latter. The answer to this question should be an emphatic '*No*', since interdependence between individuals is a basic feature of group living. Several difficulties need to be overcome, however, if we hope to extend the study of linkages beyond the macaque genus. There is no guarantee that traits between species belonging to different taxonomic groups will be homologous. The absence or addition of a single trait may induce non-linear effects, thereby modifying the relationships between traits. For example, in contrast to macaques, low dominance asymmetry is associated with high biting rates in guenons (*Cercopithecus* spp.) (Kaplan, 1987), and unidirectional aggression is associated with a weak influence of kinship in white-faced capuchins (*Cebus capucinus*) (Bergstrom and Fedigan, 2013). Comparative studies additionally require both significant levels of interspecific variation and an adequate sample of different species. The limited diversity observed in the social styles of baboon taxa (*Papio* spp.), for instance, makes it challenging to show any compelling differences between them, and although the comparison of chimpanzees (*Pan troglodytes*) and bonobos (*P. paniscus*) has highlighted an association between tolerance, affiliation, dominance asymmetry, harshness of aggression and individual reactivity similar to those found in macaques (Hare *et al.*, 2012), any deepening of these investigations is hindered by the two-species sample in the genus.

Beyond these methodological problems, a fundamental issue remains to be addressed. It is best formulated via three related questions: to what extent are individuals interdependent? To what extent are linkages between traits entrenched in social organisation? To what extent can natural selection uncouple traits? The social networks of macaques are characterised by stable membership and the occurrence of coalitions which tighten the relationships of individuals by creating multiple feedback loops between them. I would suggest that a high degree of interdependence between individuals is responsible for the strong entrenchment of behavioural traits observed in macaque societies. Conversely, weakly integrated social systems should produce lower levels of behavioural coupling. To test this hypothesis, we have to investigate social systems that present varying degrees of interdependence. Here enter the lemurs. The wide range of social systems observed in these animals, from loose associations to more complex organisations, appears to be a valuable tool to study the possible relationship between social interdependence and behavioural coupling.

Figure 4.8 *Macaca fuscata*, Japanese macaque: one of the less tolerant macaque species according to Thierry's scale (grade 1; Sueur *et al.*, 2011). Wilhelma Zoo, Germany. Photo: Elisabetta Palagi.

Box 4.2 (continued)

Box 4.2 (*cont.*)

Figure 4.9 *Macaca tonkeana*, Tonkean macaque. One of the most tolerant macaque species according to Thierry's scale (grade 4; Sueur *et al.*, 2011). Mulhouse Zoo, France. Photo: Elisabetta Palagi.

canopy can be masked by vegetation – these results support the idea that more than one hierarchy can coexist at the same time in a lemur group.

4.5 Why female dominance?

Independent of the quantitative method used to assess hierarchy, adult females turn out to have a leading role in lemur groups. Even though a weaker or unclear female dominance, needing confirmation via quantitative analyses, has been reported in a few species (e.g., *Eulemur fulvus*: Pereira and McGlynn, 1997; *Eulemur coronatus*: Pereira *et al.*, 1990; *Daubentonia madagascariensis*: Rendall, 1993), most of the studies converge in indicating that female dominance and/or priority is the norm in lemurs (e.g., *Lemur catta*: Jolly, 1966; Pereira *et al.*, 1990; Pereira and Kappeler, 1997; *Propithecus* spp.: Kubzdela *et al.*, 1992; Wright, 1993; Palagi *et al.*, 2008b; *Indri indri*: Pollock, 1979; *Varecia variegata*: Raps and White, 1995; *Eulemur rubriventer*: Tecot and Romine, 2012; *Eulemur macaco flavifrons*: Digby and Kahlenberg, 2002; *Eulemur fulvus* × *collaris*: Palagi and Norscia, 2011). There is increasing evidence that female dominance also applies to nocturnal species (*Microcebus murinus*: Radespiel and

Zimmermann, 2001; *Hapalemur griseus*: Waeber and Hemelrijk, 2003; *Avahi occidentalis*: Ramanankirahina *et al.*, 2011; *Avahi meridionalis*: later in this chapter).

The Energy Conservation Hypothesis (ECH) was put forth by Jolly (1966) to explain the female dominance in lemurs and was then extended to explain several other idiosyncrasies (*sensu* Kappeler, 2000) of lemurs (lack of sexual dimorphism, abundance of pair-living species, cathemerality, etc.; Wright, 1999). According to this hypothesis, the evolution of female dominance would be related to the highly seasonal environment of Madagascar, extremely variable and resource poor at least during several months per year. In this context, female dominance would be an adaptation to high reproductive costs and scarcity of food in comparison to other primates due to a resource-poor and highly variable island environment. In this context, females would take feeding priority over males to be able to have enough food to raise offspring successfully (Jolly, 1984; Young *et al.*, 1990; Wright, 1999). The high energetic stress imposed on both gestating lemur females (due to their low metabolic rates compared to monkeys; Richard and Nicoll, 1987) and lactating and weaning mothers was considered (Sauther, 1991, 1993; Meyers and Wright, 1993). Indeed, it was observed that in *Lemur catta*, lactating females ate more and had more aggressive interactions over food (Rasamimanana and Rafidinarivo, 1993; Sauther, 1993). However, as pointed out by Wright (1999), gestation and lactation are costly to all female primates and mammals, which makes it unclear why feeding priority evolved in lemurs, and not in other taxa. Moreover, in quantitative comparisons with anthropoid primates, no unusual reproductive or energetic costs for lemur females have been shown (Kappeler, 1996). Additionally, in non-primate taxa, species can be found with typical male dominance patterns living in environments that are substantially poorer in resources and more extreme in seasonality (e.g., Mooring *et al.*, 2003). Other predictions deriving from the ECH could not be fully confirmed, such as the fact that milk quality should be higher in lemurs than in anthropoids to allow faster infant growth in a harsh environment or the differences in the feeding competition that should be observed in more versus less seasonal habitat (and, therefore, in east versus west forests and/or wet versus dry forests) (Wright, 1999). Therefore, the ECH, although valid for *Lemur catta*, cannot convincingly explain why the phenomenon of female dominance is so widespread in lemurs.

Another hypothesis formulated to explain female dominance (and other peculiar lemur traits) is the EVolutionary Disequilibrium Hypothesis (EVDH). According to this hypothesis, female dominance (as well as the lack of sexual dimorphism) would be the result of an evolutionary disequilibrium after the recent extinction of large predatory eagles and large-bodied lemur species in Madagascar, which would have led different species to become diurnal and polygamous from nocturnal and pair-living ancestors (van Schaik and Kappeler, 1996; Kappeler, 1999). One assumption deriving from this hypothesis is that female dominance would exist in other nocturnal, non-lemur primates. However, with the exception of *Otolemur garnettii* (a species African loris; Hager and Welker, 2001) female dominance has not been reported in galagos, lorises, pottos, or tarsiers (Charles-Dominique, 1977; Bearder, 1987; Gursky, 1998). A second assumption from the hypothesis would be that female dominance is found also in pair-living primates outside Madagascar, whereas co-dominance is usually found in such species (Wright, 1984; 1993; Kappeler, 1993; Palombit, 1994, 1995). Therefore, this hypothesis also cannot actually explain female dominance in lemurs.

A more recent and convincing explanation is provided by the Cost-Asymmetry Hypothesis, which predicts the general patterns of intersexual dominance across primates and is consistent with observations in non-primate mammals and birds lacking strong sexual dimorphism (Dunham, 2008). Female dominance can be found in several non-primate mammals with similar-sized sexes, including the hyrax (*Procavia capensis*; Koren et al., 2006), spotted hyena (*Crocuta crocuta*; East et al., 1993), rufous elephant shrew (*Elephantus rufescens*; Rathbun, 1979) and velvet-furred swamp rat (*Rattus lutreolus*; Monamy, 1997), but also the brush-tailed possum (*Trichosurus vulpecula*; Jolly and Spurr, 1996) and nutria (*Myocastor coypus*; Warkentin, 1968), in which females are slightly smaller than males. This is also true for Peruvian squirrel monkeys (*Saimiri boliviensis peruviensis*; Boinski, 1999). In Columbian ground squirrels, *Spermophilus columbianus*, females become dominant over males after the mating season, until late in lactation (Murie and Harris, 1988). A similar pattern has been observed in snowshoe hares, *Lepus americanus*, the females of which become dominant over males during the summer season when breeding occurs (Graf, 1981). In giant river otters, *Pteronura brasiliensis*, the one breeding female is dominant over all other group members, including males (Duplaix, 1980). Finally, as mentioned above, female dominance has been also described in the African loris, *Otolemur garnettii* (Hager and Welker, 2001). All these species share an important feature: absence of or reduced sexual dimorphism. According to the Cost-Asymmetry Hypothesis, in monomorphic species the sex with the higher costs of reproduction will have the higher pay-off for winning a fight, resulting in an asymmetric, intersexual contest. Since females have higher reproductive costs, their dominance has been positively selected.

The lack of dimorphism between sexes in lemurs, with females bearing the highest energy costs, can indeed be a proximate factor for female dominance. The rise of female power can also be explained in the light of the biological market theory. As will be extensively discussed in Chapter 8, monomorphism tips the balance of power in favour of females because males cannot force them into mating and males depend on females for breeding opportunities and fitness success (Lewis, 2002; Norscia et al., 2009). Therefore, females become the dominant and choosing sex, because they are the limiting factor for reproduction and carry the most valuable resource: the egg ready to be turned in the male's offspring (Lewis, 2002). Trading becomes, therefore, essential and males need to offer a commodity or service in exchange for mating opportunities, because the leverage of females is higher, especially in the mating season.

4.6 Nocturnal lemurs: who rules the pair? The case of *Avahi meridionalis*

Despite the limitations of the steepness procedure, one of the advantages of its adoption is that it can be applied to 3 × 3 sociomatrices, meaning groups of three

Figure 4.10 *Microcebus murinus* from the Ankarafantsika National Park, Madagascar. Photo: Ute Radespiel.

individuals (of course paying attention to avoid observational zeros). Therefore, this method represents an interesting tool to quantitatively check the dominance relationships within familiar groups, which are generally characterised by very small sizes. For example, the dominance in nocturnal lemurs is hard to assess due to the difficult observation conditions and because there is no real social group. Nocturnal lemurs normally conduct solitary or pair-living lives, sometimes in small family groups including a pair and – more or less temporarily – their offspring of one or two years of age (Norscia and Borgognini-Tarli, 2008).

The literature reports that in captive grey mouse lemurs (*Microcebus murinus*), a cheirogaleid family, the majority of agonistic interactions (including chasing/fleeing, approach/avoidance) were decided, and were in all but one case won by females (Figure 4.10; Radespiel and Zimmermann, 2001).

In six pairs of wild western woolly lemurs (*Avahi occidentalis*) – an indriid species from the dry forest of northwestern Madagascar – it was observed that all decided agonistic conflicts (N = 15) were exclusively initiated and won by the female and that no female showed spontaneous submission towards her male partner (Ramanankirahina *et al.*, 2011). In these cases, with two animals involved and all agonistic interactions won by females, the dominance relationships are quite clear.

Figure 4.11 *Avahi meridionalis* in the forest of Sainte Luce, Madagascar. Female with offspring. Photo: Ivan Norscia.

Here we wanted to check, using a quantitative analysis, the dominance relationships in the southern woolly lemur, *Avahi meridionalis* (Zaramody *et al.*, 2006), living in the littoral rain forest, where visibility conditions are difficult and getting to spot conflicts hard. We filled in the interaction sociomatrix by using the data on *Avahi meridionalis* collected by Norscia, in 2004, in the fragment S9 (377 ha) of the forest of Sainte Luce (southeastern Madagascar; S 24.45′; E 047.11′). The small family group the data refer to (Figure 4.11) was composed of three radiocollared individuals: an adult male, an adult female, and a subadult female (6 months to 1 year of age during the observation period May–December 2004; Norscia and Borgognini-Tarli, 2008).

The interactions among the individuals were collected during 26 nights of observation (256 hr) (Norscia and Borgognini-Tarli, 2008; Norscia *et al.*, 2012). Due to the difficulty in recording agonistic interactions, the sociomatrix was filled in with bouts (N = 10) of aggression and displacement (when an individual was displaced by the other on a feeding spot).

After filling in the sociomatrix, via the freeware Steepness 1.0 (by Leiva and de Vries), we calculated the steepness, which refers to the absolute slope of the straight line fitted to the normalised David's scores (NDS; calculated on the basis of a dyadic dominance index corrected for chance) plotted against the subjects' ranks (de Vries *et al.*, 2006). We determined the NDS-based hierarchy by ranking the individuals according to their NDSs. The results (Figure 4.12) show that the hierarchy has a steepness of 1.00 and that the adult female is dominant, thus demonstrating in a quantitative way female dominance in nocturnal lemurs and providing further support to the Cost-Asymmetry Hypothesis.

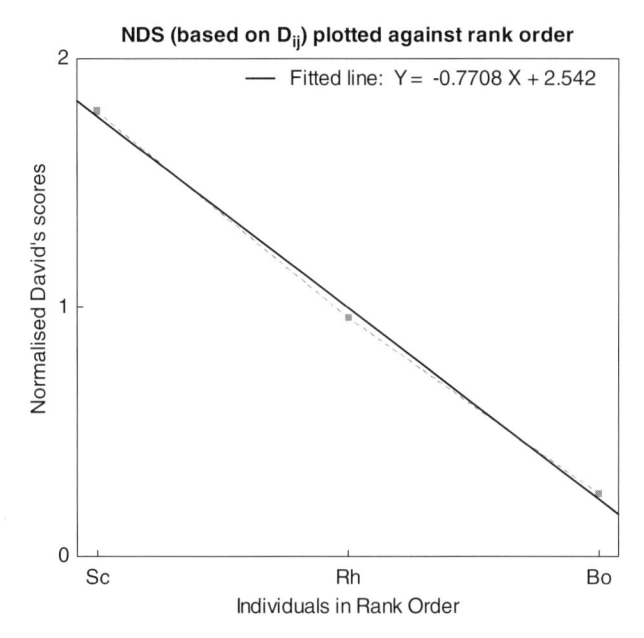

Figure 4.12 Normalised David's scores (D_{ij}, based on aggression sociomatrices) plotted against ordinal rank order (dashed black line), and the fitted line (black, solid line) for *Avahi meridionalis* from the forest of Sainte Luce, Madagascar. The Y-axis reports the normalised David's scores and the X-axis reports the individuals of each group (software: Steepness 2.2.).

References

Adams, E. S. (2005). Bayesian analysis of linear dominance hierarchies. *Animal Behaviour*, 69, 1191–1201.
Adams, M. J., Majolo, B., Ostner, J., *et al.* (2015). Personality structure and social style in macaques. *Journal of Personality and Social Psychology: Personality Processes and Social Psychology*, 109, 338–353.
Adler, N. E., Epel, E. S., Castellazzo, G. & Ickovics, J. R. (2000). Relationship of subjective and objective social status with psychological and physiological functioning: preliminary data in healthy white women. *Health Psychology*, 19, 586–592.
Adler, N. E., Singh-Manoux, A., Schwartz, J., *et al.* (2008). Social status and health: A comparison of British civil servants in Whitehall-II with European- and African-Americans in CARDIA. *Social Science and Medicine*, 66, 1034–1045.
Alfieri, V. (1996). *Della Tirannide, Del Principe e delle Lettere, La Virtù Sconosciuta*. Edizione Astense, Rizzoli BUR.
Allee, W. C. (1931). *Animal Aggregations. A Study in General Sociology*. Chicago: University of Chicago Press.
Altmann, J. (1974). Observational study of behavior: sampling methods. *Behaviour*, 49, 227–266.
Alvarez, F. (1975). Social hierarchy under different criteria in groups of squirrel monkeys, *Saimiri sciureus*. *Primates*, 16, 437–455.
Appleby, M. C. (1983). The probability of linearity in hierarchies. *Animal Behaviour*, 31, 600–608.
Balasubramaniam, K. N., Dittmar, K., Berman C. M., *et al.* (2012). Hierarchical steepness and phylogenetic models: phylogenetic signals in *Macaca*. *Animal Behaviour*, 83: 1207–1218.

Balasubramaniam, K. N., Berman, C. M., De Marco, A. et al. (2013). Consistency of dominance rank order: A comparison of David's scores with I&SI and Bayesian methods in macaques. *American Journal of Primatology*, 75, 959–971.

Bearder, S. K. (1987). Lorises, bushbabies, and tarsiers: Diverse societies in solitary foragers. In: B. B. Smuts, D. L. Cheney, R. M. Seyfarth, R. W. Wrangham & T. T. Struhsaker (eds), *Primate Societies*. Chicago: University of Chicago Press, pp. 11–24.

Bergstrom, M. L. & Fedigan, L. M. (2013). Dominance style of female white-faced capuchins. *American Journal of Physical Anthropology*, 150, 591–601.

Bernstein, I. S. (1981). Dominance: the baby and the bathwater. *Behavioral and Brain Sciences*, 4, 419–457.

Boinski, S. (1999). The social organizations of squirrel monkeys: implications for ecological models of social evolution. *Evolutionary Anthropology*, 8, 101–112.

Chance, M. R. A. (1967). Attention structure as the basis of primate rank orders. *Man*, 2, 503–518.

Charles-Dominique, P. (1977). *Ecology and Behaviour of Nocturnal Primates: Prosimians of Equatorial West Africa*. London: Duckworth.

Chase, I. D. (1980). Social process and hierarchy formation in small groups: A comparative perspective. *American Sociological Review*, 45, 905–924.

Cheney, D. L. (1977). The acquisition of rank and the development of reciprocal alliances in freeranging immature baboons. *Behavioral Ecology and Sociobiology*, 2, 303–318.

Chomsky, N., Mitchell, P. R. & Schoeffel, J. (2002). *Understanding Power: The Indispensable*, 9th edition. The New Press, New York.

Clutton-Brock, T. H. (1982). *Red Deer. Behaviour and Ecology of Two Sexes*. Edinburgh: Edinburgh University Press.

Clutton-Brock, T. H., Albon, S. D. & Guinness, F. E. (1984). Maternal dominance, breeding success and birth sex ratios in red deer. *Nature*, 308, 358–360.

de Vries, H. (1995). An improved test of linearity in dominance hierarchies containing unknown or tied relationships. *Animal Behaviour*, 50, 1375–1389.

de Vries, H. & Appleby, M. C. (2000). Finding an appropriate order for a hierarchy: a comparison of the I & SI and the BBS methods. *Animal Behaviour*, 59, 239–245.

de Vries, H., Netto, W. J. & Hanegraaf, P. L. H. (1993). MatMan: a program for the analysis of sociometric matrices and behavioural transition matrices. *Behaviour*, 125, 157–175.

de Vries H., Stevens, J. M. G. & Vervaecke, H. (2006). Measuring and testing steepness of dominance hierarchies. *Animal Behaviour*, 71, 585–592.

de Waal, F. B. M. (1982). *Chimpanzee Politics*. Baltimore: John Hopkins University Press.

de Waal, F. B. M. (1986). The integration of dominance and social bonding in primates. *Quarterly Review of Biology*, 61, 459–479.

de Waal, F. B. M. & Luttrell, L. M. (1985). The formal hierarchy of rhesus macaques (*Macaca mulatta*): an investigation of bared-teeth display. *American Journal Primatology*, 9, 73–86.

Digby, L. J. & Kahlenberg, S. M. (2002). Female dominance in blue-eyed black lemurs (*Eulemur macaco flavifrons*). *Primates*, 43, 191–199.

Donati, G., Baldi, N., Morelli, V., Ganzhorn, J. U. & Borgognini-Tarli, S. M. (2009). Proximate and ultimate determinants of cathemeral activity in brown lemurs. *Animal Behaviour*, 77, 317–325.

Drews, C. (1993). The concept and definition of dominance in animal behaviour. *Behaviour*, 125, 283–313.

Dunham, A. E. (2008). Battle of the sexes: cost asymmetry explains female dominance in lemurs. *Animal Behaviour*, 76, 1435–1439.

Duplaix, N. (1980). Observations on the ecology and behavior of the giant river otter *Pteronura brasiliensis* in Suriname. *La Terre et la Vie-Revue d'Ecologie Appliquee*, 34, 495–620.

East, M. L., Hofer, H. & Wickler, W. (1993). The erect 'penis' is a flag of submission in a female-dominated society: greetings in Serengeti spotted hyenas. *Behavioral Ecology and Sociobiology*, 33, 355–370.

Flack, J. & de Waal, F. B. M. (2004). Dominance style, social power, and conflict. In: B. Thierry, M. Singh & W. Kaumanns (eds), *Macaque Societies: A Model for the Study of Social Organization*. Cambridge University Press, pp. 157–185.

Foucault, M. (1982). The subject and power. *Critical inquiry*, 8, 777–795.

Fox, S. F., Rose, E. & Myers, R. (1981). Dominance and the acquisition of superior home ranges in the lizard *Uta stansburiana*. *Ecology*, 62, 888–893.

Fox, S. F., Rose, E. & Myers, R. (1981). Dominance and the acquisition of superior home ranges in the lizard *Uta stansburiana*. *Ecology*, 62, 888–893.

Furuichi, T. (1984). Symmetrical patterns in non-agonistic social interactions found in unprovisioned Japanese macaques. *Journal of Ethology*, 2, 109–119.

Furuichi, T. (1985). Inter-male associations in a wild Japanese macaque troop on Yakushima Island, Japan. *Primates*, 26, 219–237.

Furuichi, T. (1989). Social interactions and the life history of female *Pan paniscus* in Wamba, Zaire. *International Journal of Primatology*, 10, 173–197.

Furuichi, T. (1997). Agonistic interactions and matrifocal dominance rank of wild bonobos (*Pan paniscus*) at Wamba. *International Journal of Primatology*, 18, 855–875.

Furuichi, T. (2009). Factors underlying party size differences between chimpanzees and bonobos: a review and hypotheses for future study. *Primates*, 50, 197–209.

Furuichi, T. (2011). Female contributions to the peaceful nature of bonobo society. *Evolutionary Anthropology*, 20(4), 131–142.

Furuichi, T. & Hashimoto, C. (2002). Why female bonobos have a lower copulation rate during estrus than chimpanzees. In: C. Boesch, G. Hohmann and L. F. Marchant (eds), *Behavioural Diversity in Chimpanzees and Bonobos*. New York: Cambridge University Press, pp. 156–167.

Furuichi, T., Takasaki, H. & Sprague, D. S. (1982). Winter range utilization of a Japanese macaque troop in a snowy habitat. *Folia Primatologica*, 37, 77–94.

Furuichi, T., Idani, G., Ihobe, H., et al. (2012). Long-term studies on wild bonobos at Wamba, Luo Scientific Reserve, D. R. Congo: towards the understanding of female life history in a male-philopatric species. In: P. M. Kappeler & D. P. Watts, *Long-term Field Studies of Primates*: Berlin, Heidelberg, Springer-Verlag, pp. 413–433.

Gammel, M. P., De Vries, H., Jennings, D. J., Carlin, C. M. & Hayden, T. J. (2003). David's score: a more appropriate dominance ranking method than Clutton-Brock, *et al.*'s index. *Animal Behaviour*, 66, 601–605.

Gerloff, U., Hartung, B., Fruth, B., Hohmann, G. & Tautz, D. (1999). Intracommunity relationships, dispersal pattern and paternity success in a wild living community of bonobos (*Pan paniscus*) determined from DNA analysis of faecal samples. *Proceedings of the Royal Society of London: Biological Sciences*, 266, 1189–1195.

Gould, S. J. & Lewontin, R. C. (1979). The spandrels of San Marco and the Panglossian paradigm: a critique of the adaptationist programme. *Proceedings of the Royal Society B: Biological Sciences*, 205, 581–598.

Graf, R. P. (1981). Some aspects of snowshoe hare behavioural ecology. M.Sc. Thesis, University of British Columbia.

Grüter, C. C. & Zinner, D. (2004). Nested societies. Convergent adaptations of baboons and snub-nosed monkeys? *Primate Report*, 70, 1–98.

Guhl, A. M., Collias, N. E. & Allee, W. C. (1945). Mating behavior and social hierarchy in small flocks of white leghorns. *Physiological Zoology*, 18, 365–390.

Gursky, S. L. (1998). The effect of radio transmitter weight on a small nocturnal primate: Data on activity time budgets, prey capture rates, mobility patterns and weight loss. *American Journal of Primatology*, 46, 145–155.

Hager, R. & Welker, C. (2001). Female dominance in African lorises (*Otolemur garnettii*). *Folia Primatologica*, 72, 48–50.

Hall, C. L. & Fedigan, L. M. (1997). Spatial benefits afforded by high rank in white-faced capuchins. *Animal Behaviour*, 53, 1069–1082.

Hand, J L. (1986). Resolution of social conflicts: dominance, egalitarianism, spheres of dominance, and game theory. *Quarterly Review of Biology*, 61, 201–220.

Hare, B., Wobber, V. & Wrangham, R. (2012). The self-domestication hypothesis: evolution of bonobo psychology is due to selection against aggression. *Animal Behaviour*, 83, 573–585.

Hashimoto, C., Furuichi, T. & Takenaka, O. (1996). Matrilineal kin relationship and social behavior of wild bonobos (*Pan paniscus*): sequencing the D-loop region of mitochondrial DNA. *Primates*, 37, 305–318.

Hashimoto, C., Tashiro, Y., Hibino, E., et al. (2008). Longitudinal structure of a unit-group of bonobos: Male philopatry and possible fusion of unit-groups. In: T. Furuichi and J. Thompson. *The Bonobos: Behavior, Ecology, and Conservation*. New York, Springer, pp. 107–119.

Hawley, P. H. (1999). The ontogenesis of social dominance: a strategy-based evolutionary perspective. *Developmental Review*, 19, 97–132.

Hemelrijk, C. K. (1999). An individual-oriented model on the emergence of despotic and egalitarian societies. *Proceedings of the Royal Society of London: Biological Sciences*, 266, 361–369.

Hemelrijk, C. K., Wantia, J. & Gygax, L. (2005). The construction of dominance order: comparing performance of five methods using an individual-based model. *Behaviour*, 142, 1037–1058.

Hewitt, S. E., Macdonald, D. W. & Dugdale, H. L. (2009). Context-dependent linear dominance hierarchies in social groups of European badgers, *Meles meles*. *Animal Behaviour*, 77, 161–169.

Idani, G. (1990). Relations between unit-groups of Bonobos at Wamba, Zaire: encounters and temporary fusions. *African Study Monographs*, 11, 153–186.

Isbell, L. A. & Young, T. P. (1993). Social and ecological influences on activity budgets of vervet monkeys, and their implications for group living. *Behavioral Ecology and Sociobiology*, 32, 377–385.

Jolly, A. (1966). *Lemur Behaviour*. Chicago: University of Chicago Press.

Jolly, A. (1984). The puzzle of female feeding priority. In: M. Small (ed.), *Female Primates: Studies by Women Primatologists*. New York: A. R. Liss, pp. 197–215.

Jolly, S. E. & Spurr, E. B. (1996). Effect of ovariectomy on the social status of brushtail possums (*Trichosurus vulpecula*) in captivity. *New Zealand Journal of Zoology*, 23, 27–31.

Jolly, A., Caless, S., Cavigelli, S., et al. (2000). Infant killing, wounding, and predation in Eulemur and Lemur. *International Journal of Primatology*, 23, 327–353.

Kano, T. (1992). *The Last Ape: Pygmy Chimpanzee Behavior and Ecology*. Stanford, CA: Stanford University Press.

Kaplan, J. R. (1987). Dominance and affiliation in the *Cercopithecini* and *Papionini*: a comparative examination. In: E. L. Zucker (ed.), *Comparative Behavior of African Monkeys*, New York: Alan R. Liss, pp. 127–150.

Kappeler, P. M. (1993). Female dominance in primates and other mammals. In: P. P. G. Bateson, P. H. Klopfer & N. S. Thompson (eds), *Perspectives in Ethology Vol. 10, Behavior and Evolution*. New York: Plenum Press, pp. 143–158.

Kappeler, P. M. (1996). Causes and consequences of life-history variation among strepsirrhine primates. *American Naturalist*, 148, 868–891.

Kappeler, P. M. (1999). Lemur social structure and convergence in primate socioecology. In: P. C. Lee (ed.), *Comparative Primate Socioecology*. Cambridge University Press, pp. 273–299.

Kappeler, P. M. (2000). Causes and consequences of unusual sex ratios among lemurs. In: P. M. Kappeler (ed.), *Primate Males – Causes and Consequences of Variation in Group Composition*. Cambridge: Cambridge University Press, pp. 55–63.

Kawai, M., Dunbar, R., Ohsawa, H. & Mori, U. (1983). Social organization of gelada baboons: social units and definitions. *Primates*, 24, 13–24.

Kemp, C. & Tenenbaum, J. B. B. (2008). The discovery of structural form. *PNAS*, 105, 10687–10692.

Kendall, M. (1962). *Rank Correlation Methods*. London: Griffin.

King, A. J., Douglas, C. M. S., Huchard, E., Isaac, N. J. B. & Cowlishaw, G. (2008). Dominance and affiliation mediate despotism in a social primate. *Current Biology*, 18, 1838–1838.

Kitchen, D. M., Cheney, D. L. & Seyfarth, R. M. (2005). Contextual factor mediating contests between male chacma baboons in Botswana: effects of food, friends, and females. *International Journal of Primatology*, 26, 105–125.

Klass, K. & Cords, M. (2011). Effect of unknown relationships on linearity, steepness and rank ordering of dominance hierarchies: simulation studies based on data from wild monkeys. *Behavioural Processes*, 88, 168–176.

Koenig, A. (2000). Competitive regimes in forest-dwelling Hanuman langur females (*Semnopithecus entellus*). *Behavioral Ecology and Sociobiology*, 48, 93–109.

Koren, L., Mokady, O. & Geffen, E. (2006). Elevated testosterone levels and social ranks in female rock hyrax. *Hormones and Behavior*, 49, 470–477.

Krebs, J. R. & Davies, N. B. (1987). *An Introduction to Behavioural Ecology*. Boston: Blackwell.

Kubzdela, K. S., Richard, A. F. & Pereira, M. E. (1992). Social relations in semi-free-ranging sifakas (*Propithecus verreauxi coquereli*) and the question of female dominance. *American Journal of Primatology*, 28, 139–145.

Landau, H. G. (1951). On dominance relations and the structure of animal societies: I. Effect of inherent characteristics. *Bulletin of Mathematical Biophysics*, 13, 1–19.

Lévi-Strauss, C. (1953). Social structure. In: A. L. Kroeber (ed.), *Anthropology Today*. Chicago: University of Chicago Press, pp. 524–533.

Lewis, R. J. (2002). Beyond dominance: the importance of leverage. *Quarterly Review of Biology*, 77, 149–64.

Lewis, R. J. & van Schaik, C. P. (2007). Bimorphism in male Verreaux's sifaka in the Kirindy forest of Madagascar. *International Journal of Primatology*, 28, 159–182.

Machiavelli, N. (1961). *The Prince*. London: Penguin.

Maiya, A. S. & Berger-Wolf, T. Y. (2011). Benefits of bias: towards better characterisation of network sampling. *Proceedings of the 17th ACM SIGKDD international conference on knowledge discovery and data mining*, New York: ACM New York, pp. 105–113. ISBN: 978-1-4503-0813-7.

McGrew, W. C. (1972). *An Ethological Study of Children's Behavior*. New York: Academic Press.

Meyers, D. & Wright, P. C. (1993). Resource tracking: Food availability and *Propithecus* seasonal reproduction. In: P. Kappeler & J. Ganzhorn (eds), *Lemur Social Systems and Their Ecological Basis*. New York: Plenum Press, pp. 179–192.

Missakian, E. A. (1976). Aggression and dominance relations in peer groups of children 6–45 months of age. Paper presented at the annual meeting of the Animal Behavior Society, Boulder, Colorado.

Monamy, V. (1997). Sexual differences in habitat use by *Rattus lutreolus* (Rodentia: Muridae): the emergence of patterns in native rodent community structure. *Australian Mammalogy*, 20, 43–48.

Mooring, M. S., Fitzpatrick, T. A., Benjamin, J. E., et al. (2003). Sexual segregation in desert bighorn sheep (*Ovis canadensis mexicana*). *Behaviour*, 140, 183–207.

Murie, J. O. & Harris, M. A. (1988). Social interactions and dominance relationships between female and male Columbian ground-squirrels. *Canadian Journal of Zoology*, 66, 1414–1420.

Murray, C. M. (2007). A method for assigning categorical rank in female chimpanzees (*Pan troglodytes*) via the frequency of approaches. *International Journal of Primatology*, 28, 853–864.

Muscatell, K. A., Morelli, S. A., Falk, E. B., et al. (2012). Social status modulates neural activity in the mentalizing network. *NeuroImage*, 60, 1771–1777.

Neumann, C., Duboscq, J., Dubuc, C., et al. (2011). Assessing dominance hierarchies: validation and advantages of progressive evaluation with Elo-rating. *Animal Behaviour*, 82, 911–921.

Neumann, C., Agil, M., Widdig, A. & Engelhardt, A. (2013). Personality of wild male crested macaques (*Macaca nigra*). *PLos ONE*, 8, e69383. http://dx.doi.org/10.1371/journal.pone.0069383.

Newton-Fisher, N. E. (2004). Hierarchy and social status in Budongo chimpanzees. *Primates*, 45, 81–87.

Norscia, I. & Borgognini-Tarli, S. M. (2008). Ranging behavior and possible correlates of pair-living in Southeastern Avahis (Madagascar). *International Journal of Primatology*, 29, 153–171.

Norscia, I. & Palagi, E. (2011). Do brown lemurs reconcile? Not always. *Journal of Ethology*, 29, 181–185.

Norscia, I. & Palagi, E. (2015). The socio-matrix reloaded: from hierarchy to dominance profile in wild lemurs. *PeerJ*, 3, e729. http://dx.doi.org/10.7717/peerj.729.

Norscia, I., Antonacci, D. & Palagi, E. (2009). Mating first, mating more: biological market fluctuation in a wild prosimian. *PLoS ONE*, 4, e4679. http://dx.doi.org/10.1371/journal.pone.0004679.

Norscia, I., Ramanamanjato, J. B. & Ganzhorn, J. U. (2012). Feeding patterns and dietary profile of nocturnal southern woolly lemurs (*Avahi meridionalis*) in southeast Madagascar. *International Journal of Primatology*, 33, 150–167.

Ogola Onyango, P., Gesquiere, L. R., Wango, E. O., Alberts, S. C. & Altmann, J. (2008). Persistence of maternal effects in baboons: Mother's dominance rank at son's conception predicts stress hormone levels in subadult males. *Hormones and Behavior*, 54, 319–324.

Palagi, E. & Norscia, I. (2011). Scratching around stress: hierarchy and reconciliation make the difference in prosimians. *Stress*, 14, 93–97.

Palagi, E., Chiarugi, E. & Cordoni, G. (2008a). Peaceful post-conflict interactions between aggressors and bystanders in captive lowland gorillas (*Gorilla gorilla gorilla*). *American Journal of Primatology*, 70, 949–955.

Palagi, E., Antonacci, D. & Norscia, I. (2008b) Peacemaking on treetops: first evidence of reconciliation from a wild prosimian (*Propithecus verreauxi*). *Animal Behaviour*, 76, 737–747.

Palombit, R. A. (1994). Extra-pair copulations in a monogamous ape. *Animal Behaviour*, 47, 721–723.

Palombit, R. A. (1995). Longitudinal patterns of reproduction in wild female siamang (*Hylobates syndactylus*) and white-handed gibbons (*Hylobates lar*). *International Journal of Primatology*, 16, 739–760.

Paoli, T. & Palagi, E. (2008). What does agonistic dominance imply in bonobos? In: Takeshi, F. and Thompson, J. (eds), *Bonobos: Behaviour, Ecology, and Conservation*. New York: Springer-Verlag, pp. 35–54.

Paoli, T., Palagi, E., Tacconi, G. & Tarli, S. B. (2006). Perineal swelling, intermenstrual cycle, and female sexual behavior in bonobos (*Pan paniscus*). *American Journal of Primatology*, 68, 333–347.

Parish, A. R. (1994). Sex and food control in the uncommon chimpanzee: how bonobo females overcome a phylogenetic legacy of male dominance. *Ethology and Sociobiology*, 15, 157–179.

Parr, L. A., Matheson, M., Bernstein, I. S. & de Waal, F. B. M. (1997). Grooming down the hierarchy: allogrooming in captive brown capuchin monkeys, *Cebus apella*. *Animal Behaviour*, 54, 361–367.

Pereira, M. E. & Kappeler, P. M. (1997). Divergent systems of agonistic relationships in lemurs. *Behaviour*, 134, 225–274.

Pereira, M. E. & McGlynn, C. A. (1997). Special relationships instead of female dominance for redfronted lemurs, *Eulemur fulvus rufus*. *American Journal of Primatology*, 43, 239–258.

Pereira, M. E., Kaufmann, R., Kappeler, P. M. & Overdorff, D. J. (1990). Female dominance does not characterize all of the Lemuridae. *Folia Primatologica*, 55, 96–103.

Pinkus, S., Smith, J. N. M. & Jolly, A. (2006). Feeding competition between introduced *Eulemur fulvus* and native *Lemur catta* during the birth season at Berenty Reserve, Southern Madagascar. In: A. Jolly, R. W. Sussman, N. Koyama and H. Rasamimanana (eds), *Ringtailed Lemur Biology*. New York: Springer, pp. 119–140.

Pollock, J. I. (1979). Female dominance in *Indri indri*. *Folia Primatologica*, 31, 143–164.

Preuschoft, S. & van Schaik, C. P. (2000). Dominance and communication: conflict management in various social settings. In: F. Aureli and F. B. M. de Waal (eds), *Natural Conflict Resolution*. Berkeley, CA: University of California Press, pp. 77–105.

Pusey, A., Williams, J. & Goodall, J. (1997). The influence of dominance rank on the reproductive success of female chimpanzees. *Science*, 277, 828–831.

Radespiel, U. & Zimmermann, E. (2001). Female dominance in captive mouse lemurs (*Microcebus murinus*). *American Journal of Primatology*, 54, 181–192.

Ramanankirahina, R., Joly, M. & Zimmermann, E. (2011). Peaceful primates: affiliation, aggression, and the question of female dominance in a nocturnal pair-living lemur (*Avahi occidentalis*). *American Journal of Primatology*, 73(12), 1261–1268.

Raps, S. & White, F. J. (1995). Female social dominance in semi-free-ranging ruffed lemurs (*Varecia variegata*). *Folia Primatologica*, 65, 163–168.

Rasamimanana, H. R. & Rafidinarivo, E. (1993). Feeding behavior of *Lemur catta* females in relation to their physiological state. In: P. M. Kappeler & J. U. Ganzhorn (eds), *Lemur Social Systems and their Ecological Basis*. New York: Plenum Press, pp. 122–133.

Rathbun, G. B. (1979). The social structure and ecology of elephant-shrews. *Zeitschrift für Tierpsychologie*, 20(Supplement), 1–76.

Rendall, D. (1993). Does female social precedence characterize captive aye-ayes (*Daubentonia madagascariensis*)? *Folia Primatologica*, 14, 125–130.

Richard, A. F. (1974). Intra-specific variation in the social organization and ecology of *Propithecus verreauxi*. *Folia Primatologica*, 22, 178–207.

Richard, A. F. & Nicoll, M. E. (1987). Female social dominance and basal metabolism in a malagasy primate, *Propithecus verreauxi*. *American Journal of Primatology*, 12, 309–314.

Roberts, G. (2005). Cooperation through interdependence. *Animal Behaviour*, 70, 901–908.

Rowell, T. E. (1967). A quantitative comparison of the behaviour of a wild and a caged baboon troop. *Animal Behaviour*, 15, 499–509.

Sakamaki, T., Behncke, I., Laporte, M. et al. (2015). Intergroup transfer of females and social relationships between immigrants and residents in bonobo (*Pan paniscus*) societies. Dispersing primate females. In: T. Furuichi, J. Yamagiwa & F. Aureli (eds) *Dispersing Primate Females*. pp. 127–164. Japan, Springer.

Sapolsky, R. M. (2005). The influence of social hierarchy on primate health. *Science*, 308, 648–652.

Sapolsky, R. M., Romero, L. M. & Munck, A. U. (2000). How do glucocorticoids influence stress responses? Integrating permissive, suppressive, stimulatory, and preparative actions. *Endocrine Review*, 21, 55–89.

Sauther, M. L. (1991). Reproductive behavior of free-ranging *Lemur catta* at Beza Mahafaly special reserve, Madagascar. *American Journal of Physical Anthropology*, 84, 463–477.

Sauther, M. L. (1993). Resource competition in wild populations of ring-tailed lemurs (*Lemur catta*): Implications for female dominance. In: P. M. Kappeler and J. Ganzhorn (eds), *Lemur Social Systems and Their Ecological Basis*. New York: Plenum, pp. 135–152.

Savin-Williams, R. C. (1977). Dominance in a human adolescent group. *Animal Behavior*, 25, 400–406.

Savin-Williams, R. C. (1979). Dominance hierarchies in groups of early adolescents. *Child Development*, 50, 923–935.

Savin-Williams, R. C. (1980). Dominance hierarchies in groups of middle to late adolescent males. *Journal of Youth and Adolescence*, 9, 75–85.

Schjelderup-Ebbe, T. (1922). Beiträge zur Sozialpsychologie des Haushuhns. *Zeitschrift für Psychologie*, 88, 226–252.

Schryer, F. J. (2001). Multiple hierarchies and the duplex nature of groups. *Journal of the Royal Anthropological Institute*, 7, 705–721.

Sclafani, V., Norscia, I., Antonacci, D. & Palagi, E. (2012). Scratching around mating: factors affecting anxiety in wild *Lemur catta*. *Primates*, 53, 247–254.

Shizuka, D. & McDonald, D. B. (2012). A social network perspective on measurements of dominance hierarchies. *Animal Behaviour*, 83, 925–934.

Shizuka, D. & McDonald, D. B. (2014). Errata corrige to Shizuka, D., and McDonald, D. B. (Animal Behaviour, 83, 925–934). *Animal Behaviour*, 87, 243.

Simmen, B., Hladik, A. & Ramasiarisoa, P. (2003). Food intake and dietary overlap in native *Lemur catta* and *Propithecus verreauxi* and introduced *Eulemur fulvus* at Berenty, Southern Madagascar. *International Journal of Primatology*, 24, 949–968.

Singh-Manoux, A., Marmot, M. G. & Adler, N. E. (2005). Does subjective social status predict health and change in health status better than objective status? *Psychosomatic Medicine*, 67, 855–861.

Stevens, J. M. G., Vervaecke, H., de Vries, H. & van Elsacker L. (2007). Sex differences in the steepness of dominance hierarchies in captive bonobo groups. *International Journal of Primatology*, 28, 1417–1430.

Strayer, F. F. & Strayer, J. (1976). An ethological analysis of social agonism and dominance relations among preschool children. *Child Development*, 47, 980–989.

Sueur, C., Petit, O., De Marco, A., *et al.* (2011). A comparative network analysis of social style in macaques. *Animal Behaviour*, 82, 845–852.

Surbeck, M., Mundry, R. & Hohmann, G. (2010). Mothers matter! Maternal support, dominance status and mating success in male bonobos (*Pan paniscus*). *Proceedings of the Royal Society B: Biological Sciences*, 278, 590–598.

Tanaka, M. (2007). Habitat use and social structure of a brown lemur hybrid population in the Berenty Reserve, Madagascar. *American Journal of Primatology*, 69, 1189–1194.

Tecot, S. R. & Romine, N. K. (2012). Leading ladies: Leadership of group movements in a pair-living, co-dominant, monomorphic primate across reproductive stages and fruit availability seasons. *American Journal of Primatology*, 74, 591–601.

Thierry, B. (1986). A comparative study of aggression and response to aggression in three species of macaque. In: J. G. Else & P. C. Lee (eds), *Primate Ontogeny, Cognition and Social Behaviour* Cambridge University Press, pp. 307–313.

Thierry, B. (2000). Covariation of conflict management patterns across macaque species. In: F. Aureli & F. B. M. de Waal (eds), *Natural Conflict Resolution*, Berkeley, CA: University of California Press, pp. 106–128.

Thierry, B. (2013). Identifying constraints in the evolution of primate societies. *Philosophical Transactions of the Royal Society B*, 368, 20120342.

van Schaik, C. P. & Kappeler, P. M. (1996). The social systems of gregarious lemurs: lack of convergence with anthropoids due to evolutionary disequilibrium? *Ethology*, 102, 915–941.

Vehrencamp, S. L. (1983). A model for the evolution of despotic versus egalitarian societies. *Animal Behaviour*, 31, 667–682.

von Holst, D., Hutzelmeyer, H., Kaetzke, P., *et al.* (2002). Social rank, fecundity and lifetime reproductive success in wild European rabbits (*Oryctolagus cuniculus*). *Behavioral Ecology and Sociobiology*, 51, 245–254.

Waeber, P. O. & Hemelrijk, C. K. (2003). Female dominance and social structure in Alaotran gentle lemurs. *Behaviour*, 140, 1235–46.

Warkentin, M. J. (1968). Observations on the behavior and ecology of the nutria in Louisiana. *Tulane Studies in Zoology and Botany*, 15, 10–17.

Wasserman, S., Faust, K. & Iacobucci, D. (1994). *Social Network Analysis: Methods and Applications* (*Structural Analysis in the Social Sciences*). Cambridge University Press.

Webster, G. & Goodwin, B. C. (1982). The origin of species: a structuralist approach. *Journal of Social and Biological Structures*, 5, 15–47.

White, F. J. (1988). Party composition and dynamics in *Pan paniscus*. *International Journal of Primatology*, 9, 179–193.

Wickings, E. J. & Dixson, A. F. (1992). Testicular function, secondary sexual development, and social status in male mandrills (*Mandrillus sphinx*). *Physiology and Behavior*, 52, 909–916.

Wright, P. C. (1984). Biparental care in *Aotus trivirgatus* and *Callicebus moloch*. In: M. Small (ed.), *Female Primates: Studies by women primatologists*. New York: Alan R. Liss, Inc., pp. 59–75.

Wright, P. C. (1993). Variations in male–female dominance and offspring care in non-human primates. In: B. D. Miller (ed.), *Sex and Gender Hierachies*. Cambridge University Press, pp. 127–145.

Wright, P. C. (1999). Lemur traits and Madagascar ecology: coping with an island environment. *Yearbook of Physical Anthropology*, 42, 31–72.

Wroblewski, E. E., Murray, M. C., Keele, F. B., *et al.* (2009). Male dominance rank and reproductive success in chimpanzees, *Pan troglodytes schweinfurthii*. *Animal Behaviour*, 77(4), 873-885.

Young, A. L., Richard, A. F. & Aiello, L. C. (1990). Female dominance and maternal investment in strepsirrhine primates. *American Naturalist*, 135, 473–488.

Zaramody, A., Fausser, J. L., Roos, C., *et al.* (2006). Molecular phylogeny and taxonomic revision of the eastern woolly lemurs (*Avahi laniger*). *Primate Reports*, 74, 9–22.

Zink, C. F., Tong, Y., Chen, Q., *et al.* (2008). Know your place: neural processing of social hierarchy in humans. *Neuron*, 58, 273–283.

5 Something to make peace for: conflict management and resolution

Man is neither, by nature, peaceful nor warlike. Some conditions lead to war, others do not.
Otterbein 1997, p. 272

5.1 The end of social groups

After the death of King Alfonso XI of Castile in 1350, his eldest son Peter took control of the territory. Peter's mother, Maria of Portugal, had Alfonso's mistress, Eleanor de Guzman, killed. The death of Eleanor, who had given birth to ten children (Peter's half-siblings) split the royal descendants (legitimate and illegitimate) into two rival factions: Peter and his allies, and Alfonso's children by Eleanor, the Trastámaras, led by Henry. In the fights over the territory, Peter defeated the coalition led by his half-brother Henry in 1356 and 1360 and had two half-brothers (Henry's full brothers John and Peter) executed. Henry was forced to flee but later gained the support of Aragon, France, and many nobles of Castile. Meanwhile, Peter allied with Edward, heir to the English throne. This alliance allowed Peter to maintain the control upon the territory, until when Edward fell ill and, for political reasons, withdrew his support to Peter. Eventually, Henry took over the territory as Henry II and was responsible for the death of his half-brother Peter. John II, the great-grandson of Henry II, came to power after his mother Catherine of Lancaster, the regent, died in 1418. John II lacked authority and the territory became a battlefield for the nobles, more or less related to the ruling family, to gain power. As a result, the Trastámara family was, again, divided into two main factions: John II (with his supporting nobles) and his cousin Alfonso V of Aragon with his allies and brothers, Henry and John. In 1420, John II was kidnapped by his cousin Henry and then liberated by an ally, Álvaro de Luna. In 1429, Alfonso V ordered his brothers Henry and John to lead a joint attack (which was unsuccessful) against their cousin John II. In 1443 John II was once again captured by his cousin John of Aragon and the territory fell into near anarchy until 1445, when the group of nobles supporting John II, led by Álvaro de Luna, won a battle and Henry of Aragon was killed. When in 1454 Henry IV succeeded John II, another internal rift divided the royal family, with part of the nobles sustaining Henry IV's daughter Joan as a legitimate successor and another part sustaining Henry IV's half-brother Alfonso and, after his strange premature death in 1465, his half-sister Isabel. Isabel, who married Ferdinand (the heir to the throne of Aragon), eventually took over the territory but in the process her original family had been completely destroyed (for further information: Valdeón Baruque, 2002; Jardin, 2015).

As stressed by Otterbein (1997), 'some conditions lead to war, others do not'. War and unmanaged violence under certain circumstances may not be convenient for the preservation of social groups. For example, raiding of neighbouring groups was a common custom in the Tandroy people, traditional warrior-pastoralists living in south Madagascar and organised in clans (Jolly, 2004). However, the endless wars over cattle, land, women, tributes and succession, in which military alliances were forever changing, led to the disruption of Tandroy groups and, consequently, of their culture. Besides the cases reported above, human history is full of examples of the annihilation of entire families or groups to obtain power, such as the internal fighting among the pharaoh's descendants in ancient Egypt (e.g., Merneptah's heirs in the nineteenth dynasty), the famous War of the Roses in the England of the fifteenth century and described by Shakespeare in his first historical tetralogy, or today's Mafia blood feuds.

Lemur 'history' is also rich in episodes of internal family and group fighting to control a territory. As a matter of fact, lemurs often live in groups in which many individuals are more or less tightly related to one another. Group fission, enhanced by the death of the dominant female or by the increase of troop size, involves active targeting by the dominant clique of a subordinate clique (Jolly, 2012). Subordinates, once forced out, can fight to establish new territory by subdividing the original home range, or fight to take neighbouring troops' ranges (Koyama, 1991; Jolly et al., 1993; Hood and Jolly, 1995; Koyama et al., 2002; Takahata et al., 2005; Ichino, 2006; Ichino and Koyama, 2006; Jolly et al., 2006; Jolly, 2012).

Soma and Koyama (2013) reported that in a troop of Berenty ring-tailed lemurs, when the 'matriarch' died, a fight started among her descendants. The troop was originally composed of the descendants of one dominant female, the 'grandmother'. The grandmother and her second daughter died between December 2003 and August 2004. After the death of the grandmother, hierarchy stability crumbled and the group started reorganising. The second daughter's daughters (orphans) formed a subgroup. The fourth youngest daughter formed a subgroup with her most compliant younger sisters, ultimately depriving her older sisters of their high ranks. In fact, she became the alpha female. The oldest sister, now beta-ranked, formed another subgroup with her two sons. The middle-ranked sister associated, instead, with her subadult son. Therefore, four subgroups rose after the death of the matriarch: the dominant female and her younger sisters, the beta-ranked sister's family, the middle-ranked sister's family, and the subgroup of orphan daughters. In October 2014, the dominant female (after having offspring) and her subgroup chased away – via repeated target aggression – two of their nieces, from the orphan subgroup. The expulsion process started after one of the two nieces had delivered an infant. Eventually, the two nieces ran away, crossing the territories of two other troops, and they could never come back. After the eviction, whenever the alpha female and her younger sisters encountered the expelled nieces they pursued them aggressively. Instead, the subgroups of the older aunts did not. By September 2005, the evicted nieces had gained a sleeping site on the boundary of the territories of two other groups. However, they did not often win confrontations with other troops. One of the expelled nieces dropped her infant while retreating from a confrontation in October 2014. By the end of October the nieces had not yet succeeded in establishing their own territory and on 31 October 2005, their male consort was found dead from unidentified causes. A further example of lemur conflict is the so-called 'Civil War' reported by Alison Jolly (2004) in another troop of Berenty ring-tailed lemurs. During the 1992 drought, the group led by Diva drove the subordinate half of the group away. For several weeks the exiles ran across the ranges of at least five neighbouring troops and could feed only when no other group was around. During the repeated and exhausting confrontations with the original group, two of the exiled females lost their babies following two events of infanticide perpetrated by an immigrant male, in one case, and by a female from the original group in the other case. The evicted subgroup kept losing the contests with the original group until the exiled females changed their battle tactic. Instead of leaving just a couple of animals in the 'fire front' they started presenting a united front. The evicted subgroup started winning confrontations, taking over more and more feeding trees, thus getting to control the best and widest part of the territory. Hence the original, defeated group became marginalised and after the death of the alpha female included only a small number of individuals. In 1997, the alpha female of the fallen group, facing shrinkage of territory and food resources, chased away the daughter of the former alpha female with her little group (her daughter, her niece and their offspring). In this case the newly evicted subgroup did not succeed in conquering its own space within the original home range and its members died or disappeared one by one. By using the word 'war' Alison Jolly underlined how complex the fight tactics can be, even in lemurs, and stressed how the changing coalitions are crucial in determining the outcome of social conflicts.

In chimpanzees, the most famous case of conflict is the 'Four Year War of Gombe chimpanzees', reported by Jane Goodall in the 1970s. The seed of the conflict was planted when the so-called Kasekela community began to split. Seven adult males and three mothers and their offspring began spending most of their time in the southern part of the community's home range. By 1972, these chimpanzees had formed their own group, known as Kahama. When the males of the two communities met each other in the overlapping zone, they engaged in typical territorial fights. Then, a series of deadly brutal attacks perpetrated by the Kasekela members on the individuals of the Kahama community followed, leading to the annihilation of Kahama (Goodall *et al.*, 1979; Goodall and Berman, 2000). The coalitionary killing of chimpanzees was then observed in many study sites other than Gombe (Tai National Park, Ivory Coast: Boesch and Boesch-Achermann, 2000; Kibale, Uganda: Watts and Mitani, 2000; Muller, 2002; Watts *et al.*, 2006; Budongo, Uganda: Newton-Fischer, 2002; Loango, Gabon: Boesch *et al.*, 2006; Mahale, Tanzania: Kaburu *et al.*, 2013).

In the past, conflicts in chimpanzees have been used to make an argument in favour of the biological nature of human violence and in support of the evolutionary vision of human warfare as derived from chimpanzee-like conflicts. But when new pieces of evidence started accumulating on bonobos (*Pan paniscus*), the African ape at least as close to humans as chimpanzees, the 'warrior vision' started creaking. Bonobos have never been observed engaging in coalitionary killing and, in contrast to chimpanzees, show higher levels of cooperation in problem solving and tolerance (de Waal and Aureli, 1996; Hare *et al.*, 2012) related to more socially symmetrical relationships (de Waal, 1987; Fruth and Hohmann, 2002; Palagi and Cordoni, 2012), male–female co-dominance (Palagi *et al.*, 2006; Box 4.1), developmental delay with respect to social play (Paoli *et al.*, 2007; Palagi and Cordoni, 2012) and non-conceptive sociosexual behaviour (Koski and Sterck, 2009; Rilling *et al.*, 2012). Additionally, compared to chimpanzees, bonobos seem to possess a greater amount of grey matter in the brain regions involved in perceiving others' distress, an emotional state underpinning empathic abilities (Call *et al.*, 2002). Based on these characteristics, some scholars consider bonobos as a better model than chimpanzees to make inferences about the behaviour of extinct and extant hominids (Fry, 2012). For example, it has been reported that ancestral hominid conflict behaviour may have been more bonobo- than chimpanzee-like (de Waal, 2009), as suggested by the small, non-projecting canines and reduced sexual dimorphism of 4.4-million old *Ardipithecus ramidus* (White *et al.*, 2009; Fry 2012). The new pieces of information, added up, draw a different scenario in which humans and their predecessors were more likely prey than hunters (Hart and Sussman, 2011). Different from 'chimpanzeeists', 'bonobists' theorise that humans are not biologically deemed to violence and war.

In an interview on the BBC website, Dr Bostrom from Oxford University's Future of Humanity Institute stated that 'We're at the level of infants in moral responsibility, but with the technological capability of adults. There is a bottleneck in human history. The human condition is going to change. It could be that we end in a catastrophe or that we are transformed by taking much greater control over our biology.' (www.bbc.com/news/business-22002530). The fact that culture can somehow help us control our biology recalls Huxley's vision of the garden and the gardener, according to which ethics and human nature (the garden) need the continuous intervention of intelligence and culture (the gardener) to be maintained (Huxley, 1893). With today's knowledge, we know that culture comes from the brain, which is one of the most wonderful and marvellous products of biological evolution. Therefore, the dichotomy between culture and biology, and the idea that war is the result of thoughtless human biology whereas its control is the result of cultural intervention, have no foundation any more. Biology itself provides the tools to limit aggression and imposes the control of it because massive disruptive behaviours would prevent the survival of social groups thus proving anti-adaptive, unless group viability is jeopardised. When resources are limited the only exit strategy allowing the survival of at least part of the population may be group disruption. On one hand, the reduction of group size can restore environmental viability (with fewer animals accessing resources in a given area); on the other hand the formation of different 'factions' can increase the risk of violent intergroup fights to gain exclusive control over resources. The new social settings generated by the conflict are usually unstable and may require uneasy adjustments. Splitting can ultimately exacerbate group fragmentation, leading to the annihilation of the smallest subgroups if not of a significant part of the population. Given the costs of group disruption (see the examples of humans, chimpanzees and lemurs described above), the possible future gains resulting from engaging in a war must be extremely high and likely. This improbable condition makes the disruptive conflict the last resort to rely on.

To preserve their integrity, social groups adopt conflict management strategies that can be handled in many ways, with only a few of these actually involving physical violence (Rubin *et al.*, 1994; Fry, 2006). In humans, with variation from one culture to the next, disputants, for example, may seek the help of an impartial mediator to resolve their disagreements, negotiate a form of compensation, practice avoidance or decide to make peace without any form of intervention or compensation (Fry, 2012). These types of conflict management which for many years have been associated with the human cultural background are strategies that are in place, also in other animal species.

Box 5.1 | **by Cary J. Roseth**

Speaking of which: peacemaking in children

My interest in children's conflict resolution began as a teacher at a boarding school for 14- to 18-year-old students. Like many new educators, I started teaching with the assumption that I would have some 'good' students and some, let's say, 'less good' students. My hope was to help these less good students change their ways and become knowledgeable, well behaved, and other-oriented adults.

Of course, what I quickly realised is that students are both good and less good, and that different social contexts elicit different kinds of behaviour. These insights may seem trivial to ethologists, but most teachers rarely see their students outside of class and, as a result, tend to have a one-dimensional view of their students. As a boarding school teacher, however, I saw my students many times a day in multiple settings, and in so doing gained a much more complete understanding of their behavioural repertoire. Much to my surprise, even my good students exhibited selfish behaviours, and even 'antisocial' behaviours like aggression proved adaptive under some circumstances.

Unlike ethologists, teachers also tend to focus on improvement and reform rather than objective behavioural analysis. For peer conflict, this means that teachers are keenly aware of the costs that conflict may impose on society, especially if conflict involves physical aggression. However, at boarding school, I had the opportunity to view peer conflicts from start to finish, and in so doing learned to appreciate the variety of antecedents and consequents associated with these events. I learned that 'hitting' was not simply a 'behaviour intended to harm another', which is the typical definition of aggression in the psychological literature, nor was 'hitting' always indicative of social incompetence. Instead, hitting and other aspects of peer conflict could not be fully understood outside the social and behavioural context in which they were embedded.

In sum, what fascinated me as a young teacher about children's conflict is the same thing that continues to fascinate me today. Understanding peer conflict and its resolution demands that we reconsider common assumptions about what constitutes right and wrong behaviour and what constitutes social competence. The subject's biological roots also demand that conflict and its resolution be examined at multiple levels of analysis. Indeed, there is no end to the questions that can be asked about its immediate causation, development, function and evolution.

One of the most interesting findings from our research is the extent to which preschoolers – children 3 to 5 years old – actively balance their own self-interests with those of others. What makes this so interesting is that it contrasts so fundamentally with psychologists' historical assumption that children are developmentally self-centred and thus unable to understand others' perspectives, much less act in a way that balances their interests with those of others.

Indeed, our research has shown that preschoolers behave strategically in conflict situations, even using aggressive behaviour in functionally suitable ways. In fact, preschoolers use both *coercive* (e.g., aggression, threats) and *prosocial* behaviours (e.g., cooperation, affiliation) to control resources, and those preschoolers who use both of these strategies – so-called 'bistrategic' resource controllers – enjoy privileged access to both agentic and communal resources such as peer regard, attention and affiliation. Our research also shows that preschoolers tend to use costly coercive strategies (e.g., physical aggression) to establish resource control when peer relationships are new or unstable (Pellegrini et al., 2007; Roseth et al., 2007, 2011), less costly coercive (e.g., verbal aggression) and prosocial strategies (e.g., prosociality) to maintain resource control when peer relationships stabilise (Roseth et al., 2007), and a combination of coercive and prosocial strategies as well as reconciliation to offset potential social costs and foster affiliation (Roseth et al., 2011). Another one of our interesting findings is that the frequency of preschoolers' conflict is curvilinear, increasing then decreasing in accord with the stability of peer relationships (Roseth et al., 2007, 2011). This finding indicates that bistrategic resource control involves more than using both coercive and prosocial strategies some of the time, and instead involves careful calibration, with preschoolers matching both the form and combination of strategies to the context in which resource control bouts occur (see also Pellegrini et al., 2011a, b). Here again, these findings challenge traditional views of social

Box 5.1 (continued)

> Box 5.1 (*cont.*)
>
> competence and what 'counts' as prosocial and antisocial behaviour. These findings also challenge the common assumption that all forms of aggression are indicative of social-cognitive deficits, suggesting instead that some forms of aggression may actually encourage the development of children's conflict resolution.
>
> Perhaps most surprising, we have also found that teacher intervention during preschoolers' peer conflict may inhibit natural conflict resolution. In fact, preschoolers are more likely to stay together and resume friendly interaction when teachers do *not* intervene during coercive bouts (Roseth et al., 2008). What makes this surprising of course is that most teachers intervene in children's conflict with the intent of helping them to develop constructive means of resolving conflict. Rather than improve and reform children's conflict resolution, however, our findings suggest that teacher intervention may actually disrupt children's developing abilities.
>
> Research on children's conflict and its resolution is at its infancy now, as little is known about biological determinants, developmental trends, cognitive and sociocognitive correlates, and the way socialisation affects form and function. Future research is also needed to examine how children learn to calibrate personal and social concerns, and the extent to which such calibrations depend on their status within different social hierarchies. We also know very little about the role of children's empathic responses to others' harm and vulnerability, and whether these responses motivate reconciliation or third-party intervention.
>
> Finally, future research is needed to examine conflict resolution in other age groups and in other settings. While the preschool years offer a unique window into the emergence of conflict resolution, the setting also represents the human equivalent of captive animal research. In fact, many university preschools are described as child development 'laboratories' and are outfitted with one-way glass, observation booths and cameras. Such amenities support careful observation data collection, of course, but are hardly representative of a 'natural habitat'. Future research is therefore also needed to examine conflict resolution in the home and among older children in a variety of social settings.

5.2 Aggressive behaviour: from the individual-centred to the relational perspective

Violence is more prominent than harmony. In the media, a single case of aggression (towards family, friends, spouse, etc.) garners more attention than many incidents of cooperative behaviour enacted to prevent aggression from occurring or to buffer its consequences. The scenario drawn from the media (joined to a good dose of readers' insane curiosity) has contributed to bias people's perception towards the feeling that we live in a violence-grounded society. Violence is unquestionably present in social settings but it is hard to believe that human social life is governed by violence. What makes the news on severe violence so striking is that it is more unusual than the positive interactions normally characterising everyday life, leading to conflict prevention and resolution.

For decades after the start of the first ethological studies, aggressive behaviour has been defined as antisocial, regardless of whether it was considered an innate, pre-programmed behaviour (Lorenz, 1966: hydraulic model) or the result of both instinct and experience (Tinbergen, 1968). In such studies, the antisocial character of the aggressive behaviour was biased by the model species, the setting, or the experimental protocol selected to study the effect of agonistic events. For example, in different species, spanning rats and humans, aggression was studied in experimentally isolated individuals and was often induced through painful stimulation (Johnson, 1972). This unnatural setting strongly altered the behavioural response of animals, preventing them from engaging in any form of regulation of the aggression.

Zuckerman in his book of 1932 (*The Social Life of Monkeys and Apes*) concluded that monkey society was based on and ruled by aggression because he had observed that two-thirds of his study group of hamadryas (*Papio hamadryas*) had been exterminated by internal fights, started by males to take possession of females. De Waal (1989a) argued that this tremendous situation derived from a condition of severe crowding (more than 70 animals were kept in a 20 × 30 m enclosure) and extremely imbalanced sex ratio, since only a few females were present in the group. These conditions, which would be stressful for any social species, proved unsustainable for hamadryas, whose society is organised in one-male units (or harems: one reproductive male with several reproductive females). Again, the totally artificial and aberrant conditions prevented the possibility for Zuckerman's animals to adopt tactics of conflict management.

Studies on highly territorial and solitary animals were used as another source of data to support the antisocial nature of aggression. Indeed, some fish and bird species use conflicts to keep intruders away and secure the exclusive use of resources in an area (de Waal, 1996; de Waal *et al.*, 2000). When animals do not know each other, aggression is used to maintain distance and leads to dispersal (individual model; de Waal *et al.*, 2000). Clearly, in this context, animals compete for resources without getting back any benefit from group living, such as lower risk of predation, better ability to defend food resources, and information sharing on food, competitors or predator location. But does it really make sense talking about antisocial behaviour when the social setting is missing?

When animals coexist in social groups, conflicts of interest are common because several or many individuals share the same territory and find themselves competing for the same resources. For example, the presence of a desirable resource (food item, oestrus female, shelter, etc.) can tickle the appetite of different group members, thus leading to a conflict between the individuals interested in such resource. Yet, social living has been positively selected in many animal species. Therefore, social animals must have found a way to manage conflicts of interest, avoiding the disruption of their group. In the social context, stating that aggression is an antisocial behaviour is just not correct. First of all, a fight requires the interaction between at least two individuals, the actor and the receiver of the aggression. Moreover, aggression can be one of the very elements upon which social cohesion is grounded. The dominance hierarchy of social groups is established via aggressive events. The outcome of such events, determining winners and losers, allows animals to acquire different ranking positions. In the long run, the rise and maintenance of a hierarchy based upon previous aggression limit future aggression. By knowing their own place in the hierarchy, animals can choose to stay away from each other, with an individual, typically a subordinate, withdrawing from trying to access a resource, in order to avoid dangerous competition with a dominant. The subordinate animal has to make a decision on the basis of the costs and benefits of engaging in overt competition with a dominant. When rank differences between individuals are high, starting a fight with a powerful group member is likely to be a lost cause. Consequently, the resource automatically goes to the dominant and the risk of receiving an aggression (and being injured) is warded off. This tactic of avoidance is used to work out a conflict of interest (by simply dropping competition) in an alternative to aggression. In short, aggression regulates aggression via a negative feedback cycle.

The avoidance tactic can be also used when, by limiting social interactions, animals can reduce the probability that a conflict of interest arises in the first place. This situation has been observed in groups of apes kept under limited space conditions (Cordoni and Palagi, 2007), where no emergency exit is available and when the rise of conflicts over food, preferred locations, etc. is highly probable. The reduction of animals' activity, known as the elevator effect (de Waal *et al.*, 2000), allows avoiding severe aggression, which could have disastrous outcomes in an enclosure with no escape opportunities. Also in this case, experience of previous aggression occurring under high-density conditions helps mould avoidance tactics and, therefore, the hypothesis that past aggression may serve to limit future aggression still stands.

Besides avoidance, an individual can tolerate another over a resource, possibly sharing part of it or letting a group mate access it, in exchange for future cooperation (coalitionary support, territory defence, food search, etc.). The level of tolerance can vary according to the species considered (more or less despotic), the quality and quantity of the resource, and the behavioural mechanisms adopted by animals to buffer social tension when a conflict of interest can be foreseen (pre-conflict strategy). As will be discussed in Part III, such mechanisms can involve play, grooming, and other types of affinitive contacts, which can enhance tolerance towards others.

Sometimes aggression is inevitable but animals cannot afford to let aggression destroy their social group every time that a conflict of interest comes along. Hence, mechanisms to resolve the conflicts have been developed to fix the social relationships damaged by aggressive events. The most basic form of conflict resolution is reconciliation, occurring when former opponents directly engage in post-conflict reunion through friendly contacts, more technically defined as affinitive interactions.

It is possible to talk about reconciliation only when two individuals engage in affinitive contacts most frequently after they have fought. Otherwise, friendly contacts could be simply part of the usual interactions the former opponents engage in, and not enhanced by the conflict and used to buffer its negative consequences.

Reconciliation represents the 'formal act' promptly and unequivocally settling the hostility between former opponents and restoring their relationship. Individual recognition and memory of previous aggressive events are the only items of the 'cognitive equipment' that animals need to reconcile. Due to its low cognitive cost, reconciliation is probably the most powerful mechanism that many species of social animals can use to avoid the disruption of their group. Literature has confirmed that post-conflict reunion occurs in different species of corvids (Clayton and Emery, 2007; Fraser and Bugnyar, 2010, 2011) and across mammals, from gregarious marsupials (Cordoni and Norscia, 2014) to highly social primates (Fry, 2013). Besides reconciliation, other more sophisticated mechanisms can be adopted by social animals (including humans, see Box 5.1 above). Third parties not involved in the conflict can mediate the reunion of the two opponents by contacting either the victim or the aggressor (see Box 5.2 to further explore this aspect), also depending on the quality of the relationship binding the third party and each one of the opponents. In this case, higher relational cognitive abilities are required since animals must somehow 'understand' their social position in the group and the relationship linking the other group members to one another. However, the debate over the abilities that animals need to engage in third-party interactions and their actual significance (consolation, appeasement, protection, aggression buffering, etc.) is totally open.

Avoidance, tolerance and aggression (followed or not by reconciliation) are the three pillars of the Relational Model proposed by de Waal (2000) to explain how conflicts of interest can be managed in social animals.

Box 5.2 | **by Giada Cordoni**

Speaking of which: conflict management in non-primate mammals

Competition and aggressiveness are customarily presented as the natural state of affairs in the animal kingdom (human society included) (Koenig, 2002; Rubenstein, 2012; Thierry, 2013). Conflict management is a pool of behavioural strategies which are fundamental for the maintenance of group integrity (Aureli and de Waal, 2000). Therefore, we expect to find them wherever and whenever a social group is in place. These strategies include reconciliation between opponents (de Waal and van Roosmalen, 1979), triadic affinitive contacts directed towards victim or aggressor (Palagi *et al.*, 2004, 2006; Cordoni *et al.*, 2006; Koski and Sterck, 2007) and quadratic affiliation among bystanders (Judge and Mullen, 2005; Leone *et al.*, 2010). Starting from the pioneering study on chimpanzees by de Waal and van Roosmalen (1979),

Box 5.2 (continued)

Box 5.2 (*cont.*)

numerous studies have demonstrated the occurrence of natural conflict resolution in many primate species (Aureli *et al.*, 2002). Nevertheless, demonstrating the occurrence of post-conflict behaviours is not enough to unveil the role of such behaviours in group maintenance and social relationship establishment. Several studies have been carried out in non-primate mammals (domestic goat – Schino, 1998; spotted hyena – Wahaj *et al.*, 2001; bottlenose dolphin – Weaver, 2003; domestic dog – Cools *et al.*, 2008). By using a comparative approach, these studies underlined the differences and similarities with primates in aggression management. For example, in domestic goats, post-conflict friendly reunions can play an important role in reducing the victim's anxiety; similar findings were reported in long-tailed macaques as well (Aureli and van Schaik, 1991). Studying non-primate species can help in determining the potentially diverging roles of different post-conflict strategies.

I started investigating and comparing different features of reconciliation and third-party affiliations (e.g., frequency, timing, modality) in two non-primate species, the red-necked wallaby (*Macropus rufogriseus*) and the grey wolf (*Canis lupus lupus*).

Notwithstanding the divergence between Metatheria and Eutheria mammals that occurred some 168–178 million years ago (Luo *et al.*, 2011; dos Reis *et al.*, 2012), marsupials have evolved many morphological, behavioural and neocortical traits that are markedly comparable to those of placental mammals that occupy similar niches (Karlen and Krubitzer, 2007; Meredith *et al.*, 2008; Isler, 2011). It is possible to hypothesise that evolution has led marsupial and placental mammals to develop similar solutions to deal with similar social challenges, such as competition and aggressiveness.

Contrary to other marsupials, red-necked wallabies are gregarious and face the challenges of spatial closeness, including coexistence around feeding sites. The occurrence of reconciliation found in this species (mean group corrected conciliatory tendency: 27.40%; Cordoni and Norscia, 2014) supports the hypothesis that, like other social or gregarious placental mammals facing similar pressures, wallabies enact post-conflict strategies to reduce possible negative consequences of a conflict.

As in the great apes (Arnold and Whiten, 2001; Aureli and Schaffner, 2002; Fraser *et al.*, 2008) and two prosimian species, *Propithecus verreauxi* (Palagi *et al.*, 2008a) and *Lemur catta* (Kappeler, 1993), conciliatory contacts between wallabies can play an important role in limiting the likelihood of further attacks towards the victim by the aggressor and in reducing the rates of post-conflict scratching in both the opponents, functioning as an anxiety reliever (Arnold and Aureli, 2007) and, consequently, as a social uncertainty reductive mechanism (Uncertainty Reduction Hypothesis, Aureli and van Schaik, 1991). After an aggression, the temporary interruption of partner

5.2 Aggressive behaviour: from the individual-centred to the relational perspective

compatibility (Cords and Aureli, 2000) may endanger the ordinary social associations and the degree of interindividual tolerance in both opponents, thus causing anxiety. Red-necked wallabies maintain their interindividual relationships via continuous 'covert' interactions, such as social sniffing, social licking, feeding in contact and scent marking (Higginbottom and Croft, 1999; Jarman, 2000). Therefore, also in this species reconciliation may represent a useful tool to restore relaxed social conditions.

Moving from marsupial to placental mammals, my interest focused on a highly social and cooperative species, the grey wolf (*Canis lupus lupus*). The typical wolf pack is structured as a family group including a breeding alpha pair and its offspring (Mech and Boitani, 2003; Miklósi, 2014). In a pack, individuals generally form a well-established linear hierarchy in which all males dominate over all females (Mech, 1999; Cordoni and Palagi, 2008). My colleague Elisabetta Palagi and I demonstrated the occurrence of reconciliation (Cordoni and Palagi, 2008) and third-party affiliations (Palagi and Cordoni, 2009; Cordoni and Palagi, 2015), directed both towards the victim (victim triadic affiliation) and the aggressor (aggressor triadic affiliation), in grey wolves.

According to the Valuable Relationship Hypothesis (Kappeler and van Schaik, 1992; Cords, 1997; van Schaik and Aureli, 2000), in wolves the frequency of reconciliation was affected by the coalitionary support rates: the higher the levels of exchange support, the higher the rates of reconciliation. This result fits with the features of wolf social life. In fact, despite the strict hierarchical arrangement of individuals, the pack existence and integrity strongly rely on cooperation between fellows, which have specific roles and act in a flexible way, following environmental and social changes (Peterson *et al.*, 2002). A similar result was obtained in Assam macaques (*Macaca assamensis*) where the females reconciled more with individuals with whom they exchanged more agonistic support, which increases the probability to access resources and to maintain the social status (Cooper *et al.*, 2005).

Through the investigation of triadic post-conflict affiliations (towards either victim or aggressor), other functional similarities between primates and wolves have been demonstrated. In *Canis lupus lupus*, similar levels of victim triadic affiliation (mean group triadic contact tendency: 63.27%) and aggressor triadic affiliation (mean group triadic contact tendency: 44.78%) were found, although the two types of triadic post-conflict affiliations showed a functional dichotomy (Palagi and Cordoni, 2009; Cordoni and Palagi, 2015). Victim triadic affiliation reduces the likelihood of renewed attacks towards the victim as predicted by the Victim Protection Hypothesis, already demonstrated in bonobos (Palagi and Norscia, 2013) and chimpanzees (Palagi *et al.*, 2014). On the other hand, aggressor triadic affiliation decreases the probability of renewed attacks

Box 5.2 (continued)

> **Box 5.2** (*cont.*)
>
> towards uninvolved third parties by reducing the arousal of the aggressor as predicted by the Appeasement Hypothesis (*sensu* van Hooff, 1967). Studies on chimpanzees (Romero *et al.*, 2011) and lowland gorillas (Palagi *et al.*, 2008b) have shown a similar function of aggressor triadic affiliation.
>
> In wolves victim triadic affiliation occurred more frequently in the absence of reconciliation, functioning as a substitute for conciliatory contacts between opponents (Substitute for Reconciliation Hypothesis) as found in many primate species (e.g., chimpanzees, Palagi *et al.*, 2006; Fraser and Aureli, 2008; bonobos, Palagi *et al.* 2004; baboons, Wittig *et al.*, 2007; mandrills, Schino and Marini, 2012; monkeys, Palagi *et al.*, 2014a). Aggressor triadic affiliation tended to be more likely after the occurrence of reconciliation thus suggesting that contacting an aggressor immediately after conciliatory contacts reduced the risk for the bystander to be the object of a renewed attack. The interdependency of reconciliation and aggressor triadic affiliation in primates is still under debate. In chimpanzees, Romero *et al.* (2011) found that aggressor triadic affiliation occurred more often in the absence of reconciliation whereas in the same species Koski and Sterck (2007) found no support for the Substitute for Reconciliation Hypothesis.
>
> The study on wolves has demonstrated that victim triadic affiliation and aggressor triadic affiliation represent the two sides of the same coin: affiliation with the aggressor is distributed according to the individual ranking position and support value, thus highlighting the strict hierarchy characterising wolf society; at the same time, affiliation with the victim is affected by the strength of the relationship quality, thus underlying the cooperative aspect of the wolf social system.
>
> Contrasting victim and aggressor triadic affiliations concurrently in the same social group provides the opportunity to understand the precise role of each of the two behavioural categories. Surprisingly, in primates the victim and aggressor triadic affiliations have never been examined at the same time, although natural conflict resolution has been studied in primates more than in any other taxon. This is an interesting line of research that I would like seen addressed in the future.

5.3 Give peace a chance: reconciliation and lemurs

Pre-conflict mechanisms

Different behavioural mechanisms can be used to reduce conflict probability in social groups. The affinitive contacts adopted by group mates to reduce the likelihood of aggression in stressful conditions, such as the contacts occurring right before food distribution or in crowding conditions, depend on the species behavioural repertoire. Humans can use language, songs, symbolic gestures and friendly contacts such as handshakes, embraces, nose rubbing and kisses to increase tolerance (Firth, 1972; Kendon and Ferber, 1973; de Waal, 1989a; Floyd 1999, 2001; Field, 2014; daily news and…look around). Gorillas – which are not renowned for their proclivity to engage in intense affiliative interactions (Harcourt, 1979; Watts, 1995, 1996) – have been observed touching each other more often in response to greater density (Cordoni and Palagi, 2007). In this case, touch is a rapid affiliative item selectively used across the different sex–class combinations as appeasement behaviour (Cordoni and Palagi, 2007). The exuberant bonobos, which contrary to gorillas enjoy the pleasure of vigorous social contacts, cope with stressful conditions by selectively increasing grooming, play and the frequency of non-reproductive sexual interactions (Palagi et al., 2006; Paoli et al., 2007; Tacconi and Palagi, 2009). Chimpanzees, other than increasing grooming, use greetings such as kissing and submissive bowing as an aggression disclaimer (de Waal, 2000; Palagi et al., 2004).

In monkeys, grooming is the most relevant behaviour to increase tolerance (for further information on grooming see also Chapter 8). The highly despotic rhesus macaques seem to cope with crowded conditions by grooming each other more frequently. Males groom other males and females more frequently under limited space conditions, with a consequent reduction of aggression rates (Judge and de Waal, 1997; de Waal, 2000).

Spider monkeys (*Ateles geoffroyi*) show aggressive escalation during subgroup fusion events. During these events, animals embrace each other more often thus mitigating aggression. In this case, embraces (but not grooming) appear to be effective in appeasing or reassuring others (Aureli and Schaffner, 2007). In capuchin monkeys, grooming has been found to serve as a conflict prevention mechanism. Before scheduled feeding, a predictable competitive situation, grooming rates increase and the risk of aggressive escalation decreases (Polizzi di Sorrentino et al., 2010).

Grooming also seems to work in reducing the probability of aggression in strepsirrhines, even if the data available in this respect are meagre.

In red-fronted lemurs, low-ranking females exchange grooming for the tolerance of dominant females. In fact, subordinate females have been observed giving the largest amount of grooming (relative to what they received) to high-ranking females probably to reduce the risk of being evicted from their group (Port et al., 2009). In general, the higher the level of grooming asymmetry (grooming given versus grooming received), the higher the frequency of aggression observed. This is consistent with previous findings on wedge-capped capuchin monkeys showing that equal reciprocation of grooming time between partners characterises affiliative grooming relationships to strengthen social bonds and reduce the probability of aggression (O'Brien, 1993). These results highlight the long-term relationship between grooming reciprocity within dyads and the probability of agonistic events (Port et al., 2009).

Overall, the mechanisms of conflict prevention in lemurs have been largely neglected, if not completely ignored. Indeed, finding appropriate literature to draft this chapter has not been easy at all. This is why at this point we decided to stop writing and start analysing unprocessed data in order to fill part of the gap.

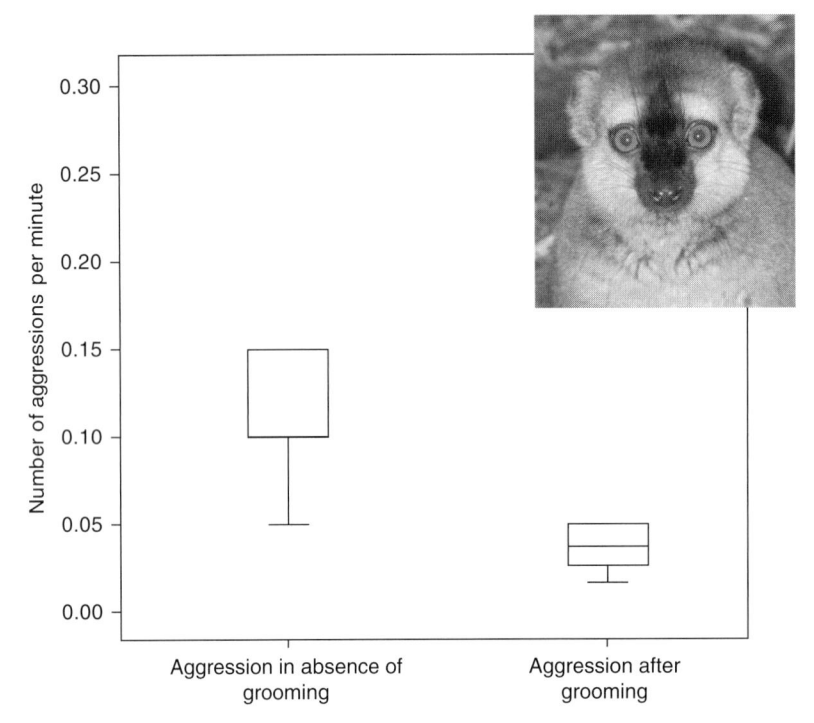

Figure 5.1 Frequency of aggressive events that occurred in presence and absence of previous grooming interactions in *Eulemur rufus* × *collaris* from the Berenty forest, Madagascar. Exact Wilcoxon's test: T = 0, ties = 0, n = 8, $p < 0.05$. Photo: Elisabetta Palagi.

By scrutinising our records on lemurs from the Berenty forest, collected from 2006 to 2011, we verified if it was less likely that two individuals engaged in aggressive interactions after being involved in a grooming session. We focused on the possible cause–effect relationship existing between the occurrence of grooming and subsequent aggression. Within focal sampling periods of ten minutes, we observed that grooming significantly reduced the probability of aggression involving the former groomer and groomee in *Eulemur rufus* x *collaris*, *Propithecus verreauxi* and *Lemur catta* (Figures 5.1, 5.2 and 5.3). These results confirm that also in the short term, grooming can be a pre-conflict tool employed by lemurs to limit the occurrence of aggression. This is a coping strategy in that tolerance levels are raised by increasing affiliation rates.

Lemur catta seems able to deal with the possible devastating effect of violent conflicts under space reduction. To verify the effect of reduced space availability in lemurs, we analysed the data collected in the period 2004–2005 in captive lemurs hosted at the Pistoia Zoo (Tuscany, Italy). Here, the lemurs could stay in a grassy yard outside (100 m^2) or in an inside enclosure (20 m^2). To reduce the biases related to seasonality we restricted the analyses to those months when both conditions (inside and outside) were present. Data were gathered over 119 hours inside and 67 hours outside. When comparing the agonistic events (normalised over the hours of observation) occurring in the

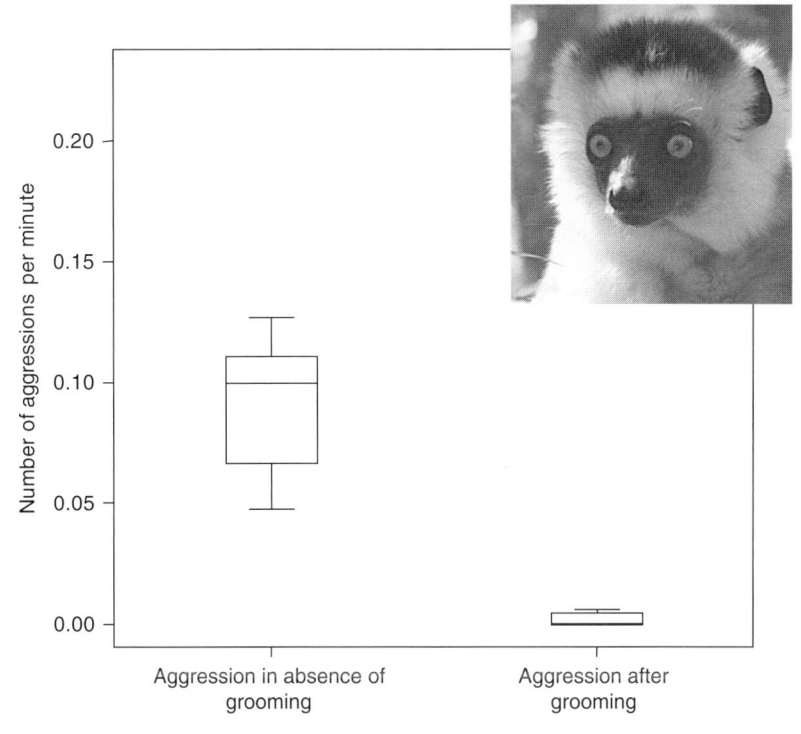

Figure 5.2 Frequency of aggressive events that occurred in presence and absence of previous grooming interactions in *Propithecus verreauxi* from the Berenty forest, Madagascar. Exact Wilcoxon's test: T = 1.00, ties = 1, n = 14, $p < 0.001$. Photo: Ivan Norscia.

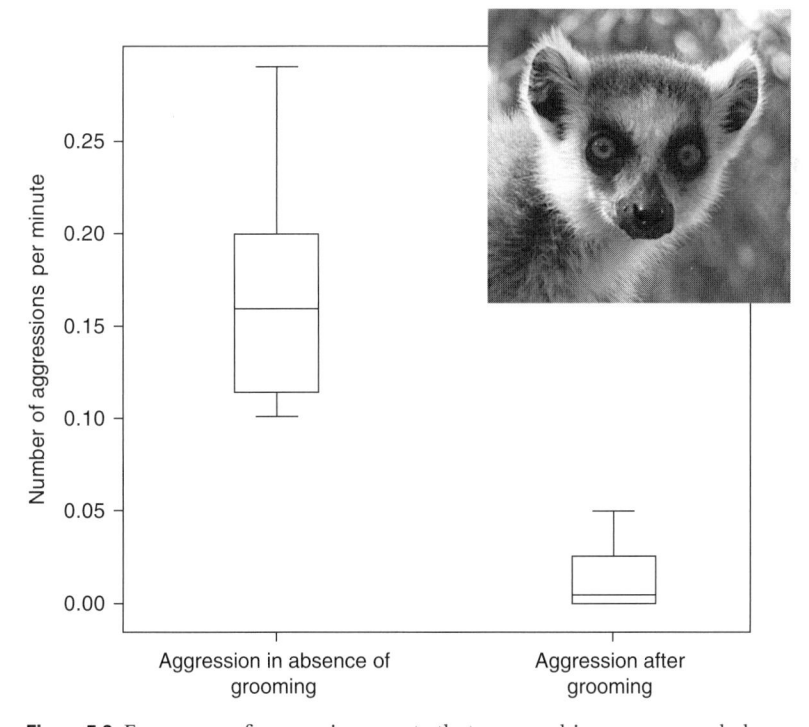

Figure 5.3 Frequency of aggressive events that occurred in presence and absence of previous grooming interactions in *Lemur catta* from the Berenty forest, Madagascar. Exact Wilcoxon's test: T = 0, ties = 0, n = 10, $p < 0.01$. Photo: Elisabetta Palagi.

outside enclosure with those occurring in the inside room, we found that the overall level of agonistic events did not differ between the two conditions;[1] nor did grooming.[2] Yet, lemurs were able to limit aggressive incidents of high intensity, involving chasing, bite and grab bouts (*sensu* Palagi et al., 2005). In fact, the proportion of severe conflicts over the total number of aggressive events was significantly reduced when animals were hosted inside, under short-term crowding conditions.[3] Therefore, under brief crowding periods, dominants do not appear to control the quantity of their attacks but their modality, making them less intense and therefore less dangerous. Even if we considered the lemur species characterised by the most despotic social structure (*Lemur catta*, see also Chapter 4), our results do not support the density–aggression model, predicting that high-density conditions lead to higher levels of aggression (Calhoun, 1962). Instead, the tactic put in place by ring-tailed lemurs to avoid the escalation of aggressive encounters is consistent with the coping model, predicting that animals can respond to crowded conditions by modifying their behaviour to reduce the number and/or severity of aggressive encounters (de Waal, 1989b; Aureli and de Waal, 1997; Judge and de Waal, 1997; de Waal et al., 2000; Judge, 2000).

The 'elevator effect' described by de Waal and coworkers (2000) predicts that individuals finding themselves in confined spaces tend to inhibit their activity in order to avoid conflicting interactions. This paradigm is consistent with our results on captive lemurs which, once again, express tactic abilities also observed in monkeys and apes. For example, capuchin monkeys significantly reduce intense aggression, play and social grooming when spatially confined, thus limiting social encounters. A clear signal of the elevator effect is that under acute crowding capuchins increase self-grooming, which reduces arousal and does not implicate any social interaction (van Wolkenten et al., 2006). Adult gorillas have been observed avoiding interactions under space reduction by staying more spatially dispersed and increasing the levels of sitting alone, avoidance, and dismissing behaviours (Cordoni and Palagi, 2007).

Another behaviour used by lemurs to prevent conflicts from occurring is play. The linkage between social play and aggressive contacts is particularly evident in captive ring-tailed lemurs. For example, those dyads showing low aggression rates can engage most frequently in play fighting (Palagi, 2009). While analysing the behavioural sequences of play fighting in juvenile ring-tailed lemurs, Pellis and Pellis (1997) found that play dynamics strongly resembled those of real aggression. In many primate and non-primate species, the roughness of play fighting can likely lead to the escalation into serious fighting (Fagen, 1981; Pellis, 2002a). However, the low levels of aggression found in ring-tailed lemurs under playful contexts (Palagi, 2009) suggest that animals are able to cope with possible ambiguous and dangerous situations. In this view, play can be considered as a tool to increase tolerance and, at the same time, to assess own and others' physical and social skills

---

[1] Overall aggression rates did not differ between outdoor and indoor conditions (Exact Wilcoxon's test: n = 9, T = 19.00, p = 0.734).

[2] Overall grooming frequencies did not differ between outdoor and indoor conditions (Exact Wilcoxon's test: n = 9, T = 10.00, p = 0.164).

[3] The proportion of high-intensity conflicts over the total number of aggressive occurrences was lower indoors than outdoors (Exact Wilcoxon's test: n = 9, T = 3.00, p = 0.039).

in a safe context. The potential of play in relaxing social interactions and increasing tolerance is reinforced by playful signals such as relaxed open mouth (Palagi *et al.*, 2014a), body postures and movements (Palagi, 2009; Yanagi and Berman, 2014), which can modulate the intensity of the interaction and increase the fairness between the playmates.

In a species characterised by a high level of social tolerance, the Verreaux's sifaka, play is used by adult males as an ice-breaker mechanism, thus establishing the conditions to accept new adult males in the social group during the mating season, a period of high conflict of interest (Antonacci *et al.*, 2010). Play is used not only by lemurs but also by many other primate taxa (and not exclusively to increase tolerance), as extensively explained in Chapter 7.

Post-conflict mechanisms

As already discussed earlier in this chapter, despite the existence of several mechanisms to avoid aggression, sometimes the conflict is inevitable because the wanted resource is crucial and inalienable or because there are not the conditions to employ pre-conflict buffering measures. Once the aggression has occurred it is necessary to engage in post-conflict affiliative interactions, such as reconciliation (Figure 5.4), to restore the relationship between former opponents and avoid negative consequences (e.g., conflict spreading) for the entire social group. The occurrence of reconciliation – defined as the first exchange of affiliative contact between opponents soon after a conflict (de Waal and van Roosmalen, 1979) – has been found in humans (Fujisawa *et al.*, 2005; Box 5.1) and other social or gregarious mammals (e.g., wallabies, Cordoni and Norscia, 2014; domestic goats, Schino, 1998; horses, Cozzi *et al.*, 2010, spotted hyenas, Wahaj *et al.*, 2001; wolves, Cordoni and Palagi, 2008; domestic dogs, Cools *et al.*, 2008; dolphins, Weaver, 2003; primates, Aureli *et al.*, 2002; see also Box 5.2).

Within primates, this kind of conflict resolution has been extensively studied in monkeys and apes but much more rarely in strepsirrhines (Figure 5.5). Yet, comparing social strepshirrhines with the best-known anthropoids is crucial for a better understanding of the evolution of conflict resolution mechanisms. Alison Jolly stressed the importance of comparative studies as the key to understand in depth the dynamics of lemur social behaviour. Based on early studies she pointed out that 'ringtailed lemur interactions are much more black and white than in many anthropoids – either affiliative or aggressive – between any two animals, with minimal ambiguity, and no reconciliation after quarrels...Reconciliation has recently been asserted for sifaka, ringtails and brown lemurs, using different measurements....Complexity of social relations would be worth revisiting by someone very familiar with behavior of both monkeys and prosimians' (Jolly, 2012, p. 33).

As explained in the preface to this volume, lemurs (which retain ancestral traits such as a small brain and communication highly based on smell) contrast with anthropoids in various behavioural features, including female dominance, lack of pronounced sexual dimorphism and strict seasonal breeding (Martin, 1990; Wright, 1999). However, group-living lemurs share basic features with anthropoids such as

Figure 5.4 Tail anointing in *Lemur catta* male. His ears are flattened, which preludes a ritualised or real fight (top). *Propithecus verreauxi* during grooming, one of the most common affinitive patterns used by primates to reconcile (bottom). Berenty forest, Madagascar. Photos: Ivan Norscia and Elisabetta Palagi.

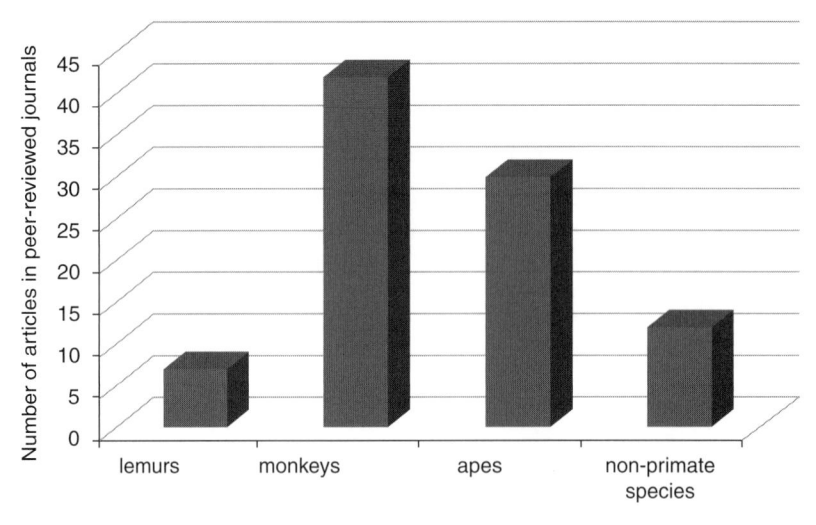

Figure 5.5 Scientific articles on natural conflict resolutions published in peer-reviewed journals up to February 2013.

cohesive multimale/multifemale societies, female philopatry (Pereira and Kappeler, 1997) and individual recognition (Palagi and Dapporto, 2006, 2007), a prerequisite for reconciliation (Aureli *et al.*, 2002).

Despite the importance of extending conflict management studies to strepsirrhines, until 2008 post-conflict behaviour had been investigated only in a handful captive groups of lemurs: *Eulemur fulvus*, *Eulemur macaco*, *Lemur catta* (Kappeler, 1993; Rolland and Roeder, 2000; Roeder *et al.*, 2002; Palagi *et al.*, 2005). Later, the investigation was extended to wild lemurs (*Propithecus verreauxi*, Palagi *et al.*, 2008a; *Eulemur rufus* × *collaris*, Norscia and Palagi, 2011; Palagi and Norscia, 2015). In 1993, Kappeler had predicted that the occurrence of reconciliation could be linked to the levels of tolerance of social groups because he had found reconciliation to be present in *E. fulvus* but not in *L. catta*, with the latter being characterised by the highest levels of despotism. The subsequent studies have confirmed this hypothesis. In fact, reconciliation was not found in *Eulemur macaco* (Roeder *et al.*, 2002) and in the majority of groups of *Lemur catta* in which the phenomenon had been investigated (Kappeler, 1993; Palagi *et al.*, 2005). *Eulemur macaco*, compared to *Eulemur fulvus*, shows despotic female dominance (Roeder *et al.*, 2002; Hemelrijk *et al.*, 2008). Studies in the wild have confirmed that in more tolerant species, namely *Propithecus verreauxi* and *Eulemur rufus* × *collaris,* reconciliation is present. Therefore, the presence of reconciliation in *P. verreauxi* suggests that this species, characterised by relaxed interindividual relationships, is more similar to *E. fulvus* (also showing looser hierarchical relationships) than to *L. catta* and *E. macaco*, at least in terms of dominance style. In social strepsirrhines, the occurrence of reconciliation, scattered across different taxonomic groups, depends on the dominance style of the societies, more than on the phylogenetic closeness of the species. Something similar has been observed in 20 different *Macaca* species, which share similar social organisation (multimale-multifemale) but which largely diverge in the dominance style. In macaques, a positive correlation was found between the degree of group tolerance and the level of reconciliation (Thierry, 1986, 2000; de Waal and Luttrell, 1989). For example, reconciliation rates are higher in the tolerant Tonkean (*Macaca tonkeana*; Thierry, 1985a, b; Petit and Thierry, 1994; Demaria and Thierry, 2001; Ciani *et al.*, 2012; Palagi *et al.*, 2014a) and crested macaques (*Macaca nigra*; Petit *et al.*, 1997) than in the despotic Japanese macaques (*Macaca fuscata*; Chaffin *et al.*, 1995; Schino *et al.*, 2004).

According to the Social Constraints Hypothesis (de Waal and Aureli, 1996) the differences in primate social styles (de Waal and Luttrell, 1989), already present in infancy (Thierry, 1985a), influence a wide range of behaviours including aggression, affiliation, dominance and nepotism (Thierry, 1985b, 1990; Aureli *et al.*, 1997; Petit *et al.*, 1997; Balasubramaniam *et al.*, 2012). The studies on lemurs have revealed that in strepsirrhines, as in haplorrhines, peacemaking is not only possible but also shaped by similar social variables. Hence, the Social Constraints Hypothesis largely confirmed in monkeys and apes (de Waal and Aureli, 1996) applies to strepsirrhines, as well.

It is hard to determine whether the similar distribution of reconciliation according to the social style found in both strepsirrhines and haplorrhines is an evolutionary analogy or homology. The forces driving the reconciliation phenomenon have probably been in place since the origin of the primate group but the entity of the phenomenon likely depends on whether social groups have acquired (or not acquired) a certain dominance style. The cost-benefit ratio, as predicted by the basic sociobiology theories (Wilson, 2000), determines if taking the risk of reconciling is worth it or not. Considering the presence of reconciliation as the simple aftermath of the dominance style would be reductive. Beyond primates, high rates of reconciliation have been found in wolves (*Canis lupus*; mean conciliatory tendency, 44.1% in the wild: Baan *et al.*, 2014; 53.3% in captivity: Cordoni and Palagi, 2008). Within a pack, every subject knows its social standing with every other individual and each group defends its own territory as a unit. Yet, even if the alpha male normally guides the movements of the pack and initiates aggressions against intruders (Mech, 1977), the subordinate members can sometimes oppose their leader's actions. According to Zimen (1981), no subject decides alone the carrying out of activities that are vital to the group cohesion. In short, wolves are highly despotic but also extremely cooperative, to the point that bystanders can conveniently affiliate with either aggressors or their victims depending on the situations (Cordoni and Palagi, 2015; Box 5.2). The existence of an extremely cooperative pack has presumably to do not only with hunting but also with the collective rearing of offspring and, consequently, with reproductive success (Mech and Boitani, 2003). It is clear that in wolves the benefit of reconciling and preserving the social bonds outweighs the cost deriving from pack disruption, which would be detrimental for both dominants and subordinates. Thus, reconciliation can be found in despotic groups provided that they are cooperative. Further evidence of this assumption is the presence of reconciliation in spotted hyenas (*Crocuta crocuta*). Hyenas are despotic but often depend on help from other group members during hunts, defence of ungulate carcasses against competitors, and coalition formation that is important in both the acquisition and maintenance of social rank (Wahaj *et al.*, 2001). In hyenas, as in wolves, the necessity to cooperate overcomes the competition between dominants and subordinates, which explains the presence of reconciliation. The lower levels of reconciliation observed in hyenas (mean conciliatory tendency: 11.3%) may be due to the fact that, contrary to wolves, spotted hyenas live in a fission–fusion society allowing dispersal – other than reconciliation – as an exit strategy.

Cooperation and despotism are two opposite forces that contribute in shaping reconciliation patterns, in a number of species, including lemurs. When the balance between these forces strongly fluctuates according to seasonal variations, long-term studies are needed in order to unveil reconciliation, especially if occurring at low frequencies. This is the case of *Lemur catta*, a species considered as virtually unable to reconcile. However despotic *Lemur catta* may be, female–female coalitionary support is present, especially during aggressive interactions. Therefore, reconciliation can play a role also in this species. Consistently, the phenomenon was present in one captive group hosted at the Pistoia Zoo, Italy and two wild groups from the Berenty forest, Madagascar (Palagi and Norscia, 2015). The presence of reconciliation could be detected thanks to the extensive database available, which allowed

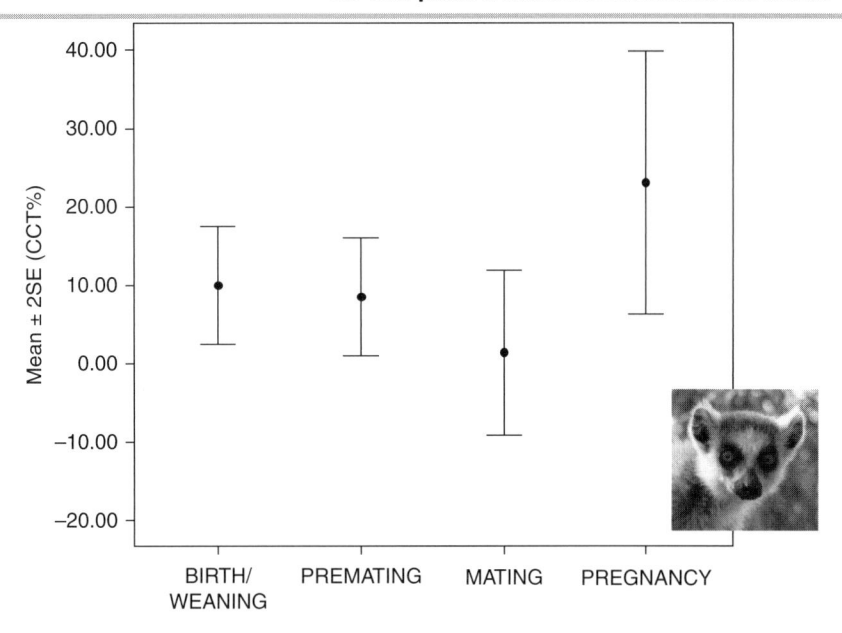

5.3 Give peace a chance: reconciliation and lemurs   133

Figure 5.6 Seasonal fluctuation of corrected conciliatory tendencies (CCTs) in *Lemur catta*. Reconciliation rates are minimum during the mating period (Palagi and Norscia, 2015), thus the period in which data are collected is crucial to highlight the presence of the phenomenon. Photo: Elisabetta Palagi.

analysis of 2339 post-conflict (PCs) and matched control (MCs) focal observations collected from eight groups, five in the wild and three in captivity.[4] We found that the season, more than other variables (wild/captivity setting, rank or individual features), influenced the reconciliation levels, which were lowest during the mating period (Figure 5.6). This result is consistent with the strict reproductive seasonality of *Lemur catta*, in which female oestrus lasts around one day per year. Overall, it is confirmed that reconciliation can be present in a despotic species but not when the advantages of intragroup cooperation are annihilated by intragroup competition, as it occurs in seasonal breeders when reproduction is at stake.

In lemur species characterised by lower levels of despotism, the phenomenon of reconciliation seems to have a more prominent role in the maintenance of group homeostasis. In *Propithecus verreauxi* and in *Eulemur rufus* × *collaris*, the conciliatory tendency can reach high levels (Verreaux's sifaka: mean 44.72% ±6.51 SE; brown lemurs: mean 26.62% ±8.34 SE) and reconciliation could be detected over

---

[4] After the last aggressive pattern of any given agonistic event, the loser of the interaction was followed as the focal individual for a ten minute post-conflict period (PC). Matched control observations (ten minute long MCs) took place during the next possible day at the same time, context (feeding, resting or travelling) and physiological season (lactation, pre-mating, mating, and pregnancy) as the original PC. The MC was conducted on the same focal animal, in the absence of agonistic interactions during the ten minutes before the beginning of the MC and when the opponents had the opportunity to interact. For further details on the methods used for this study case see Palagi and Norscia, 2015. For details on the general methodology see for example: de Waal and Yoshihara, 1983 (original article) and Aureli and de Waal, 2000 (extensive review).

short observation periods. Nevertheless, also in these species reconciliation remains linked to cooperation/competition balance. In fact, reconciliation did not reach any significant level when the aggression occurred in the feeding context (Palagi *et al.*, 2008a; Norscia and Palagi, 2011), when (again) a crucial resource is at stake and competition higher. Therefore, in lemurs as in other primates, reconciliation can be influenced by season or context and be subject to the evaluation of costs and benefits, which in turn depends on whether cooperating is more rewarding than competing. This theoretical concept found an empiric support in *Propithecus verreauxi*. In this species, victims were found more likely to engage in post-conflict affiliation with the former aggressor after low-intensity agonistic encounters occurred outside the feeding context. Moreover, lemurs were more likely to reconcile with valuable partners: post-conflict affiliations were preferentially initiated by subordinates with top-ranking individuals, and occurred more frequently between animals sharing good relationships. Hence, lemurs can evaluate possible risks and benefits before engaging in post-conflict reunions, in order to gain long-term benefits such as future cooperation. In *P. verreauxi*, reconciliation was also found to reduce – in the short term – the probability of further attacks on the victim by the same aggressor (Palagi *et al.*, 2008a). In this respect, reconciliation may be seen as an *hic-et-nunc* mechanism,[5] needed to avoid conflict spreading across group members, possibly leading to social disruption. While a similar short-term function of reconciliation was found in *Eulemur rufus × collaris*, the strategic use of reconciliation as a tool to gain possible long-term benefits was not detected in this species. Here, reconciliation was not biased towards valuable or high-ranking group mates (Norscia and Palagi, 2011).

Reconciliation offers a further example of the importance of investigating phenomena by applying a comparative approach not only to strepsirrhines and haplorrhines but also within strepsirrhines themselves. Lemur species (and groups!) characterised by differences in the interindividual relationships, social looseness and hierarchical steepness (see Chapter 4 for reference) can provide different elements to understand the proximate and ultimate factors underlying the evolution of important behavioural phenomena, such as conflict management and resolution.

Box 5.3 | **by Peter Verbeek**

Speaking of which: the contribution of peace ethology to life science

I came to the study of aggressive and peaceful behaviour from a broad interest in social behaviour. Social behaviour is a fascinating topic of study to me because it involves interactions between actors with potentially diverging interests. How will social actors work out inevitable conflicts of interest? Do they opt for dealing with it aggressively or peacefully? And what factors

---

[5] *Hic et nunc* means 'here and now', a mechanism that works in the same moment in which the action is enacted.

affect their choice of strategy? Seeking answers to these kinds of questions got me started in this field and still motivates my work today.

The main purpose of life on Earth as we understand it is to sustain and propagate itself. I want to know the role that peaceful behaviour plays in sustaining and propagating life across the entire spectrum of life, including the human manifestation of it. The notion that human beings occupy a pinnacle position on this Earth, somehow removed in behaviour and mind from nature, has never made sense to me. I agree with psychiatrist and environmentalist Ian McCallum that 'strictly speaking, there is no such thing as human nature. There is only nature and the very human expression of it' (McCallum, 2012). Doing comparative research on aggressive and peaceful behaviour in multiple species, including my own, feels natural to me. Moreover, I am convinced that the comparative method is indispensable if we want to uncover the very human expression of aggression and peace (cf. Tinbergen, 1968; see also Verbeek, 2008, 2013).

Most valuable to me are findings or insights that help me plot a new course for my work. The findings from my own work and that of my colleagues that show that across cultures young children can make peace with peers without adult intervention is a good example of that (Butovskaya et al., 2000; Verbeek and de Waal, 2001; cf. Kempes, 2008). The fact that in timing, form and function, early peacemaking in our own species resembles not only that of our primate cousins (Silk, 2002; Verbeek 2008), but also that of several other mammal species (e.g., Cordoni and Palagi, 2008), and even some birds (e.g., Fraser and Bugnyar, 2011), is shaping my research and thoughts about the future of the field as a whole. My research on aggression is teaching me the importance of the distinction between species-typical and species-atypical aggression (Verbeek et al., 2007; cf. Haller and Kruk, 2006). Recent findings that show that peaceful behaviour can also be species-typical or species-atypical (e.g., Sapolksy, 2006, 2013) have inspired me to pursue the distinction between species-typical and species-atypical in both aspects of my research.

I foresee an exciting and productive future for this research domain as it evolves from the study of natural conflict resolution (Aureli and de Waal, 2000) into a comprehensive study of peaceful behaviour. Now that we are finally looking at it systematically and comprehensively, it turns out that peaceful behaviour, such as friendly cooperation, helping and sharing, and behaviours that keep aggression in check, or re-establish non-violent relations and tolerance following conflict, appears in a wide range of species. Intriguing new findings such as sharing and peacekeeping in colonial orb web spiders (Wenseleers et al., 2013), increased cooperation among more distant relatives in species ranging from eusocial insects (van Zweden et al., 2012) to cichlid fish (Stiver et al., 2005), and the policing of selfish behaviour in

Box 5.3 (*cont.*)

genetically identical ants (Oldroyd, 2013), all clearly illustrate that explaining how and why peaceful behaviour has evolved and persists across a wide range of species now counts among the greatest challenges for behavioural biology. I see three interrelated ways in which the research domain will most likely develop and expand. First, as mentioned earlier, I expect that the field of natural conflict resolution will develop into a comprehensive peace ethology (Verbeek, 2008, 2013). Second, I foresee an important role for such a behavioural biology of peace in the future development of life science as a whole. Third, I expect that peace ethology will become one of the key disciplines within an emerging multidisciplinary behavioural science of peace. I discuss each of these three future directions next.

Peace ethology. In the final chapter of their epoch-making volume *Natural Conflict Resolution*, Filippo Aureli and Frans de Waal list eight research questions for which they felt answers were specifically needed (Aureli and de Waal, 2000, pp. 375–379). Four of these questions address ultimate concerns (species-comparisons on natural conflict resolution, including human vs non-human, functional questions, and links to the evolution of morality and justice), while four address proximate concerns (health, development, peace-keeping and social structure) (cf. Tinbergen, 1963).

Natural Conflict Resolution effectively translated Niko Tinbergen's ideas about an ethological approach to 'War and peace in animals and man' (Tinbergen, 1968) into testable research questions and integrative goals. However, at the time the volume was published, behavioural biology had become conceptually and institutionally divided (Thierry, 2007). Ethologists were mainly interested in behavioural mechanisms and behavioural development (i.e., proximate concerns; Kappeler *et al.*, 2013), while the study of the evolution and function of behaviour had been all but claimed by behavioural ecologists (i.e., ultimate concerns; Kappeler *et al.*, 2013). Now, 13 years after the publication of *Natural Conflict Resolution*, and 50 years after Tinbergen's seminal 'On Aims and Methods of Ethology', (Tinbergen, 1963) the ethological approach to social behaviour that effectively integrates ultimate and proximate considerations is making a strong comeback (Kappeler *et al.*, 2013; cf. Blumstein *et al.*, 2010). The stage appears set for a behavioural biology of peace that does full justice to the vision and expectations of its pioneers.

Peace ethology's expected contributions to life science. I believe that understanding the interplay between aggressive and peaceful behaviour is fundamental to understanding life. I predict that following contemporary life science's recent quantitative successes (e.g., genomics; molecular biology) new breakthroughs in our understanding of life will be more qualitative in nature. I expect that new holistic approaches will shed light on mutually beneficial interdependencies in life on Earth (Scofield and Margulis, 2012), including those between microbiomes and genomes, as expressed, for example, through behaviour (Ezenwa *et al.*, 2012) and development (Pennisi, 2013). New findings on mutualisms and interdependencies in nature will revitalise life science, and I predict that research on obstacles and catalysts of peaceful behaviour will make significant contributions to this, in particular through the coming of age of a genuine peace ethology.

Peace ethology's expected contributions to a multidisciplinary behavioural science of peace. There are compelling reasons for a translational multidisciplinary behavioural science of peace. While seeking sustainable development solutions will be a key goal for present and future generations, 'The most important public good is peace' (Sustainable Development Solutions Network, 2013, pp. 26). Like health, peace is a requirement for human happiness and well-being. Unlike health, of which we understand many of the physical and mental processes, our understanding of the processes of peace and what sustains them are still limited (cf. Coleman and Deutsch, 2012). Our knowledge of health derives for a great part from the fact that we routinely look beyond the human condition towards the rest of nature for an understanding of what impedes or optimises our health. A mature peace ethology can lead the way in doing the same for peace.

References

Antonacci, D., Norscia, I. & Palagi, E. (2010). Stranger to familiar: wild strepsirrhines manage xenophobia by playing. *PLos ONE*, 5(10) e13218. http://dx.doi.org/10.1371/journal.pone.0013218.

Arnold, K. & Aureli, F. (2007). Postconflict reconciliation. In: C. J. Campbell, A. Fuentes, A. C. MacKinnon, M. Panger & S. Bearder (eds.), *Primates in Perspective*. Oxford University Press, pp. 592–608.

Arnold, K. & Whiten, A. (2001). Post-conflict behaviour of wild chimpanzees (*Pan troglodytes schweinfurthii*) in the Budongo forest, Uganda. *Behaviour*, 138, 649–690.

Aureli, F. & de Waal, F. B. M. (1997). Inhibition of social behavior in chimpanzees under high-density conditions. *American Journal of Primatology*, 41, 213–228.

Aureli, F. & de Waal, F. B. M. (2000). *Natural Conflict Resolution*. California: Regents of the University of California.

Aureli, F. & Schaffner, C. M. (2002). Empathy as a special case of emotional mediation of social behavior. *Behavioural and Brain Science*, 25, 23–24.

Aureli, F. & Schaffner, C. M. (2007). Aggression and conflict management at fusion in spider monkeys. *Biology Letters*, 3, 147–149.

Aureli, F. & van Schaik, C. P. (1991). Post-conflict behaviour in long-tailed macaques (*Macaca fascicularis*): coping with the uncertainty. *Ethology*, 89, 101–114.

Aureli, F., Das, M. & Veenema, H. C. (1997). Differential kinship effect on reconciliation in three species of macaques (*Macaca fascicularis, M. fuscata*, and *M. sylvanus*). *Journal of Comparative Psychology*, 111(1), 91–99.

Aureli, F., Cords, M. & van Schaik, C. P. (2002). Conflict resolution following aggression in gregarious animals: a predictive framework. *Animal Behaviour*, 64, 325–343.

Baan, C., Bergmüller, R., Smith, D. W. & Molnar, B. (2014). Conflict management in free-ranging wolves, *Canis lupus*. *Animal Behaviour*, 90, 327–334.

Balasubramaniam, K. N., Dittmar, K., Berman, C. M., et al. (2012). Hierarchical steepness, counter-aggression, and macaque social style scale. *American Journal of Primatology*, 74, 915–925.

Blumstein, D. T., Ebensperger, L. A., Hayes, L. D., et al. (2010). Toward an integrative understanding of social behavior: new models and new opportunities. *Frontiers in Behavioral Neuroscience*, 4, 1–9.

Boesch, C. & Boesch-Achermann, H. (2000). *The Chimpanzees of the Taï Forest: Behavioural Ecology and Evolution*. Oxford: Oxford University Press.

Boesch, C., Head, J., Tagg, N., et al. (2006). Fatal chimpanzee attack in Loango National Park, Gabon. *International Journal of Primatology*, 28, 1025–1034.

Butovskaya, M., Verbeek, P., Ljungberg, T. & Lunardini, A. (2000). A multi-cultural view of peacemaking among young children. In: F. Aureli and F. B. M. de Waal (eds), *Natural Conflict Resolution*. Berkeley, CA: University of California Press, pp. 243–258.

Calhoun, J. B. (1962). Population density and social pathology. *Scientific American*, 206, 139–148.

Call, J., Aureli, F. & de Waal, F. B. M. (2002). Postconflict third party affiliation in stumptailed macaques. *Animal Behaviour*, 63, 209–216.

Chaffin, C. L., Friedlen, K. & de Waal, F. B. M., (1995). Dominance style of Japanese macaques compared with rhesus and stumptail macaques. *International Journal of Primatology*, 25, 1283–1312.

Ciani, F., Dall'Olio, S., Stanyon, R. & Palagi, E. (2012). Social tolerance and adult play in macaque societies: a comparison with different human cultures. *Animal Behaviour*, 84, 1313–1322.

Clayton, N. S. & Emery, N. J. (2007). The social life of corvids. *Current Biology*, 17, R652–R656.

Deutsch, M. & Coleman, P. T. (2012). Psychological components of sustainable peace: an introduction. In: *Psychological Components of Sustainable Peace*. New York: Springer.

Cools, A. K. A., van Hout, A. J. M. & Nelissen, M. H. J. (2008). Canine reconciliation and third-party-initiated postconflict affiliation: do peacemaking social mechanisms in dogs rival those of higher primates? *Ethology*, 114, 53–63.

Cooper, M. A., Bernstein, I. S. & Hemelrijk, C. K. (2005). Reconciliation and relationship quality in Assamese macaques (*Macaca assamensis*). *American Journal of Primatology*, 65, 269–282.

Cordoni, G. & Norscia, I. (2014). Peace-making in marsupials: the first study in the red-necked wallaby (*Macropus rufogriseus*). *PLoS ONE*, 9(1), e86859. http://dx/doi.org/10.1371/journal.pone.0086859.

Cordoni, G. & Palagi, E. (2007). Response of captive lowland gorillas (*Gorilla gorilla gorilla*) to different housing conditions: testing the aggression/density and coping models. *Journal of Comparative Psychology*, 121, 171–180.

Cordoni, G. & Palagi, E. (2008). Reconciliation in wolves (*Canis lupus*): new evidence for a comparative perspective. *Ethology*, 114, 298–308.

Cordoni, G. & Palagi, E. (2015). Being a victim or an aggressor: Different functions of triadic post-conflict interactions in wolves (*Canis lupus lupus*). *Aggressive Behavior*, http://dx.doi.org/10.1002/ab.21590.

Cordoni, G., Palagi, E. and Borgognini-Tarli, S. M. (2006). Reconciliation and consolation in captive western gorillas. *International Journal of Primatology*, 27, 1365–1382.

Cords, M. (1997). Friendship, alliances, reciprocity and repair. In: A. Whiten & R. W. Byrne, (eds), *Machiavellian Intelligence II*. Cambridge University Press, pp. 24–49.

Cords, M. & Aureli, F. (2000). Reconciliation and relationship qualities. In: F. Aureli & F. B. M. de Waal (eds), *Natural Conflict Resolution*. Berkeley, CA: University of California Press, pp. 177–198.

Cozzi, A., Sighieri, C., Gazzano, A., Nicol, C. J. & Baragli, P. (2010). Post-conflict friendly reunion in a permanent group of horses (*Equus caballus*). *Behavioural Processes*, 85, 185–190.

de Waal, F. B. M. (1987). Tension regulation and nonreproductive functions of sex in captive bonobos (*Pan paniscus*). *National Geographic Research*, 3, 318–335.

de Waal, F. B. M. (1989a). *Peacemaking Among Primates*. Cambridge, MA: Harvard University Press.

de Waal, F. B. M. (1989b). The myth of a simple relation between space and aggression in captive primates. *Zoo Biology Suppl.*, 1, 141–148.

de Waal, F. B. M. (1996). *Good Natured*. Cambridge, Massachusetts: Harvard University Press.

de Waal, F. B. M. (2000). Primates—a natural heritage of conflict resolution. *Science*, 289, 586–590.

de Waal, F. B. M. (2009). *The Age of Empathy*. New York: Harmony Books.

de Waal, F. B. M. & Aureli, F. (1996). Consolation, reconciliation and a possible cognitive difference between macaques and chimpanzees. In: A. E. Russon, K. A. Bard, S. T. Parker (eds), *Reaching into Thought: The Minds of Great Apes*. Cambridge University Press, pp. 80–110.

de Waal, F. B. M. & Luttrell, L. M. (1989). Toward a comparative socioecology of the genus *Macaca*: different dominance style in rhesus and stumptailed monkeys. *American Journal of Primatology*, 19, 83–109.

de Waal, F. B. M. & van Roosmalen, A. (1979). Reconciliation and consolation among chimpanzees. *Behavioral Ecology and Sociobiology*, 5, 55–66.

de Waal F. B. M. & Yoshihara, D. (1983). Reconciliation and redirected affection in rhesus monkeys. *Behaviour*, 85, 224–241.

de Waal, F. B. M., Aureli, F. & Judge, P. G. (2000). Coping with crowding. *Scientific American*, 282, 76–81.

Demaria, C. & Thierry, B. (2001). A comparative study of reconciliation in rhesus and Tonkean macaques. *Behaviour*, 138, 397–410.

dos Reis, M., Inoue, J., Hasegawa, M., *et al.* (2012). Phylogenomic datasets provide both precision and accuracy in estimating the timescale of placental mammal phylogeny. *Proceedings of the Royal Society B: Biological Sciences*, 279, 3491–3500.

Ezenwa, V. O., Gerardo, N. M., Inouye, D. W., Medina, M. & Xavier, J. B. (2012). Animal behavior and the microbiome. *Science*, 338, 198–199.

Fagen, R. (1981). *Animal Play Behavior*. New York: Oxford University Press, p. 684.

Field, T. (2014). *Touch*. MIT Press.

Firth, R. (1972). Verbal and bodily rituals of greeting and parting. In: J. S. La Fontaine (ed.), *The Interpretation of Ritual* London, UK: Routledge, pp. 1–38.

Floyd, K. (1999). All touches are not created equal: effects of form and duration on observers' interpretations of an embrace. *Journal of Nonverbal Behavior*, 23, 283–299.

Floyd, K. (2001). Human affection exchange: I. Reproductive probability as a predictor of men's affection with their sons. *The Journal of Men's Studies*, 10, 39–50.

Fraser, O. N. & Aureli, F. (2008). Reconciliation, consolation and postconflict behavioral specificity in chimpanzees. *American Journal of Primatology*, 70, 1–10.

Fraser, O. N. & Bugnyar, T. (2010). Do ravens show consolation? Responses to distressed others. *PLoS ONE*, 5(5), e10605, http://dx.doi.org/10.1371/journal.pone.0010605.

Fraser, O. N, & Bugnyar, T. (2011). Ravens reconcile after aggressive conflicts with valuable partners. *PLoS ONE*, 6(3), e18118, http://dx.doi.org/10.1371/journal.pone.0018118.

Fraser, O. N., Schino, G. & Aureli, F. (2008). Components of relationship quality in chimpanzees. *Ethology*, 114, 834–843.

Fruth, B. & Hohmann, G. (2002). How bonobos handle hunts and harvests: why share food? In: Boesch, C., Hohmann, G. & Marchant, L. F. (eds), *Behavioural Diversity in Chimpanzees and Bonobos*. New York: Cambridge University Press, pp. 231–243.

Fry, D. P. (2006). *The Human Potential for Peace: An Anthropological Challenge to Assumptions About War and Violence*. New York: Oxford University Press.

Fry, D. P. (2012). Life without war. *Science*, 336, 879–884.
Fry, D. P. (ed.) (2013). *War, Peace, and Human Nature: the convergence of evolutionary and cultural views*. Oxford University Press.
Fujisawa, K. K., Kutsukake, N. & Hasegawa, T. (2005). Reconciliation pattern after aggression among Japanese preschool children. *Aggressive Behavior*, 31, 138–152.
Goodall, J. & Berman, P. (2000). *Reason for Hope: A Spiritual Journey*. Grand Central Publishing.
Goodall, J., Bandura, A., Bergmann, E., *et al.* (1979). Inter-community interactions in the chimpanzee populations of the Gombe National Park. In: D. Hamburg and E. McCown (eds), *The Great Apes*. Menlo Park, CA: Benjamin/Cummings, pp. 13–53.
Haller, J. & Kruk, M. R. (2006). Normal and abnormal aggression: human disorders and novel laboratory models. *Neuroscience and Biobehavioral Reviews*, 30, 292–303.
Harcourt, A. H. (1979). Social relationships between adult male and female mountain gorillas in the wild. *Animal Behaviour*, 27, 325–342.
Hare, B., Wobber, V. & Wrangham, R. (2012). The self-domestication hypothesis: evolution of bonobo psychology is due to selection against aggression. *Animal Behaviour*, 83, 573–585.
Hart, D. & Sussman, R. W. (2011). In: R. W. Sussman & C. R. Cloninger (eds), *Origins of Altruism and Cooperation*. New York: Springer, pp. 19–40.
Hemelrijk, C. K., Wantia, J. & Isler, K. (2008). Female dominance over males in primates: Self-organisation and sexual dimorphism. *PLoS ONE*, 3(7), e2678, http://dx.doi.org/10.1371/journal.pone.0002678.
Higginbottom, K. & Croft, D. B. (1999). Social learning in marsupials. In: Box, H. O. and Gibson, K. R. (eds), *Mammalian Social Learning – Comparative and Ecological Perspectives*. Cambridge University Press, pp. 80–101.
Hood, L. C. & Jolly, A. (1995). Troop fission in female *Lemur catta* at Berenty Reserve, Madagascar. *International Journal of Primatology*, 16, 997–1015.
Huxley, T. H. (1893). Evolution and Ethics. Prolegomena. http://aleph0.clarku.edu/huxley/CE9/E-EProl.html
Ichino, S. (2006). Troop fission in wild ring-tailed lemurs (*Lemur catta*) at Berenty, Madagascar. *American Journal of Primatology*, 68, 97–102.
Ichino, S. & Koyama, N. (2006). Social changes in a wild population of ringtailed lemurs (*Lemur catta*) at Berenty, Madagascar. In: Jolly, A. Sussman, R. W., Koyama, N. & Rasamimanana, H. R. (eds), *Ringtailed Lemur Biology: Lemur catta in Madagascar*. New York: Springer, pp. 233–244.
Isler, K. (2011). Energetic trade-offs between brain size and offspring production: marsupials confirm a general mammalian pattern. *BioEssays*, 33(3), 173–179.
Jardin, J. P. (2015). La reina María de Portugal, entre padre, marido, hijo e hijastros: la mediación imposible. *e-Spania*, 20, http://dx.doi.org/10.4000/e-spania.24140.
Jarman, P. J. (2000). Males in macropod society. In: Kappeler, P. (ed), *Primate Males: Causes and Consequences of Variation in Group Composition*. Cambridge University Press, pp. 21–33.
Johnson, R. N. (1972). *Aggression in Man and Animals*. Philadelphia: W. B. Saunders.
Jolly, A. (2004). *Lords and Lemurs: Mad Scientists, Kings with Spears, and the Survival of Diversity in Madagascar*. Houghton Mifflin Harcourt.
Jolly, A. (2012). Berenty Reserve, Madagascar: A long time in a small space. In: P. M. Kappeler & D. P. Watts (eds), *Long-Term Field Studies of Primates*. Berlin, Heidelberg: Springer-Verlag, pp. 21–44.
Jolly, A. Rasamimanana, H. R., Kinnaird, M. F., *et al.* (1993). Territoriality in *Lemur catta* groups during the birth season at Berenty, Madagascar. In: Kappeler, P. M. & Ganzhorn, J. U. (eds), *Lemur Social Systems and their Ecological Basis*. New York: Plenum, pp. 85–109.
Jolly, A. Rasamimanana, H. R., Braun, M. A., *et al.* (2006) Territory as bet-hedging: *Lemur catta* in a rich forest and an erratic climate. In: Jolly, A. Sussman, R. W., Koyama, N. & Rasamimanana, H. R. (eds), *Ringtailed Lemur Biology: Lemur catta in Madagascar*. New York: Springer, pp. 187–207.

Judge, P. G. (2000). Coping with crowded conditions. In: Aureli, F. & de Waal, F. B. M. (eds), *Natural Conflict Resolution*. Berkeley: University of California Press, pp. 129–154.

Judge, P. G. & de Waal, F. B. M. (1997). Rhesus monkey behaviour under diverse population densities: coping with long-term crowding. *Animal Behaviour*, 54, 643–662.

Judge, P. G. & Mullen, S. H. (2005). Quadratic postconflict affiliation among bystanders in a hamadyras baboon group. *Animal Behaviour*, 69, 1345–1355.

Kaburu, S. S. K., Inoue, S. & Newton-Fisher, N. E. (2013). Death of the alpha: Within-community lethal violence among chimpanzees of the Mahale Mountains National Park. *American Journal of Primatology*, 75, 789–797.

Kappeler, P. M. (1993). Reconciliation and post-conflict behaviour in ringtailed lemurs, *Lemur catta* and redfronted lemurs, *Eulemur fulvus rufus*. *Animal Behaviour*, 45(5), 901–915.

Kappeler, P. M. & van Schaik, C. P. (1992). Methodological and evolutionary aspects of reconciliation among primates. *Ethology*, 92, 51–69.

Kappeler, P. M., Barrett, L., Blumstein, D. T. & Clutton-Brock, T. H. (2013). Constraints and flexibility in mammalian social behavior: introduction and synthesis. *Philosophical Transactions of the Royal Society: Biological Sciences*, 368, 1–10.

Karlen, S. J. & Krubitzer, L. (2007). The functional and anatomical organization of marsupial neocortex: evidence for parallel evolution across mammals. *Progress in Neurobiology*, 82, 122–141.

Kempes, A. (2008). Preface to special issue on natural conflict resolution in humans. *Behaviour*, 145, 1493–1496.

Kendon, A. & Ferber, A. (1973). A description of some human greetings In: R. P. Michael & J. H. Crook, *Comparative Ecology and Behaviour of Primates*. London, UK: Academic Press, pp. 591–668.

Koenig, A. (2002). Competition for resources and its behavioral consequences among female primates. *International Journal of Primatology*, 23, 759–783.

Koski, S. E. & Sterck, E. H. M. (2007). Triadic post-conflict affiliation in captive chimpanzees: does consolation console? *Animal Behaviour*, 73, 133–142.

Koski, S. E. & Sterck, E. H. (2009). Post-conflict third-party affiliation in chimpanzees: what's in it for the third party? *American Journal of Primatology*, 71, 409–418.

Koyama, N. (1991). Troop division and inter-troop relationships of ring-tailed lemurs (*Lemur catta*) at Berenty, Madagascar. In: Ehara, A., Kimura, T., Takenaka, O., Iwamoto, M. (eds), *Primatology Today*. Amsterdam: Elsevier, pp. 173–176.

Koyama, N., Nakamichi, M., Ichino, S., & Takahata, Y. (2002). Population and social dynamics changes in ring-tailed lemur troops at Berenty, Madagascar between 1989–1999. *Primates*, 43, 291–314.

Leone, A., Mignini, M., Mancini, G. & Palagi, E. (2010). Aggression does not increase friendly contacts among bystanders in geladas (*Theropithecus gelada*). *Primates*, 51, 299–305.

Lorenz, K. (1966). *On Aggression*, trans. Marjorie Latzke. London: Methuen.

Luo, Z-X., Yuan, C-X., Meng, Q-J. and Ji, Q. (2011). A Jurassic eutherian mammal and divergence of marsupials and placentals. *Nature*, 476, 442–445.

Martin, R. D. (1990). *Primate Origins and Evolution*. Princeton, NJ: Princeton University Press.

McCallum, I. (2012). A wild psychology. In: P. H. Kahn, Jr. and P. H. Hasbach (eds), *Ecopsychology. Science, Totems, and the Technological Species*. Cambridge, MA: The MIT Press, pp. 139–156.

Mech, L. D. (1977). Wolf-pack buffer zones as prey reservoirs. *Science*, 198, 320–321.

Mech, L. D. (1999). Alpha status, dominance, and division of labor in wolf packs. *Canadian Journal of Zoology*, 77, 1196–1203.

Mech, L. D. & Boitani, L. (2003). Wolf social ecology. In: Mech, D. & Boitani, L. (eds), *Behavior, Ecology, and Conservation*. Chicago: University of Chicago Press, pp. 1–34.

Meredith, R. W., Westerman, M., Case, J. A. & Springer, M. S. (2008). A phylogeny and timescale for marsupial evolution based on sequences for five nuclear genes. *Journal of Mammal Evolution*, 15, 1–26.

Miklósi, Á. (2014). *Dog Behaviour, Evolution, and Cognition.* Oxford: Oxford University Press.

Muller, M. N. (2002). Agonistic relations among Kanyawara chimpanzees. In: C. Boesch, Hohmann, G. & Marchant, L. F. (eds), *Behavioural Diversity in Chimpanzees and Bonobos.* Cambridge University Press, pp. 112–124.

Newton-Fisher, N. E. (2002). Relationships of male chimpanzees in the Budongo Forest, Uganda. In: C. Boesch, Hohmann, G. & Marchant, L. F. (eds), *Behavioural Diversity in Chimpanzees and Bonobos.* Cambridge University Press, pp. 125–137.

Norscia, I. & Palagi, E. (2011). Do wild brown lemurs reconcile? Not always. *Journal of Ethology*, 29(1), 181–185.

O'Brien, T. G. (1993). Allogrooming behaviour among adult female wedge-capped capuchin monkeys. *Animal Behaviour*, 46, 499–510.

Oldroyd, B. P. (2013). Social evolution: policing without genetic conflict. *Current Biology*, 23(5), R208-R210.

Otterbein, K. F. (1997). The origins of war. *Critical Review*, 11, 251–277.

Palagi, E. (2009). Adult play fighting in a prosimian (*Lemur catta*): modalities and roles of tail signals. *Journal of Comparative Psychology*, 123, 1–9.

Palagi, E. & Cordoni, G. (2009). Postconflict third-party affiliation in *Canis lupus*: do wolves share similarities with the great apes? *Animal Behaviour*, 78, 97–986.

Palagi, E. & Cordoni, G. (2012). The right time to happen: play developmental divergence in the two *Pan* species. *PLoS ONE*, 7(12), e52767. http://dx.doi.org/10.1371/journal.pone.0052767.

Palagi, E. & Dapporto, L. (2006). Beyond odor discrimination: demonstrating individual recognition by scent in *Lemur catta*. *Chemical Senses*, 31, 437–443.

Palagi, E. & Dapporto, L. (2007). Females do it better. Individual recognition experiments reveal sexual dimorphism in *Lemur catta* (Linnaeus 1758) olfactory motivation and territorial defence. *Journal of Experimental Biology*, 210, 2700–2705.

Palagi, E. & Norscia, I. (2013). Bonobos protect and console friends and kin. *PLoS ONE*, 8, e79290. http://dx.doi.org/10.1371/journal.pone.0079290.

Palagi, E. & Norscia, I. (2015). The season for peace: reconciliation in a despotic species (*Lemur catta*). *PLoS ONE*, 10(11), e0142150. http://dx.doi.org/10.1371/ journal.pone.0142150.

Palagi, E., Paoli, T. & Borgonini-Tarli, S. (2004). Reconciliation and consolation in captive bonobos (*Pan paniscus*). *American Journal of Primatology*, 62, 15–30.

Palagi, E., Cordoni, G. & Borgognini-Tarli, S. M. (2004). Immediate and delayed benefits of play behaviour: New evidence from chimpanzees (*Pan troglodytes*). *Ethology*, 110, 949–962.

Palagi, E., Paoli, T. & Tarli, S. B. (2005). Aggression and reconciliation in two captive groups of *Lemur catta*. *International Journal of Primatology*, 26, 279–294.

Palagi, E., Paoli, T. & Borgognini-Tarli, S. M. (2006). Immediate and delayed benefits of play behaviour: new evidence from chimpanzees (*Pan troglodytes*). *International Journal of Primatology*, 27, 1257–1270.

Palagi, E., Cordoni, G. & Borgonini-Tarli, S. (2006). Possible roles of consolation in captive chimpanzees (*Pan troglodytes*). *American Journal of Physical Anthropology*, 129, 105–111.

Paoli, T., Tacconi, G., Borgognini-Tarli, S. & Palagi, E. (2007). Influence of feeding and short-term crowding on the sexual repertoire of captive bonobos (*Pan paniscus*). *Annales Zoologici Fennici*, 44, 84–88.

Palagi, E., Antonacci, D. & Norscia, I. (2008a). Peacemaking on treetops: first evidence of reconciliation from a wild prosimian (*Propithecus verreauxi*). *Animal Behaviour*, 76, 737–747.

Palagi, E., Chiarugi, E. & Cordoni, G. (2008b). Peaceful post-conflict interactions between aggressors and bystanders in captive lowland gorillas (*Gorilla gorilla gorilla*). *American Journal of Primatology*, 70, 949–955.

Palagi E, Dall'Olio S, Demuru E, & Stanyon, R. R. (2014a). Exploring the evolutionary foundations of empathy: consolation in monkeys. *Evolution and Human Behavior*.

Palagi E, Norscia I, & Spada G (2014b) Relaxed open mouth as a playful signal in wild ring-tailed lemurs. *American Journal of Primatology*, 76, 1074–1083.

Pellegrini, A. D., Roseth, C. J., Mliner, S., *et al.* (2007). Social dominance in preschool classrooms. *Journal of Comparative Psychology*, 121, 54–64.

Pellegrini, A. D., Van Ryzin, M. J., Roseth, C. J., *et al.* (2011a). Behavioral and social cognitive processes in preschool children's social dominance. *Aggressive Behavior*, 35, 1–10.

Pellegrini, A. D., Bohn-Gettler, C. M., Dupuis, D., *et al.* (2011b). An empirical examination of sex differences in scoring preschool children's aggression. *Journal of Experimental Child Psychology*, 109, 232–238.

Pellis, S. M. (2002a). Sex differences in play fighting revisited: traditional and nontraditional mechanisms of sexual differentiation in rats. *Archives of Sexual Behavior*, 31, 17–26.

Pellis, S. M. & Pellis, V. C. (1997). Targets, tactics, and the open mouth face during play fighting in three species of primates. *Aggressive Behavior*, 23, 41–57.

Pennisi, E. (2013). How do microbes shape animal development? *Science*, 340, 1159–1160.

Pereira, M. E. & Kappeler, P. M. (1997). Divergent systems of agonistic behaviour in lemurid primates. *Behaviour*, 134, 225–274.

Peterson, R. O., Jacobs, A. K., Drummer, T. D., Mech, L. D. & Smith, D. W. (2002). Leadership behaviour in relation to dominance and reproductive status in grey wolves, *Canis lupus*. *Canadian Journal of Zoology*, 80, 1405–1412.

Petit, O. & Thierry, B. (1994). Aggressive and peaceful interventions in conflicts in Tonkean macaques. *Animal Behaviour*, 48, 1427–1436.

Petit O, Abegg C, Thierry B (1997). A comparative study of aggression and conciliation in three ercopithecine monkeys (*Macaca fuscata, Macaca nigra, Papio papio*). *Behaviour*, 134, 415–432.

Polizzi di Sorrentino, E. P., Schino, G., Visalberghi, E. & Aureli, F. (2010). What time is it? Coping with expected feeding time in capuchin monkeys. *Animal Behaviour*, 80, 117–123.

Port, M., Clough, D. & Kappeler, P. M. (2009). Market effects offset the reciprocation of grooming in free-ranging redfronted lemurs, Eulemur fulvus rufus. *Animal Behaviour*, 77, 29–36.

Rilling, J. K., Scholz, J., Preuss, T. M., *et al.* (2012). Differences between chimpanzees and bonobos in neural systems supporting social cognition. *Social cognitive and affective neuroscience*, 7, 369–379.

Roeder, J. J., Fornasieri, I. & Gosset, D. (2002). Conflict and postconflict behaviour in two lemur species with different social organizations (*Eulemur fulvus* and *Eulemur macaco*): a study on captive groups. *Aggressive Behavior*, 28, 62–74.

Rolland, N. & Roeder, J. J. (2000). Do ringtailed lemurs (*Lemur catta*) reconcile in the hour post-conflict?: a pilot study. *Primates*, 41, 223–227.

Romero, T., Castellanos, M. A. and de Waal, F. B. M. (2011). Post-conflict affiliation by chimpanzees with aggressors: other-oriented versus selfish political strategy. *PLoSONE*, 6(7), e22173. http://dx.doi.org/10.1371/journal.pone.0022173.

Roseth, C. J., Pellegrini, A. D., Bohn, C. M., Van Ryzin, M. & Vance, N. (2007). Preschoolers' aggression, affiliation, and social dominance relationships: An observational, longitudinal study. *Journal of School Psychology*, 45, 479–497.

Roseth, C. J., Pellegrini, A. D., Dupuis, D. N., *et al.* (2008). Teacher intervention and U.S. preschoolers' natural conflict resolution after aggressive competition. *Behaviour*, 145, 1601–1626.

Roseth, C. J., Pellegrini, A. D., Dupuis, D. N., *et al.* (2011). Preschoolers' bistrategic resource control, reconciliation, and peer regard. *Social Development*, 1, 185–211.

Rubenstein, D. R. (2012). Family feuds: social competition and sexual conflict in complex societies. *Philosophical Transactions of the Royal Society B*, 367, 2304–2313.

Rubin, J. Z., Pruitt, D. G. & Kim, S. H. (1994). *Social Conflict: Escalation, Stalemate, and Settlement*, McGraw-Hill, New York.

Sapolsky, R. M. (2006). Social cultures among nonhuman primates. *Current Anthropology*, 47, 641–656.

Sapolsky, R. M. (2013). Rousseau with a tail. Maintaining a tradition of peace among baboons. In: D. P. Fry (ed.), *War, Peace, and Human Nature. The Convergence of Evolutionary and Cultural Views*. New York: Oxford University Press, pp. 421–438.

Schino, G. (1998). Reconciliation in domestic goats. *Behaviour*, 135, 343–356.

Schino, G. & Marini, C. (2012). Self-protective function of post-conflict bystander affiliation in mandrills. *PLoS ONE*, 7(6), e38936. http://dx.doi.org/10.1371/journal.-pone.0038936.

Schino, G., Geminiani, S., Rosati, L. & Aureli, F. (2004). Behavioral and emotional response of Japanese macaque (*Macaca fuscata*) mothers after their offspring receive an aggression. *Journal of Comparative Psychology*, 118, 340–346.

Scofield, B. & Margulis, L. (2012). Psychological discontent: Self and science on our symbiotic planet. In, P. H. Kahn, Jr. and P. H. Hasbach (eds), *Ecopsychology. Science, Totems, and the Technological Species*. Cambridge, MA: The MIT Press pp. 219–240.

Silk, J. B. (2002). The form and function of reconciliation in primates. *Annual Reviews of Anthropology*, 31, 21–44.

Soma, T. & Koyama, N. (2013). Eviction and troop reconstruction in a single matriline of ring-tailed lemurs (*Lemur catta*): what happened when 'grandmother' died? In: J. Masters, Gamba M., Génin F. & Tuttle R. (eds), *Leaping Ahead: Advances in Prosimian Biology*. New York: Springer, pp. 137–146.

Stiver, K. A., Dierkes, P., Taborsky, M., Gibbs, H. L. & Balshine, S. (2005). Relatedness and helping in fish: examining the theoretical predictions. *Proceedings of the Royal Society B: Biological Sciences*, 272, 1593–1599.

Sustainable Development Solutions Network (2013). *An action agenda for sustainable development*. Report for the UN Secretary Geneneral.

Tacconi, G. & Palagi, E. (2009). Play behavioural tactics under space reduction: social challenges in bonobos, Pan paniscus. *Animal Behaviour*, 78, 469–476.

Takahata Y, Koyama N, Ichino S, & Miyamoto N (2005). Inter- and within-troop competition of female ring-tailed lemurs: a preliminary report. *African Study Monographs*, 26, 1–14.

Thierry, B. (1985a). Social development in three species of macaque (*Macaca mulatta, M. fascicularis, M. tonkeana*): A preliminary report on the first ten weeks of life. *Behavioural Processes*, 11, 89–95.

Thierry, B. (1985b). Patterns of agonistic interactions in three species of macaque (*Macaca mulatta, Macaca fascicularis, Macaca tonkeana*). *Aggressive Behavior*, 11, 223–233.

Thierry, B. (1986). A comparative study of aggression and response to aggression in three species of macaque. In: J. G. Else & P. C. Lee (eds), *Primate Ontogeny, Cognition, and Social Behavior*. Cambridge University Press, pp. 307–313.

Thierry, B. (1990). Feedback loop between kinship and dominance: the macaque model. *Journal of theoretical Biology*, 145(4), 511–522.

Thierry, B. (2000). Covariation and conflict management patterns across macaque species. In: F. Aureli & F. B. M. de Waal (eds), *Natural conflict resolution*. Berkeley, CA: University of California Press, pp. 106–128.

Thierry, B. (2007). Behaviourology divided: shall we continue? *Behaviour*, 144, 861–878.

Thierry, B. (2013). Identifying constraints in the evolution of primate societies. *Philosophical Transactions of the Royal Society B*, 368, 20120342. http://dx.doi.org/10.1098/rstb.2012.0342.

Tinbergen, N. (1963). On aims and methods of ethology. *Zeitschrift für Tierpsychologie*, 20, 410–433.

Tinbergen, N. (1968). On war and peace in animals and man. An ethologist's approach to the biology of aggression. *Science*, 160, 1411–1418.

Valdeón Baruque, J. (2002). *Pedro I el Cruel y Enrique de Trastámara: ¿la primera guerra civil española?* Barcelona: Aguilar, 2003.

van Hooff, J. A. R. A. M. (1967). The facial displays of catarrhine monkeys and apes. In Morris, D. (Ed.). *Primate Ethology* (pp. 7–68). New York: Aldine.

van Schaik, C. P. & Aureli, F. (2000). The natural history of valuable relationships in primates. In Aureli, F. and de Waal, F. B. M. (eds), *Natural Conflict Resolution* (pp. 307–333). Berkeley: University of California Press.

van Wolkenten, M. L., Davis, J. M., Gong, M. L. & de Waal, F. B. M. (2006). Coping with acute crowding by *Cebus apella*. *International Journal of Primatology*, 27, 1241–1256.

van Zweden, J. S., Cardoen, D. & Wenseleers, T. (2012). Social evolution: when promiscuity breeds cooperation. *Current Biology*, 22, R922–R924.

Verbeek, P. (2008). Peace ethology. *Behaviour*, 145, 1497–1524.

Verbeek, P. (2013). An ethological perspective on war and peace. In: D. P. Fry (ed.), *War, Peace, and Human Nature: The Convergence of Evolutionary and Cultural Views*. New York: Oxford University Press.

Verbeek, P. and de Waal, F. B. M. (2001). Peacemaking among preschool children. *Peace and Conflict: Journal of Peace Psychology*, 7, 5–28.

Verbeek, P., Iwamoto, T. & Murakami, N. (2007). Differences in aggression among wild type and domesticated fighting fish are context dependent. *Animal Behaviour*, 73, 75–83.

Wahaj, S. A., Guse, K. R. & Holekamp, K. E. (2001). Reconciliation in spotted hyenaa (*Crocuta crocuta*). *Ethology*, 107, 1057–1074.

Watts, D. (1995). Post-conflict social events in wild mountain gorillas (Mammalia, Hominoidea). I. Social interactions between opponents. *Ethology*, 100, 139–157.

Watts, D. P. (1996). Comparative socioecology of mountain gorillas. In: W. C. McGrew, L. F. Marchant & T. Nishida (eds), *Great Ape Society*: Cambridge University Press, pp. 16–28.

Watts, D, & Mitani, J. (2000). Infanticide and cannibalism by male chimpanzees at Ngogo, Kibale National Park, Uganda. *Primates*, 41, 357–365.

Watts, D., Muller, M., Amsler, S., Mbabazi, G. & Mitani, J. (2006). Lethal intergroup aggression by chimpanzees in the Kibale National Park, Uganda. *American Journal of Primatology*, 68, 161–180.

Weaver, A. (2003). Conflict and reconciliation in captive bottlenose dolphins, *Tursiops truncatus*. *Marine Mammal Science*, 19, 836–846.

Wenseleers, T., Bacon, J. P., Alves, D. A., et al. (2013). Bourgeois behavior and freeloading in the colonial orb web spider *Parawixia bistriata* (Araneae, Araneidae). *The American Naturalist*, 182, pp. 120–129.

White, T. D., Asfaw, B., Beyene, Y., et al. (2009). Ardipithecus ramidus and the paleobiology of early hominids. *Science*, 326(5949), 64–86.

Wilson, E. O. (2000). *Sociobiology: The New Synthesis*. Cambridge, MA: Belknap Press of Harvard University Press.

Wittig, R. M., Crockford, C., Wikberg, E., Seyfarth, R. M. & Cheney, D. L. (2007). Kin mediated reconciliation substitutes for direct reconciliation in female baboons. *Proceedings of the Royal Society B: Biological Sciences*, 274, 1109–1115.

Wright, P. C. (1999). Lemur traits and Madagascar ecology: coping with an island environment. *American Journal of Physical Anthropology*, 110, 31–72.

Yanagi, A & Berman, C. M. (2014). Body signals during social play in free-ranging rhesus macaques (*Macaca mulatta*): a systematic analysis. *American Journal of Primatology*, 76, 168–179.

Zimen, E. (1981). *The Wolf, a Species in Danger*. Delacorte Press.

Zuckerman, S. (1932). *The Social Life of Monkeys and Apes*. London: Routledge and Kegan Paul.

6 Anxiety...from scratch: emotional response to tense situations

You have probably noticed the ambiguity and vagueness in the use of the word 'anxiety.' Generally one means a subjective condition, caused by the perception that an 'evolution of fear' has been consummated. Such a condition may be called an emotion. What is an emotion in the dynamic sense? Certainly something very complex.

Sigmund Freud, 1969

6.1 Stress, anxiety and...evolution

Back in 1872, Darwin, in *The Expression of the Emotions in Man and Animals*, wrote about emotion expression as the possible result of 'the direct action of the excited nervous system on the body independently of the will'. In the whole animal world, bodily movements are able to reveal and unveil emotions not apparent otherwise. In humans, gestures, facial expressions and, more in general, non-verbal behaviours are an integral part of how we express our emotions. Emotions include anxiety, which can be linked to fear, motivational conflict, or uncertainty in decision-making (Morgan, 2006).

Defining anxiety is not easy, as also recognised by Sigmund Freud back in 1920 (see the 1969 edition). Linguistically speaking, anxiety is defined as a '...feeling of nervousness or worry about something that is happening or might happen in the future' or, similarly, 'a feeling of worry, nervousness, or unease about something with an uncertain outcome'. Stress is defined, instead, as 'great worry caused by a difficult situation, or something that causes this condition' or 'a state of mental or emotional strain or tension resulting from adverse or demanding circumstances' (definitions by Cambridge and Oxford online dictionaries, respectively). From these non-technical definitions, it is possible to infer intuitively that anxiety can be related to stress but also that the difference between stress and anxiety is not clear-cut. Indeed, a plethora of models, theories and hypotheses have been put forth to integrate stress and anxiety into a single framework. However, the behavioural, physiological, psychological and clinical implications of anxiety and stress responses have proven difficult to untangle.

In *Homo sapiens*, non-human primates and other mammals, anxiety is an emotional state involving tension and/or agitation, with both physiological and behavioural implications (van Riezen and Segal, 1988; Craig *et al.*, 1995; Barros and Tomas, 2002; Bourin *et al.*, 2007). Indeed, anxiety is associated with a characteristic set of behavioural responses, including avoidance, vigilance and arousal, which evolved to protect the individual from danger (Gross and Hen, 2004).

Stress is an integrated neuroendocrine response that can be measured by quantifying glucocorticoid or its metabolites in blood, saliva, excreta and integumentary structures (hair and feathers) (Sapolsky, 2005; Sheriff et al., 2011). Stress has been also defined as a non-specific general adaptation syndrome, which involves hormonal and metabolic changes following a traumatic event and leading to the fight-or-flight response (Weissman, 1990). Anxiety – which plays an important role in enhancing distress buffering behaviours – is one of the main emotional components of the stress experience, being itself connected to escalation (fight) or de-escalation (flight) responses. As a matter of fact, in psychology research one approach to the measurement of stress is through the assessment of anxiety (e.g., Shulman and Jones, 1996).

Following the psychological model proposed by Price (2003), the decision to escalate (fighting) or de-escalate (avoid conflict/flee) can take place at the rational, emotional and instinctive levels. These levels correspond to the so-called 'triune forebrain' (MacLean, 1990) composed, respectively, by the neomammalian brain located in the neocortex, the palaeomammalian brain situated in the limbic system, and the reptilian brain occupying the basal ganglia or corpus striatum. At the rational level, the decision of fighting or backing off is made consciously based on a cost-benefit assessment, which includes the evaluation of possible risks and advantages related to each option. At the emotional or limbic level, escalation can take the form of anger and combat excitement, associated with bodily changes. De-escalation, instead, can recruit the dysphoric emotion of anxiety. Finally, at the instinctive level, escalation may take the form of elevated mood, whereas de-escalation can result in depressed mood and unfocused anxiety (Price, 2003). Recent studies have incorporated the three levels of the MacLean's triune forebrain into a dual process model, including reflective and reflexive systems (Hecht et al., 2012). The reflective system (neomammalian complex) includes conscious, controlled and cognitive responses, whereas the reflexive system (reptilian and paleomammalian complexes) includes unconscious or preconscious, automatic and perceptually driven responses, such as anxiety-related behaviours.

Whatever the model considered, the different parts of the forebrain tell different 'chapters' of the evolutionary history of the mammalian brain and, to use a computer science metaphor, work as different, but interacting processors (let's say from the 1980s to the 1990s and 2000!) integrated into a single, modern processing unit.

The responses related to anxiety are, therefore, part of the most ancient mechanisms regulating basic spheres of the social life of an individual, such as courtship, intrasexual competition, and anti-predatory behaviours (Price, 2003). In short, anxiety seems to be hard-wired into the mammalian brain. Surprisingly, lemurs' response to stressful events has been neglected in the field of anxiety research. The time has come to fill the gap.

6.2 Measuring anxiety: from physiology to behaviour

The endocrine response to stress can be emotionally translated into anxiety which, as occurs for any emotional state, is behaviourally translated into changes of the motor patterns. What are the behavioural indicators that can be used to measure anxiety and, indirectly, provide hints on the stress experienced by an individual?

Figure 6.1 Self-scratching in *Gorilla beringei*, Bwindi Impenetrable National Park, Uganda. Photo: Ivan Norscia.

Researchers have adopted a range of tools to measure anxiety (Troisi, 2002; Balon, 2005). In humans, self-report of mental state, mood, behaviour and symptom experience have been extensively employed, using different measure scales (Craig *et al.*, 1995; Balon, 2005). In human and non-human primates, there is evidence that self-directed behaviours, namely self-scratching (Figures 6.1 and 6.2), self-grooming, yawning and body shaking can provide an index of anxiety (Maestripieri *et al.*, 1992; Schino *et al.*, 1996; Aureli *et al.*, 2002), even though they do not always predict stress levels revealed by hormonal analyses (Troisi, 2002; Higham *et al.*, 2009).

In *Homo sapiens* tension and stressful situations increase rates of self-directed behaviours, including body scratching (Morris, 1977; Fried, 1994; Troisi *et al.*, 2000). For example, during interviews with helping professionals, depressed and anxious outpatients with difficulties in describing their emotional states showed higher frequencies of self-directed behaviours (such as self-touching, -scratching, and -grooming) than similar outpatients without such difficulties (Troisi *et al.*, 2000). In non-human apes and in cercopithecines, increased levels of self-directed behaviours are associated with different stressful situations. In chimpanzees, *Pan troglodytes*, scratching levels were particularly high in cases of crowded conditions (Aureli and de Waal, 1997), in subordinate males and in females when a non-affiliative group member was in their proximity (Kutsukake, 2003), and after aggression (Fraser *et al.*, 2008). In cercopithecines, self-directed behaviours have been found to increase in the presence of potentially threatening dominant neighbours (olive baboons,

Figure 6.2 *Chlorocebus aethiops* self-scratching. Entebbe botanical garden, Uganda. Photo: Ivan Norscia.

Papio anubis: Castles *et al.*, 1999) and after agonistic episodes (macaques, *Macaca sylvanus* and *Macaca fascicularis*: Aureli, 1997; vervet monkeys, *Chlorocebus aethiops*: Daniel *et al.*, 2008). Moreover, post-conflict scratching has been observed to decrease to baseline following former opponents' reunion, when the main source of stress (relationship disruption and/or fear of renewed attacks) is removed (*Macaca* spp.: Aureli and van Schaik, 1991; *Papio anubis*: Castles and Whiten, 1998; *Pan troglodytes*: Kutsukake and Castles, 2001; Fraser *et al.*, 2010).

Among self-directed behaviours, self-scratching, in particular, appears to be one of the most reliable behavioural tools to measure anxiety. Indeed, anxiety states can share common biochemical origins with the physiological sensation of pruritus (that is, severe itching; *sensu* Rothman, 1941), leading to the itch-scratch cycle (Shankly, 1988; Stangier *et al.*, 2003; Tran *et al.*, 2010).

Behavioural observations combined with pharmacological data have clearly shown the tight linkage between stress-induced hormones and self-directed behaviours, or stereotypies, in human and non-human primates (Tuinier, *et al.*, 1996; Troisi, 2002). Studies on group-living monkeys revealed that self-scratching is sensitive to pharmacological manipulation of mood through anxiolytic and anxiogenic substances (Schino *et al.*, 1996).

In *Homo sapiens*, anxiolytic substances (e.g., nitrazepam and diazepam) reduce pruritus and scratching (Krause and Shuster, 1983; van Moffaert, 2003). Schino and coworkers (1991) found that in female long-tailed macaques (*Macaca fascicularis*) scratching rates decreased after the pharmacological manipulation of mood through the anxiolytic lorazepam. Maestripieri *et al.* (1992) observed that the acute administration of the anxiolytic midazolam reduced the rate of scratching in infants of rhesus macaques (*Macaca mulatta*). Similarly, in common marmosets (*Callithrix jacchus*) and black-tufted marmosets (*Callithrix penicillata*), the administration of the anxiolytic diazepam induced a significant reduction in scratching (Cilia and Piper, 1997; Barros *et al.*, 2000). In mice showing increased anxiety due to the genetic deletion of Sapap3, rough self-grooming (leading to facial hair loss) was alleviated by a selective serotonin reuptake inhibitor (Welch *et al.*, 2007).

In conclusion, self-scratching is tightly and directly connected to anxiety in that it can be manipulated via anxiolytic/anxiogenic substances and it is unquestionably related to tense, demanding or uncomfortable situations leading to the disruption of the subject's internal homeostasis. Therefore, self-scratching is a reliable tool to detect and measure anxiety.

Box 6.1 | **by Filippo Aureli**

Speaking of which: anxiety-related behaviour

It is a common, almost trivial behaviour, but it is a powerful tool to study emotionality. I am talking about a simple scratch, a type of self-directed behaviour. I started to study self-directed behaviour almost by chance, being exposed to its possible link to anxiety by Alfonso Troisi and Gabriele Schino during my final undergraduate year. Then, I 'exported' this knowledge from Italy to Holland when I started to do my PhD on post-conflict behaviour at the University of Utrecht. I surprised Carel van Schaik and Jan van Hooff, my PhD supervisors, when I showed them the power of self-directed behaviour. I found that after receiving aggression long-tail macaques (*Macaca fascicularis*) increased the rate of scratching and other self-directed behaviour, such as self-grooming and body shake, compared to control periods. After reconciliation (i.e., a friendly reunion between former opponents: de Waal and van Roosmalen, 1979) took place, the rates of self-directed behaviour returned to baseline levels. These results allowed us to infer that aggression increased uncertainty in the recipient of aggression and that reconciliation functioned in restoring emotional balance (Aureli *et al.*, 1989; Aureli and van Schaik, 1991).

While I was making some progress in using self-directed behaviour to understand my study subjects' post-conflict emotionality, my Italian colleagues were doing the same in other behavioural aspects. Thus, we teamed up under Dario Maestripieri's lead and reviewed behavioural, physiological and pharmacological evidence for self-directed behaviour being linked to emotionality and wrote 'a modest proposal' about its value as an indicator of emotions in primates (Maestripieri *et al.*, 1992). This review pointed out that in the empirical articles published until 1992 the emotion for which self-directed behaviour was used as an indicator was ambiguously labelled as stress, motivational conflict, uncertainty, tension or anxiety. The advent of elegant ethopharmacological experiments helped to clarify that self-directed behaviour is a reliable indicator of anxiety-like emotions at least in some primates as scratching selectively increased after administration of anxiety inducing drugs and decreased after administration of anxiety reducing drugs (Schino et al., 1996; Barros and Tomaz, 2002). Self-directed behaviour has clearly a hygienic function. Thus, the anxiety-like emotion has to be inferred when rates of self-directed behaviour are above baseline levels.

Although Maestripieri et al. (1992) has been a successful article (cited over 250 times so far), for many years the use of self-directed behaviour as an indicator of anxiety was linked to personal contacts. For example, it became 'a must' in the behavioural studies on primates at Utrecht University, but its use did not spread to other research teams for several years. Apart from the common difficulty for new perspectives to be accepted, the early period of the use of self-directed behaviour as an indicator of anxiety was hindered by two main factors. First, the study of emotions in animals has been controversial because of the conscious feeling that humans associate with them, and only recently has it been at least partially accepted as a valid area of research (cf. Aureli and Whiten, 2003). Second, there was ambivalence in early studies between interpreting the same type of results as a validation of self-directed behaviour being an anxiety indicator and as a demonstration of the individual experiencing anxiety by using its self-directed behaviour. The pharmacological validation made a big difference, and self-directed behaviour has become a standard element in data collection on primate behaviour.

This was a long answer to the simple question put by the book's authors: 'What makes the study of anxiety-related behaviour interesting?' I hope that by now the reader would know that my answer to this question is that I am interested in self-directed behaviour because it opens a window to animal emotionality in a simple, non-invasive way. This is a wonderful opportunity, but we should not misuse it (see below).

One of the most interesting outcomes of my research was the opportunity to be more specific about the emotions experienced by other animals. Until the advent of using self-directed behaviour, scientists were struggling to pinpoint animal emotions on the basis of only behavioural cues. I remember that in more than one oral conference presentation the speaker tried to illustrate tension, stress, and motivation ambiguity by showing and interpreting images of the study subjects. While their descriptions were fascinating, the evidence for the behaviours being indicators of such internal states was lacking. Interestingly, most times the images shown revealed the study subjects

Box 6.1 (continued)

Box 6.1 (*cont.*)

engaging in self-directed behaviour. Several times this was followed by people in the audience turning to me and commenting that it was obvious in which emotion the study subjects were engaged because of their self-directed behaviour. The message was slowly getting through. If we use self-directed behaviour as an indicator we can be more specific about which emotion our study subjects experience. For example, I could confidently publish about post-conflict anxiety in macaques (Aureli, 1997; Aureli and Smucny, 2000; Aureli *et al.*, 2002).

To associate the behavioural responses to anxiety with pharmacological data requires that animals are clinically manipulated. This is difficult in most species and situations. Still, demonstrating such association is needed before using self-directed behaviour as an anxiety indicator in any species, and researchers need to find the appropriate conditions for behavioural, rather than pharmacological, validation tests. Another outcome of my research was to show various examples of how to rise to such a challenge. For example, we used the response of captive chimpanzees to vocalisation produced by neighbouring groups, which was demonstrated to increase the likelihood of intragroup aggression (Baker and Aureli, 1997). In socially housed chimpanzees self-directed behaviour, such as rough and gentle scratching, was more common after neighbouring individuals vocalised than before, whereas this was not the case for single-caged chimpanzees, for whom neighbour vocalisation carried no risk of aggression (Baker and Aureli, 1997). Thus, our validation of chimpanzees' self-directed behaviour based on the risk of intragroup aggression following neighbour vocalisation has allowed us and others to use it as a reliable anxiety indicator in other contexts (Arnold and Whiten, 2001; Leavens *et al.*, 2001; Baker, 2004; Koski *et al.*, 2007; Fraser *et al.*, 2008).

A third outcome of research on self-directed behaviour was the possibility to provide evidence for the calming effect of grooming that has been viewed as one of its main benefits since the 1970s (Terry, 1970). Although a reduction in heart rate can provide such evidence (Boccia *et al.*, 1989; Aureli *et al.*, 1999), this approach is complicated by the need to manipulate the animal and the pronounced effect of general physical activity on heart rate (Smith *et al.*, 1993; Aureli *et al.*, 1999). The use of self-directed behaviour allowed Schino *et al.* (1988) to demonstrate that long-tail macaques relax after receiving grooming. However, recipients are not the only individuals to benefit from anxiety reduction. Crested black macaques (*Macaca nigra*) scratch themselves less after giving grooming (Aureli and Yates, 2010). However, an increase in self-directed behaviour of both the groomer and the recipient was found in Barbary macaques (*Macaca sylvanus*; Molesti and Majolo, 2013). Thus, more research on the emotional responses to giving and receiving grooming is needed in macaques and other species.

Before using a behavioural indicator a critical aspect is its validation to be a reliable measure of a given phenomenon. Self-directed behaviour has been used too often without such a validation. Thus, an important investigation domain for the future is the systematic validation of self-directed behaviour before its use as an anxiety indicator (i.e., validation as an anxiety indicator in one context, so it can be used as such in other contexts). As there is still much confusion and misuse of self-directed behaviour, the validation should be done for any species, because we do not know if it works as an anxiety indicator in all species.

In this respect, one challenge is to find appropriate contexts for its validation. In many species, the approach of and being in proximity to a more dominant individual is perceived as an uncertain situation that provokes an increase in heart rate (Aureli *et al.*, 1999) and self-directed behaviour (Pavani *et al.*, 1991; Castles *et al.*, 1999). This context can be then easily used for the validation of self-directed behaviour in a variety of species. However, there are species, such as spider monkeys (*Ateles* spp.: Aureli and Schaffner, 2008), for which this context cannot be used, as dominance relationships are not pronounced, if they exist at all. Thus, in some species finding the appropriate context for the validation of self-directed behaviour as an anxiety indicator can be challenging and will require a considerable dose of researcher creativity.

Self-directed behaviour is usually linked to uncertainty (Maestripieri *et al.* 1992), and it has been used as an indicator of anxiety (Schino *et al.*, 1996) and stress (Troisi, 2002). However, anxiety and stress are different phenomena, but very few studies have attempted to empirically distinguish them with the use of self-directed behaviour and other indicators (see Higham *et al.*, 2009). Thus, an interesting avenue of research is the integrated study of stress and anxiety with the goal of highlighting their differences and similarities.

All the above points are relevant to the relatively new research area of emotional mediation of social interactions. The window into animal emotions afforded by self-directed behaviour was at the basis of the proposal for variation in social interactions being mediated by variation in emotional responses (Aureli, 1997). Because emotional responses are so variable, it makes them suitable candidates to mediate the behavioural flexibility typical of human and non-human animals (Aureli and Whiten, 2003). A key issue is to understand the mechanisms that underlie an individual's ability to modify its behaviour depending on the quality of its relationship with a potential partner in a particular context. This can be accomplished by relationship assessment based on emotional mediation as the emotional experience of an individual is affected by the frequency and quality of previous interactions with group members. Thus, the emotional experience can be functionally equivalent to

Box 6.1 (continued)

> Box 6.1 (*cont.*)
>
> the processes of bookkeeping, frequency computation and quality conversion of the interactions with a partner needed for updated relationship assessment. The resulting emotional experience is partner-dependent. Thus, emotional differences can be at the core of the observed variation in social interactions reflecting the variation in relationship quality across partners (Aureli and Schaffner, 2002; Aureli and Whiten, 2003). Emotional mediation may be the mechanism for many cases of reciprocity (Schino and Aureli, 2009, 2010) and has been the focus of recent models explaining the occurrence of social interactions (Campennì and Schino, 2014; Evers *et al.*, 2014). Therefore, self-directed behaviour has been critical for the first steps and can further support the progress of this intriguing area of research.

6.3 Self-directed behaviours and anxiety in lemurs

In lemurs, as in other primates, self-scratching has proven a useful tool not only to detect, but also to measure anxiety, depending on the source and event perturbing the subject's internal homeostasis (Figures 6.3 and 6.4).

Various perturbing events can occur in the forest including the presence of competitors, predators and unfamiliar stimuli (Figure 6.5). Here we present three case studies, in which we used self-scratching to measure anxiety upraise in the lemurs of the Berenty forest (south Madagascar), following predation attempts (in *Eulemur rufus* × *collaris*), aggression (in *Eulemur rufus* × *collaris* and *Propithecus verreauxi*), and during mating (in *Lemur catta*).

We collected the scratching bouts in the *Eulemur rufus* × *collaris* (brown lemurs) in the 15 minutes following predation attempts by *Polyboroides radiatus*, Madagascar's

Figure 6.3 Self-scratching in *Lemur catta* at Parco Natura Viva (Bussolengo, Italy). Photo: Tommaso Ragaini.

Figure 6.4 Self-scratching in *Propithecus verreauxi*. Berenty forest, Madagascar. Photo: Ivan Norscia.

harrier-hawk, a predator present in the Berenty area. During such attempts hawks flew overhead, close to forest canopy, sometimes targeting lemurs while they were feeding or resting on tree tops, and almost always inducing lemurs to produce alarm calls and/or a flee response. Immediately after predation attempts scratching levels peaked, indicating an increase of anxiety (Palagi and Norscia, 2011).

Indeed, a predatory attack is always a source of acute anxiety, and probably stress, inducing the typical fight-or-flight response, part of a general adaptation syndrome (see Sapolsky, 2005 for a review).

Another context that we examined was intragroup aggression, which also determined an increase of scratching in the victims. Such increase was probably related to the uncertainty in decision-making: withdrawing for fear of renewed attacks and approaching to reconcile (Aureli, 1997). Therefore, attacks by both conspecifics and hawks determined an increase of scratching but the response was higher after predation attempts, because life was in jeopardy.

Categorising potentially stressful events according to their dangerousness allowed assessing the differing response of scratching according to the attack's severity. The change in scratching rates, as it happens with the fluctuation in the cortisol levels in the presence of stressors (Sapolsky, 2005), is not an all-or-nothing response to

Figure 6.5 *Lemur catta* coming across the Malagasy ground boa (*Acrantophis madagascariensis*) at Berenty, Madagascar. Photo: Ivan Norscia.

anxiety. While the scratching variation itself is considered to indicate a change in the animal's emotional state (is the animal anxious/not anxious?), the entity of such variation can provide an estimate of the relative amount of the anxiety accumulated after different events (is the animal more or less anxious?).

After showing that the brown lemur's hierarchy was linear, we tested the relationship between ranking position and scratching. We found that baseline scratching decreased as an animal's ranking position increased. Nevertheless, the negative correlation vanished when we replaced baseline scratching with scratching performed by victims after both predation attempts and aggressions.

As a further step in the analysis, we calculated a scratching variation index for each lemur. This index was determined as the difference between scratching levels after either predatory or conspecific's attacks and baseline conditions, over the total scratching levels performed by each subject. There was a negative correlation between the scratching variation index and the animals' ranking position, with the dominant individuals showing the highest differential increase of scratching, after predatory attempts (Figure 6.6) (Palagi and Norscia, 2011).

Why do dominants seem to cope worse than subordinates with uncontrolled dangerous situations? The result can be interpreted following a theoretical framework derived from studies using plasma and salivary cortisol analyses, in primate and

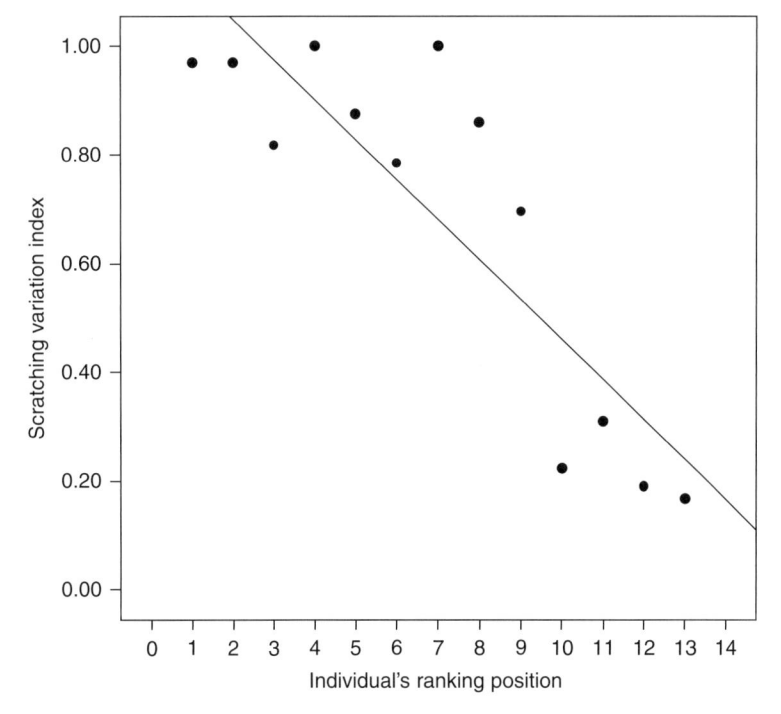

Figure 6.6 Negative Spearman's correlation between the scratching variation index (Y-axis) and rank in *Eulemur fulvus* × *collaris*. The scratching variation index was calculated as the difference between scratching levels after either predatory or conspecific's attacks and baseline conditions, over the total scratching levels performed by each subject.

non-primate mammals. Interestingly, the anxiety-related scratching behaviour of brown lemurs and the hormonal data from other investigations draw a similar profile shaped by specific social network features.

According to the hormonal framework, subordinates will exhibit the highest baseline stress levels in cases of social stability; that is, when hierarchy is stable and maintained by reiterated aggressions over subordinates, which receive low social support by group members (Abbott et al., 2003). The wild brown lemurs that we studied showed linear hierarchy (Palagi and Norscia, 2011), informing social stability (Sapolsky, 1992; Abbott et al., 2003) and, consistently, the highest levels of anxiety-related scratching were shown by low-ranking individuals. Indeed, subordinate lemurs are normally threatened by dominants via agonistic displays and aggression and receive the lowest coalitionary support, mostly supplied to dominants (Pereira and Kappeler, 1997).

Additionally, hormonal literature suggests that dominants and subordinates can be affected to a similar extent when events perturb individual safety and social stability, which in group-living mammals induces an increase in corticosteroid concentrations (Sapolsky, 1990; Alexander and Irvine, 1998). In brown lemurs immediately after perturbing events (predation attempts and aggression), both dominants and subordinates experienced similar, high levels of anxiety revealed by their comparable scratching rates (Palagi and Norscia, 2011). These external events leading to temporary social uncertainty are 'blind' agents impacting the anxiety levels – and probably stress – of all group members.

Analogously, in captive lowland gorillas (*Gorilla gorilla gorilla*), another species showing a linear hierarchy, Cordoni and Palagi (2007) found that scratching – more frequent when animals were kept in indoor facilities – was higher in subordinates than in dominants in the absence of space constraints (outdoor facilities). Instead, subordinates and dominants showed similar scratching levels when confined indoors. According to the authors, this finding indicated that all group members reacted to the high-density condition in terms of emotional response independently of one's own ranking position.

Interestingly, in an egalitarian society, with non-linear hierarchy, such as that of *Macaca tonkeana* (Thierry, 2004), Palagi and colleagues (2014) found that all group members scratched to a similar extent in baseline conditions, whereas dominants experienced higher levels of anxiety (and scratching) in the minutes immediately following an aggressive event. This is the flip side of the coin of the framework presented above. When the starting point is not a rigidly structured society the anxiety response by group members is just different. In these societies, dominants are not used to harassing subordinates in order to repeatedly re-establish dominance, the social environment is relaxed and baseline scratching levels are expected to be low for all group members. On the other hand, in these societies subordinates, or coalitions of subordinates, can counterattack high-ranking fellows and possibly reverse – more or less temporarily – dominance relationships. Hence, dominants have more to lose than subordinates from aggressive events. The dominants' emotional uncertainty during conflicting situations is probably translated into higher anxiety levels.

The highest differential increase in the scratching levels after predation attempts in dominant brown lemurs (Palagi and Norscia 2011; Figure 6.1 of this chapter) is in line with reports on stress neuroendocrinology, indicating that environmental threats can induce stress increase (linked to adrenocortical reactivity), especially in dominants (Manogue *et al.*, 1975; Hellhammer *et al.*, 1997).

Preliminary analyses on the relationship between scratching and stressful events have been also investigated in *Propithecus verreauxi* (sifaka) and have confirmed that also in this species scratching works in detecting anxiety increase (Figure 6.7).

During the oestrus period, in *Lemur catta*, anxiety can increase in both females, as a result of increased harassment by males, and males as a consequence of higher competition levels. Indeed, in *Lemur catta* reproduction is not an easy affair. Males have to seize the day by catching the opportunity to mate in a narrow time window, in that female receptivity lasts 10–24 hours per year (Jolly, 1966; Evans and Goy, 1968; Koyama, 1988; Cavigelli and Pereira, 2000). Obtaining access to females is a very demanding task for males both because they have to play off rivals they are competing with for female mounting and because females have got mate choice. As a result, males are subject to aggression from both other male competitors and females during mating attempts. On the other hand, females are

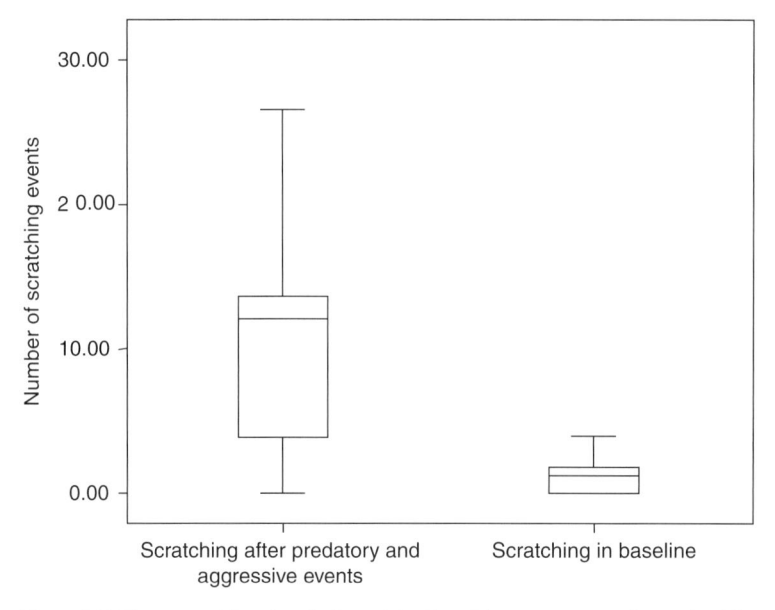

6.3 Self-directed behaviours and anxiety in lemurs

Figure 6.7 The scratching levels found in *Propithecus verreauxi* are significantly higher after potentially stressful events than in baseline condition (Wilcoxon's test: T = 0, ties = 0, n = 17, $p = 0.0001$).

busy attempting to repel particular males from attempting to copulate or to separate a male from another female during mounting (Sussman and Richard, 1974; Sauther, 1991). Things may be even more complicated than this. A female-female mounting incident was seen in Berenty (secondary forest Ankoba) in April 2008 (Daniela Antonacci, pers. comm.). Dominant females target subordinate ones to prevent them from conceiving because subordinate females with vital offspring can potentially acquire a higher ranking status in the social group and subtract resources from the dominant females (Vick and Pereira, 1989). The sole presence of a female showing a full perineal swelling, signalling the beginning of receptivity, determines a sharp increase in the aggression levels between group members, accompanied by a strong increase in self-scratching rates in both males and females, as observed by our study group (Sclafani et al., 2012). Even if both males and subordinate females are subject to attacks, aggressions towards males are particularly dangerous: some forms of conflict, exclusive of the mating season (such as jump fights), can escalate, causing serious injury, such as wounds 4–5 cm long on thighs, arms, or sides, and bleeding (Jolly, 1966; Sussman and Richard, 1974). Considering the 10-min time slot following an aggressive event during females' receptive period, we found that males tended to scratch more than females,[1] with male victims scratching significantly more when the aggressor was a female (Figure 6.8). The exclusive dominance of females over males and the high despotic relationships of *Lemur catta* society (Jolly, 1966; Pereira and

---

[1] Mann-Whitney's test $U = 33.00$, $n_{males} = 9$, $n_{females} = 14$, $p = 0.058$.

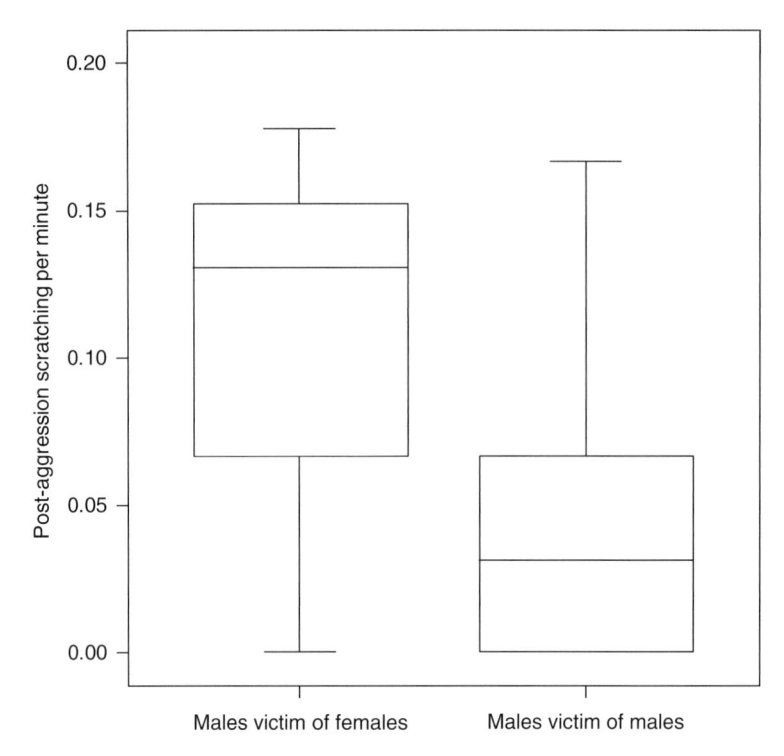

Figure 6.8 The scratching levels of *Lemur catta* males are higher after receiving aggression from females than from other males (Wilcoxon's test T = 1.00, ties = 2, n_{males} = 9, p = 0.031).

Kappeler, 1997; Chapter 4 of this book) explain the high anxiety experienced by males, especially when attacked by females.

We used scratching levels in *Lemur catta* to detect anxiety in aggressors and aggressees in presence or absence of receptive females. The anxiety-related scratching levels were not affected by frequency of aggression that lemurs performed, regardless of the presence of receptive females.[2] Instead, the presence of receptive females in the group influenced victim's anxiety because the individuals that received more aggression also showed higher levels of scratching.[3] This result indicates that during the copulation days, anxiety levels are more strictly related to aggressions probably because they are more dangerous and because reproduction – and not ordinary matters of conflict, such as food access – is involved. Reproduction is indeed a rare event, triggering a major conflict of interest between group members of both sexes (as we specified above). A further confirmation that lemurs' anxiety is particularly sensitive to reproduction-related conflicts of interest comes from the variation observed in the scratching levels immediately after an aggression. Results show that in the ten

---

[2] Scratching levels and aggressions performed – absence of receptive females: Spearman's test, r = -0.166, n = 23, p = 0.448; presence of receptive females: Spearman's test, r = -0.370, n = 23, p = 0.1.

[3] Scratching levels and aggression received – absence of receptive females: Spearman's test, r = 0.001, n = 23, p = 0.997; presence of receptive females: Spearman's test, r = 0.481, n = 23, p = 0.020.

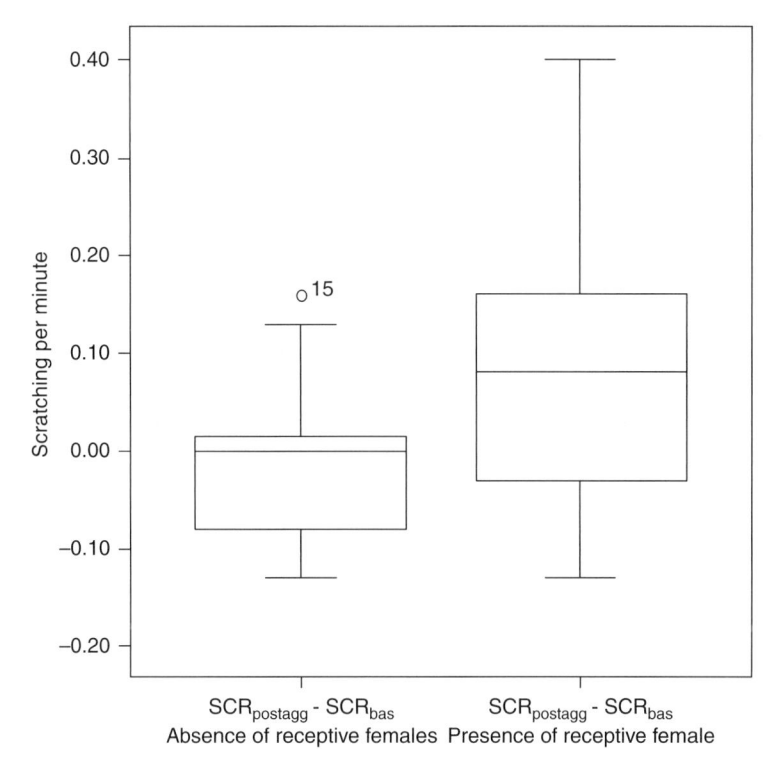

6.3 Self-directed behaviours and anxiety in lemurs

Figure 6.9 After aggression, the scratching rates in the *Lemur catta* victims (compared to baseline: Scratching$_{post-aggression}$ - Scratching$_{baseline}$) raise more in the presence than in the absence of receptive females (Wilcoxon's test T = 60.50, ties = 0, n = 23, *p* = 0.018).

minutes following an agonistic event, the victim's scratching levels increased both in the absence and in the presence of females ready to mate. However, compared to baseline, the scratching raise provoked by an aggression was higher during the copulation days than in the days when no female was receptive (Figure 6.9).

Box 6.2 | **by Lisa Gould**

Speaking of which: sex and hormones in wild lemurs

At the International Primatological Society Congress in 1996, I attended an excellent plenary talk by Dr Robert Sapolsky, relating to his pioneering work on the endocrine stress response, social status and behaviour of wild adult male baboons in Kenya (e.g., Sapolsky, 1982, 1983, 1992). Sapolsky's presentation fascinated me, as I had studied male affiliative patterns and sociality in male *Lemur catta* for my PhD dissertation research in 1992–93 (Gould, 1994), and the idea of examining hormonal correlates of male behaviour in wild ring-tailed lemurs was truly exciting. *L. catta* are characterised as

Box 6.2 (continued)

Box 6.2 (*cont.*)

exhibiting the most extreme form of female dominance (Jolly *et al.*, 1993) – all adult females in a group are dominant to males in all contexts (Jolly, 1966; Kappeler, 1990; Jolly *et al.*, 1993; Sauther, 1993; Sauther and Sussman, 1993; Gould, 1994; Sauther *et al.*, 1999). Furthermore, males within a group form a dominance hierarchy (Jolly, 1966; Sussman, 1977), but this hierarchy is highly unstable and changes frequently (Gould, 1994, 1997), particularly during the very short (~4 week) annual mating period, which is accompanied by marked intermale physical aggression (Jolly, 1966; Koyama, 1988; Sauther, 1991; Gould and Ziegler, 2007). Given all of these variables, the possibility of examining potential hormonal correlates of male behaviour during both mating and post-mating seasons seemed like a great project to undertake.

At the time of Sapolsky's 1996 presentation, few researchers had looked at the relationships between stress hormones (glucocorticoids), sex hormone levels and behavioural patterns in wild lemurs. Beginning in the early 1990s, Brockman and colleagues (1996, 1998, 2001) examined reproductive endocrinology and patterns of mating and aggression in both female and male *Propithecus verreauxi*, and in the mid-1990s, Cavigelli (1999) studied the effects of dominance rank, predation threat and food accessibility on stress hormone levels in wild female *Lemur catta*. But no behavioural endocrinology study of wild male *L. catta* had occurred. I was fortunate to receive funding from three different agencies to conduct the research, so I set out in 2001 (post-mating season) and 2003 (mating and post-mating seasons) to collect faecal and behavioural samples from individually identified adult male *L. catta* residing in three social groups at the Beza Mahafaly Reserve in southwestern Madagascar.

During Sapolsky's male baboon study in the late 1970s, the only way to measure hormonal levels was via blood samples, which necessitated actual capturing of the study animals. By the mid-to-late 1990s, new and far less invasive techniques of hormonal extraction from faecal samples had been developed (Whitten *et al.*, 1998), which made this area of research much more feasible. In the field, I was able to easily collect and preserve many faecal samples for later lab analysis.

When a female becomes receptive to mating, the resident males in her group, as well as males from other groups, engage in both marked physical combat with each other and ritual (stink-fight) displays during the female's receptive period (Jolly, 1966; Sauther, 1991; Sauther *et al.*, 1999). Since the mating season lasts approximately four weeks, males are in a constant state of high alert, and face potentially deadly encounters with rival males throughout this period. Males are often injured, sometimes fatally, from attacks and falls (Jolly, 1966; Gould *et al.*, 2005; Gould and Ziegler, 2007). One of my predictions was that males would exhibit significantly higher glucocorticoid (stress) levels during the short mating season compared to the post-mating period, because in order to achieve mating and potential reproductive success, a ring-tailed lemur male must be prepared to engage in physical combat during female receptivity periods. Most males in a social group will attempt to mate with the resident females, and will fight with each other for this opportunity, but many males also enter other nearby groups containing receptive females and must engage in male–male competition with the resident males of these groups as well. Furthermore, females decide with whom they will mate, so even if a male wins a stink fight or physical fight with a rival male, the receptive female may reject him as a mating partner. If a male's mating overtures (primarily tail-waving and squealing at the female) are accepted by the receptive female and he is able to mount and copulate, there are usually a number of rival males in close proximity who attack and push the male away from the female in order to disrupt copulation. Mating season is not an easy time for ring-tailed lemur males!

Glucocorticoids are adrenal hormones that are secreted during periods of stress (Sapolsky, 1982, 1983; Genuth, 1993; Balm, 1999), and prolonged release of glucocorticoids can result in serious physiological issues such as immunosuppression (Wingfield and Ramenofsky, 1999; Abbott et al., 2003). Given the level of intermale aggression that occurs during the *Lemur catta* mating season, I assumed that males would exhibit significantly higher glucocorticoid levels at this time, compared with the post-mating period, when testosterone levels drop dramatically and males spend much time resting and recuperating. But surprisingly, I found no statistical difference in mean fGc (faecal glucocorticoid) levels between mating and post-mating periods (Gould et al., 2005). The lack of marked differences in fGc levels between these seasons corresponded with Wingfield and Ramenofsky's (1999) argument that our expectation of 'reproductive stress' can be inaccurate, as the mating season is a predictable event, and these lemurs may not perceive the accompanying heightened male–male aggression as particularly stressful.

Mean faecal testosterone (fT) levels, on the other hand, *did* differ significantly between seasons, with mean fT levels rising dramatically in the early weeks of the mating period, and then dropping to baseline levels as the end of the season approached (Gould and Ziegler, 2007). This pattern of testosterone secretion was an exact fit with Wingfield et al.'s (1990) 'challenge hypothesis'. So it seems that in male *Lemur catta* the testosterone production increases rapidly during the early mating period, but such levels just as rapidly decrease towards the end of this time, preventing any long-term negative health effects of sustaining high testosterone levels.

In the future, it would be fascinating to investigate and compare stress hormone profiles across an annual cycle in male *Lemur catta*. A 'lemur year'

Box 6.2 (continued)

> Box 6.2 (*cont.*)
>
> in terms of reproduction consists of the mating season, a post-mating period, and birth and infant rearing season. The male migration season begins just after birth season, and continues for several months (Sussman, 1992; Gould, 1997, 2006). Male ring-tailed lemurs disperse, on average, every 3.5 years, and normally switch groups several times during their lives (Sussman, 1992; Koyama *et al.*, 2002; Gould, 2006). Is the migration period physiologically stressful for a dispersing male, who must, at times, travel alone and face increased predation risk and aggression from resident males in his search for a new group? Do resident males experience an increase in stress hormones during an immigration attempt by an unknown male? Or is the dispersal/immigration process another instance of a predictable life-history event? Many questions arise in relation to male and female *Lemur catta* life histories and the role that hormones play in the evolution of behavioural patterns. Hopefully some of these questions will be addressed by a new generation of lemur researchers in the years to come.
>
> Another potentially very useful avenue of research relates to the current state of lemur conservation in Madagascar. There are currently more than 100 recognised lemur species, and most of these are now threatened with rapid extinction due to massive deforestation and over-hunting for the illegal bushmeat trade throughout Madagascar (Schwitzer *et al.*, 2014). Lemurs are now (2014) considered the most threatened mammalian group on Earth. Studies focusing on stress responses to rapid habitat change could provide crucial information in terms of the development of conservation plans for these highly threatened and fascinating animals.

6.4 Anxiety reducing mechanisms in lemurs

Defeating anxiety is crucial, not only for lemurs but for any mammal, because it can impair the basic domains of social life. Reproduction, cognition and sociality, among other things, are all negatively affected by states of severe anxiety. Well-controlled studies on *Homo sapiens* indicate, for example, that pregnant women suffering high levels of anxiety are at increased risk for spontaneous abortion and preterm labour (Mulder *et al.*, 2002). In general, stressful conditions strongly and negatively affect vertebrates' reproduction potential by inhibiting mating behaviour in both sexes. In males the main stress hormone (cortisol) inhibits the release of testosterone, which promotes mating behaviours (Valenstein and Young, 1955; Beach and Inman, 1965; Sapolsky, 1987; Lord *et al.*, 2009); in females, glucocorticoids inhibit gonadotropin secretion and disrupt ovarian activity (Saketos *et al.*, 1993; Tilbrook *et al.*, 2002; Breen and Karsch, 2004; Blaustein, 2010). Since anxiety is a proxy for stress, the negative effect of stress-related anxiety on mating can be detrimental to reproduction. It has also been reported that in a wide array of mammalian species, spanning rodents to humans, environmental stressors (segregation, unstable housing, removal of environmental enrichment, etc.) causing anxiety are associated with negative cognitive biases (Brydges *et al.*, 2012). In dogs, for example, anxiety-related behaviours, including those in response to separation, are negatively correlated with motivation and learning ability measures (Mendl *et al.*, 2010).

Besides reproduction and cognition, the domain that is probably most sensitive to anxiety effects is sociality. It has been observed, in humans, that individuals suffering social phobia are less able to manage and control high anxiety social situations (Coles *et al.*, 2001). In children, the behaviour of social withdrawal can reflect underlying thoughts and feelings of social anxiety. According to Rubin and Burgess (2001), the relation between withdrawal and anxiety is 'dialectic' and 'cyclical'. Anxiety may be reinforced by frequent avoidance of interaction with peers but, concurrently, avoidance negatively affects the development of social skills. Taking an example from the aquatic world, it is interesting to observe that in marine mammals, the anxiety provoked by noise pollution (causing an impairment in intraspecific acoustic communication) can reduce social cohesion (e.g., with individuals deviating from group migration paths) with possible subsequent negative effects on survivorship and reproductive success (Wright *et al.*, 2007).

Social animals recover better from experiences of distress when group members are together and exchange positive interactions, such as embracing, grooming, touching and body contact. This phenomenon, known as 'social buffering' (see Kikusui *et al.*, 2006 for an extensive review), is widespread in mammals (Davitz and Mason, 1955; Coe *et al.*, 1978; Levine *et al.*, 1978; Mendoza *et al.*, 1978; Thorsteinsson *et al.*, 1998; Hennessy *et al.*, 2000). Primates make no exception: when they experience anxiety they can buffer it by engaging in social interactions (de Waal, 1987; Palagi *et al.*, 2004, 2006; van Wolkenten *et al.*, 2006; Aureli and Yates, 2010; Norscia and Palagi, 2011). The most typical and powerful behaviour used to reduce anxiety is allogrooming (hereafter, simply indicated as grooming), a form of fur cleaning performed using hands, lips or teeth depending on the species considered (see also Chapter 8). Besides toileting, grooming is involved in the establishment and maintenance of social bonds (Dunbar, 1991). This social function seems to be mediated by the release of brain opioids such as beta-endorphins causing an effect of relaxation (Keverne *et al.*, 1989). Grooming also appears to decrease the heart rate (Boccia *et al.*, 1989; Aureli *et al.*, 1999). Interestingly, in humans, body massage seems to have similar effects as grooming, leading to the decrease of heart rate and blood pressure and probably inducing relaxation (Kaada and Torsteinbø, 1989; Hayes and Cox, 1999). However (and incredibly) in humans, the available literature does not allow stating with certainty that massage induces relaxation (Smith and O'Driscoll, 2003), although this function is often assumed in scientific reports (e.g., Cherkin *et al.*, 2009). In this case, quite surprisingly, objective data on the role of social contacts in reducing anxiety comes from quantitative investigation on non-human primates.

In haplorrhines, the beneficial effect of grooming in reducing anxiety has been reported since the 1980s (Schino *et al.*, 1988; Aureli and Yates, 2010). Only recently has this issue been addressed in strepsirrhines, whose anxiety seems to be sensitive to grooming as well. In species with diurnal activity (Jolly, 1966; Barton, 1987) grooming is almost always reciprocal, with two lemurs grooming each other at the same time. Consequently, the benefits deriving from fur cleaning are mutual.

A study conducted on *Lemur catta* during the mating period showed that in this species grooming has a short-term effect on anxiety, with scratching rates decreasing within the 15 minutes following the grooming session (Sclafani et al., 2012). Interestingly, in the same study group and period, grooming negatively correlated with scratching when no female was receptive but not during the days in which at least one female was ready to mate,[4] the most stressful period. In addition, during the copulation days a negative correlation trend was detected between scratching and aggression performed whereas no correlation existed in the absence of receptive females.[5] How can these additional results be interpreted? Probably, single grooming sessions are not sufficient to buffer the effect of anxiogenic factors (aggressive events), which are highly recurring, severe and relevant for the acquisition of a rare and inalienable resource (the eggs to be fertilised).

On the other hand, the negative correlation trend between scratching levels and aggression performed by animals allows us to hypothesise that in *L. catta* aggression per se might be effective as an energy releaser and stress reliever mechanism. Our interpretation is supported by previous reports on olive baboons (*Papio anubis*), showing that animals that initiated or redirected aggression against others had significant lower basal stress levels (measured via glucocorticoid concentration) and a better response to acute challenge (Sapolsky and Ray, 1989; Virgin and Sapolsky, 1997). In addition, in *Homo sapiens* several studies have found that aggressive adults have lower levels of stress hormones (van Goozen, 2005).

Among positive social interactions, grooming is not the only behaviour working as an anxiety reliever; social play can work too. In our two study groups of sifaka (observed in the Berenty forest), for example, the levels of scratching performed after a stressful event (aggression, noise disturbance, presence of observers the animals were not habituated to) decreased after the animals had engaged in playful interactions.[6] Similar results were found in a family group of captive marmosets (*Callithrix jacchus*). In this species, we observed that scratching levels, increased after aggressive events, were significantly reduced after social play sessions (Norscia and Palagi, 2011; the same methods were applied to obtain the results on sifaka reported here). The linkage between play and anxiety reduction is consistent with previous reports in other mammals. In a study on young bears, Fagen and Fagen (2004) speculated that play experience could relieve past anxiety and

---

[4] Scratching and grooming were not correlated in the absence of receptive females (Spearman's test: n = –0.006, n = 23, p = 0.977) but they were negatively correlated when at least one female was receptive (Spearman's test: r = –0.480, n = 23, p = 0.020).

[5] Scratching and aggressions performed were not correlated in the absence of receptive females (Spearman's test: r = –0.166, n = 23, p = 0.448) but they tended to be correlated when at least one female was receptive (Spearman's test: r = –0.370, n = 23, p = 0.082).

[6] In the 15 minutes following a stressful event, social play interactions significantly reduced scratching levels (Wilcoxon's test: n = 12, T = 12.00, p = 0.034).

build resistance to future stressful events. In juvenile rats and in the New World's squirrel monkey, stress hormones appeared to be negatively correlated with play (Biben and Champoux, 1999; Wilson, 2001). Interestingly, Klein *et al.* (2010) found that, in male rats, acute restraint stress suppressed play, whereas repeated restraint stress affected all social interactions but left play largely unaffected. This framework suggests that while acute stress can suppress play, mild or repeated stress can probably be buffered by play (Pellis and Pellis, 2009).

Finally, peace restoration between two opponents – one of the conflict resolution mechanisms adopted by social groups – works in reducing post-conflict anxiety, in primates and other mammals. This phenomenon, known as reconciliation (see Chapter 5), is based on a wide array of affiliative interactions (which depend on the behavioural repertoire of a species) between the aggressor and the victim. Such interactions can include grooming and play (which – as we have already discussed above – can work per se as anxiety relievers) but also many other positive social contacts performed in single or combined patterns.

What is important here is the conciliatory value of the social contact, and its implications for conflict resolution, and not the pattern used. Unresolved conflicts can lead to social disruption and have serious consequences for the entire community, such as fragmentation, isolation, animal wounding or killing (see also Chapter 5, focussed on post-conflict management).

In *Homo sapiens*, a cross-sectional study using self-rated and validated anxiety and stress assessment, general stressor and sociodemographic questionnaires showed that conflict resolution between colleagues contributes to work-related anxiety reduction in house officers (Tan *et al.*, 2013). In wild chimpanzees, Arnold and Whiten (2001) found that despite the low rates of self-directed behaviours during post-conflict periods (lower than baseline), mean rates of scratching were higher in unreconciled than in reconciled conflicts. This result was confirmed in a captive group of chimpanzees, in which Fraser *et al.* (2010) showed that post-conflict levels of self-directed behaviour, including scratching, were lower after reconciliation than when reconciliation did not occur, thus backing up the stress-alleviating function for reconciliation. Thus, despite a controversial result (Koski *et al.*, 2007), reconciliation also appears to work as an anxiety reliever in non-human apes. It is worth highlighting, however, that literature showing the direct response of anxiety-related behaviours to conflicts is meagre, and that most of the reports have focused on the effect of third-party affiliation (consolation) on anxiety levels (e.g., Romero *et al.*, 2010).

Restoring the relationship between victim and aggressor also works in reducing anxiety in Old World monkeys (Aureli *et al.*, 2002). In olive baboons (*Papio anubis*), and bonnet (*Macaca radiata*) and long-tailed (*Macaca fascicularis*) macaques, self-directed behaviour (including scratching) in the former opponents increased after aggression and decreased following reconciliation (Aureli and van Schaik, 1991; Castles and Whiten, 1998; Das *et al.*, 1998). In Barbary (*Macaca sylvanus*) and Japanese (*Macaca fuscata*) macaques the rate of post-conflict self-scratching in the victims of aggression was found to be significantly lower following reconciled conflicts compared to non-reconciled conflicts (Kutsukake and Castles, 2001; McFarland

and Majolo, 2011). It is interesting to notice that reconciliation can have a calming effect in conflict observers other than in the opponents. In this respect, anxiety can be reduced not only in the animals directly engaging in reconciliation (victim and aggressor) but also in those group members witnessing post-conflict reunions between the former opponents. This phenomenon was recently described in *Papio hamadryas* by Judge and Bachman (2013), who observed that witnessing a fight increased the self-directed behaviour whereas witnessing the post-conflict reunion reduced it. In their elegant study, the authors also showed, as a control, that anxiety reduction in the bystanders occurred after observing the affiliation of former combatants and not when the affiliation was engaged by other group members.

In a captive group of New World monkeys, brown capuchins (*Cebus apella*), Weaver and de Waal (2003) found that reconciliation worked in reducing anxiety-related scratching in the youngsters having a weak bond with their mothers and being particularly sensitive to emotional arousal after a conflict.

If reconciliation has been neglected in the studies on strepsirrhines, its possible role in reducing anxiety had been ignored until data were collected on the hybrid lemur species *Eulemur rufus × collaris* at Berenty (south Madagascar) (Palagi and Norscia 2011). Besides observing that the phenomenon of reconciliation was present in the brown lemurs (Chapter 5), we examined how scratching levels in the victim of aggression fluctuated after a conflict in the presence and in the absence of conciliatory contacts with the former opponent. We found that after post-conflict affiliation, the victim's scratching decreased to the baseline levels. This finding indicates that not only are lemurs capable of reconciling (Norscia and Palagi, 2011) but also that reconciliation is able to reduce anxiety – ultimately double-bonded to a stress physiological response – regardless of the cognitive abilities characterising the primate species considered. This is because anxiety – when uncontrolled – is a disruptive phenomenon for all.

To conclude, we would like to point out that a further mechanism used for anxiety reduction may be scratching itself. This hypothesis is backed up by research on a variety of animals, including humans, where changes in plasma endorphin levels were associated with self-directed behaviours (stereotypies), suggesting a role of these behaviours in self-medication for distress (see Morgan, 2006 for an extensive review). However, only a combined approach, including both hormonal analyses and a behavioural investigation, can allow drawing final conclusions in this respect.

6.5 Yawning and motivational conflict: why do lemurs yawn?

From lemurs to humans, primates yawn. Yawning is an involuntary and stereotyped behaviour, or fixed action pattern (*sensu* Tinbergen, 1951), present in many vertebrate species, including humans, from foetal stages to old age (Baenninger, 1997). In mammals, the yawning patterns include mouth opening, deep inspiration, brief apnoea and slow expiration. Yet, despite the clear pattern and ubiquity of yawning, controversy has surrounded this behaviour on its possible functions (Guggisberg *et al.*, 2010). Blood and brain oxygenation, vigilance arousal, thermoregulation and

6.5 Yawning and motivational conflict: why do lemurs yawn?

sleep–awake rhythm regulation are the main functions put forth to explain the presence of yawning. Some of them have been discarded and some others, such as thermoregulation, are still under debate in the scientific world (Baenninger, 1987; Gallup et al., 2009; Guggisberg et al., 2010; Gallup and Eldakar, 2013). Until recently yawning had been considered as a purely physiological behaviour, of relevance only to the yawner, and thus little attention was paid to the phenomenon. The interest in yawning has risen as it is now considered as a possible emotional signal. This convinced researchers that the phenomenon deserved better attention and motivated them to put more effort in investigating it (Guggisberg et al., 2010). Indeed, other than in homeostatic processes, yawning may be linked to social contexts, for example as a signal of aggressiveness, hierarchical dominance, frustration, sexual excitement, or as a means of synchronising activities within the group (Hadidian, 1980; Troisi et al., 1990; Deputte, 1994; Palagi et al., 2009; Vick and Paukner, 2009; Guggisberg et al., 2010; Norscia and Palagi, 2011; Demuru and Palagi, 2012).

In human and non-human primates, two different types of yawn are generally distinguished according to the physiological state and social context: rest yawns related to sleepiness or boredom (Provine, 1986; Provine et al., 1987; Zilli et al., 2008) and tension yawns (Guggisberg et al., 2007; Leone et al., 2014) related to conflict situations and possibly indicating arousal (Bertrand, 1969; Hadidian, 1980; Maestripieri et al., 1992; Deputte, 1994; Baenninger et al., 1996). Prior to these studies, Altmann (1967) had pointed out that both types of yawns may indicate levels of physiological arousal and, therefore, it is extremely difficult to disentangle the original stimuli that trigger them.

Consistently, other studies on primates have not adopted a dichotomous view – rest versus tension yawns – choosing, instead, to relate yawning variation, if any, to specific contexts. For example, macaques, geladas and mangabeys are known to exhibit 'emotion yawns' or 'social yawns' during antagonistic social encounters (Deputte, 1994; Smith, 1999; Paukner and Anderson, 2006; Leone et al., 2014). Among the great apes, chimpanzees were found to yawn more in response to human proximity (Goodall, 1968) and during conditions of social tension (Baker and Aureli, 1997).

Finally, spontaneous yawning is also described as a form of self-directed behaviour linked to anxiety states in primates (Dunbar and Dunbar, 1975; Maestripieri et al., 1992; Troisi, 2002). Bearing this possibility in mind, we decided to collect – besides scratching bouts – data on yawning in the sifaka and the ring-tailed lemurs at Berenty (Madagascar) (Figure 6.10). Thanks to the fieldwork of three students, episodes of yawning were gathered during observations on focal animals and – since yawning overall is an infrequent behaviour – also via all occurrences (Altmann, 1974). Two main, different conditions were recognised: baseline condition, in the absence of potentially stressful events, and disturbance condition, when an environmental or social stressor was detected (intragroup aggressions, presence of other groups or of unfamiliar observers within 10 m). Yawning levels were measured in both conditions and were then compared, at the individual level, through a paired sample test. We found that yawning was sensitive to the environmental

Figure 6.10 Yawning in *Lemur catta*. Berenty Reserve, Madagascar. Photo: Elisabetta Palagi.

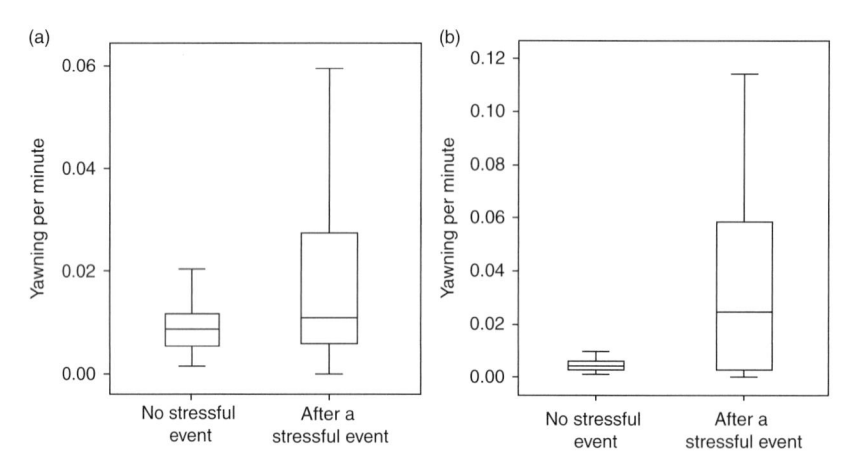

Figure 6.11 Yawning levels after a perturbing event are higher than in baseline in both *Propithecus verreauxi* (a; Exact Wilcoxon's test: n = 12; T = 11; p = 0.027) and *Lemur catta* (b; Exact Wilcoxon's test: n = 17; T = 33; p = 0.040).

perturbation provoked by the disturbing event, in both the sifaka and the ring-tailed lemur (Figure 6.11). These findings, consistent with Zannella *et al.* (2015), suggest that yawning in lemurs is modulated by emotional arousal and can indicate anxiety levels, similarly to scratching. Can we therefore suggest that yawning is suitable to be used as a measure of anxiety in the same way scratching can? As a rather infrequent behaviour, it is unlikely that yawning can replace scratching as a behavioural indicator of anxiety in the wild. In fact, the general low number of yawns during the day make it difficult, for example, to split analyses by context (e.g., before and

6.5 Yawning and motivational conflict: why do lemurs yawn?

Figure 6.12 Yawning in *Propithecus verreauxi* during a sleep–awake state transition. The sequence was shot on 27 February 2011 between 10:57:42 and 11:06:30 am in the Berenty forest, Madagascar. Photo: Ivan Norscia.

after conciliatory contacts to check whether they reduce anxiety) or to perform in-depth tests to compare different situations (aggression, predation attempts, presence of unfamiliar observers) due to the small sample size. Moreover, other than to anxiety, yawning is also linked to sleep–awake state transition (Figure 6.12). To better understand the linkage between anxiety and yawning, it would be useful to move observations to controlled conditions, such as those provided by captivity. Controlled conditions would allow measuring, for example, the effect of crowding on yawning frequencies (indoor vs outdoor facilities). It would also be easier to provide the animals with artificial disturbing stimuli (e.g., dummies to simulate competitors, predator vocalisations, etc.) to evaluate if there is an all-or-nothing or a graded yawning response (with this last case applying to scratching; Palagi and Norscia, 2011), according to the different stimulus intensity. Controlled conditions, finally, would permit an accurate comparison among different species, with respect to how yawning responds to different stressors.

The yawn, as a signal to others, has been linked to emotional arousal (Lehmann, 1979) and can be 'transmitted' to other subjects. This phenomenon, known as contagion, was once thought to take place only in extant hominids (*Homo sapiens*) and was associated with highly sophisticated cognitive abilities, such as self-awareness and mental state attribution (e.g., Platek *et al.*, 2003). Then, a study found it present also in chimpanzees (*Pan troglodytes*: Anderson *et al.*, 2004), which opened the door to the investigation of contagion in other primate species. Yet, based on the

Mind-State Attribution Model, it was hard – at the time – to predict that contagious yawning could be present in non-ape primates. Instead, the Perception–Action Model (Preston and de Waal, 2002) predicted that contagious yawning might occur in other social primates, as it postulates a more basic and widespread mechanism for behavioural coordination. Specifically, during the observation of a facial expression, the observer can involuntarily re-enact the same motor pattern by recruiting neural mechanisms – grounded in so-called mirror neurons or homologous brain areas – that concurrently activate the same affective state associated with that specific facial expression (Preston and de Waal, 2002; Gallese et al., 2004).

Since 2004, yawn contagion has been better investigated in humans (Anderson and Meno, 2003; Giganti and Zilli, 2011; Norscia and Palagi, 2011; Norscia et al., 2016), in chimpanzees (Campbell et al., 2009; Massen et al., 2012) and found in bonobos (*Pan paniscus*: Demuru and Palagi, 2012) but it has also been detected in non-ape primates such as geladas (*Theropithecus gelada*: Palagi et al., 2009) and macaques (*Macaca arctoides*: Paukner and Anderson, 2006). Yawn infectiveness has been associated with emotional contagion, a basic form of empathy (e.g., Palagi et al., 2009; Campbell and de Waal, 2011; Norscia and Palagi, 2011; Demuru and Palagi, 2012).

At a more basic level, it has been hypothesised that the social yawn may be a non-verbal form of communication used to synchronise the behaviour of a group (Guggisberg et al., 2010).

Interestingly, an unpublished study by Gervais, Pokorny and de Waal[7] found that in *Cebus apella* – a species with a documented propensity for behavioural coordination (e.g., Galloway et al., 2005) – the number of observed yawns was significantly greater than predicted by chance during the projection of a yawn video (Chi-square test, $p < 0.05$). Now, it is possible that studies on yawn contagion in platyrrhines have been discouraged by negative results, but since negative results often do not get to be published we cannot know that for sure. We simply believe that wider efforts are needed to investigate yawning as a social signal across the primate order, also considering that the studies addressing this issue are few even for catarrhines.

What about the strepsirrhines? Group-living lemurs are certainly characterised by strong behavioural coordination, necessary to maintain cohesion. Indeed, they displace in the forest as 'foraging units' (e.g., *Propithecus verreauxi*: Richard, 1978; Norscia et al., 2006), cohesively move in troops (e.g., *Lemur catta*: Jolly, 1966), split and reunite during night activities using acoustic communication (*Avahi meridionalis*: Norscia and Borgognini-Tarli, 2008), and coordinate through vocal calls (*Indri indri* and *Varecia variegata*: Geissmann and Mutschler, 2006; Maretti et al., 2010). Movement coordination of group members can also vary according to the group leader and the season, as revealed by the observation of which individuals initiate and lead the displacement, how many group members follow, and the

---

[7] Gervais, M., Pokorny, J. J. and de Waal, F. B. M. (August, 2005). Contagious yawning and scratching in brown capuchin monkeys (*Cebus apella*). Poster presented at the Emory Summer Undergraduate Research Experience, Emory University, Atlanta, GA. Link: www.cse.emory.edu/sciencenet/undergrad/SURE/Posters/2005_Gervais.cfm

termination of group movements (e.g., in *Eulemur rufifrons*: Pyritz *et al.*, 2011). Behavioural coordination has been confirmed in foraging trials (in *Eulemur rufifrons*: Pyritz *et al.*, 2013). It is therefore clear that group synchronisation is a key aspect of lemur behaviour.

Several questions can be addressed by future research. Which ecological stressors impact lemurs the most? Scratching can provide a real-time measure on the anxiety, and indirectly on the stress, of lemurs after being exposed to socioecological stressors. Therefore – besides population size, density, sex ratio, fur conditions, etc. – measuring anxiety can be an additional, useful tool to assess the impact of different environmental conditions on the lemurs' welfare. Why not trying to investigate it in different lemur species, during ecological studies?

Is the yawn a communication signal? Yawning can indicate emotional arousal in lemurs. But it might be used also as a signal to synchronise – to a certain extent – group activity. It would be interesting to have yawning bouts recorded across group members to assess if yawns are more likely to occur after some individuals have started yawning, in certain periods of the day or when group activity changes. May yawn synchronisation, if present, be distributed differently according to differing social systems (e.g., family versus large multimale-multifemale groups) or to differing habits (diurnal versus cathemeral)?

References

Abbott, D. H., Keverne, E. B., Bercovitch, F. B., *et al.* (2003). Are subordinates always stressed? A comparative analysis of rank differences in cortisol levels among primates. *Hormones and Behavior*, 43, 67–82.

Alexander, S. L. & Irvine, C. H. G (1998). The effect of social stress on adrenal axis activity in horses: The importance of monitoring corticosteroid-binding globulin capacity. *Journal of Endocrinology*, 157, 425–432.

Altmann, J. (1974). Observational study of behaviour sampling methods. *Behaviour*, 49, 227–265.

Altmann, S. A. (1967). The structure of primate social communication. In: Altmann, S. A (ed.), *Social Communication among Primates*. Chicago: University of Chicago Press, pp. 325–362 .

Anderson, J. R. & Meno, P. (2003). Psychological influences on yawning in children. *Current Psychology Letters*, 11, 1–7.

Anderson, J. R., Myowa-Yamakoshi, M. & Matsuzawa, T. (2004). Contagious yawning in chimpanzees. *Proceedings of the Royal Society B: Biological Sciences*, 271, 468–470.

Arnold, K. & Whiten, A. (2001). Post-conflict behaviour of wild chimpanzees (*Pan troglodytes schweinfurthii*) in the Budongo Forest, Uganda. *Behaviour*, 138, 649–690.

Aureli, F. (1997). Post-conflict anxiety in nonhuman primates: The mediating role of emotion in conflict resolution. *Aggressive Behaviour*, 23, 315–328.

Aureli, F. & de Waal, F. B. M. (1997). Inhibition of social behavior in chimpanzees under high-density conditions. *American Journal of Primatology*, 41, 213–228.

Aureli, F. & Schaffner, C. M. (2008). Social interactions, social relationships and the social system of spider monkeys. In: *Spider Monkeys: Behavior, ecology and evolution of the genus Ateles*. Cambridge University Press, pp. 236–265.

Aureli, F. & Smucny, D. A. (2000). The role of emotion in conflict and conflict resolution. In: Aureli, F. & de Waal, F. B. M. (eds), *Natural Conflict Resolution*. Berkeley, CA: University of California Press, pp. 199–224.

Aureli, F. & van Schaik, C. P. (1991). Post-conflict behaviour in long-tailed macaques (*Macaca fascicularis*): II Coping with the uncertainty. *Ethology*, 89, 101–114.

Aureli, F. & Whiten, A. (2003). Emotions and behavioral flexibility. In: D. Maestripieri (ed.), *Primate Psychology – The Mind and Behavior of Human and Nonhuman Primates*. Cambridge, MA: Harvard University Press, pp. 289–323.

Aureli, F. & Yates, K. (2010). Distress prevention by grooming others in crested black macaques. *Biology Letters*, 6, 27–29.

Aureli, F., van Schaik, C. P. & van Hooff, J. A. R. A. M. (1989). Functional aspects of reconciliation among captive long-tailed macaques (*Macaca fascicularis*). *American Journal of Primatology* 19, 38–51.

Aureli, F., Preston, S. D. & de Waal, F. (1999). Heart rate responses to social interactions in free-moving rhesus macaques (*Macaca mulatta*): a pilot study. *Journal of Comparative Psychology*, 113, 59–65.

Aureli, F., Cords, M. & van Schaik, C. P. (2002). Conflict resolution following aggression in gregarious animals: a predictive framework. *Animal Behaviour*, 64, 325–343.

Baenninger, R. (1987). Some comparative aspects of yawning in *Betta splendens*, *Homo sapiens*, *Panthera leo* and *Papio sphinx*. *Journal of Comparative Psychology*, 101, 349–354.

Baenninger, R. (1997). On yawning and its functions. *Psychonomic Bulletin & Review*, 4, 198–207.

Baenninger, R., Binkley, S. & Baenninger, M. (1996). Field observations of yawning and activity in humans. *Physiology and Behavior*, 59, 421–425.

Baker, K. C. (2004). Benefits of positive human interaction for socially-housed chimpanzees. *Animal Welfare* (South Mimms, England), 13, 239–245.

Baker, K. C. & Aureli, F. (1997). Behavioural indicators of anxiety: an empirical test in chimpanzees. *Behaviour*, 134, 1031–1050.

Balm, P. H. M. (1999). *Stress Physiology in Animals*. Sheffield: Sheffield Academic Press.

Balon, R. (2005). Measuring anxiety: are we getting what we need? *Depress Anxiety*, 22, 1–10.

Barros, M. & Tomaz, C. (2002). Non-human primate models for investigating fear and anxiety. *Neuroscince & Biobehavioral Reviews*, 26, 187–201.

Barros, M., Boere, V., Huston, J. P. & Tomaz, C. (2000). Measuring fear and anxiety in the marmoset (*Callithrix penicillata*) with a novel predator confrontation model: effects of diazepam. *Behavioural Brain Research*, 108, 205–211.

Barton, R. A. (1987). Allogrooming as mutualism in diurnal lemurs. *Primates*, 28, 539–542.

Beach, F. A. & Inman, N. G. (1965). Effects of castration and androgen replacement on mating in male quail. *PNAS*, 54, 1426–1431.

Bertrand, M. (1969). *The Behavioural Repertoire of the Stumptail Macaque*. Basel: Karger.

Biben, M. & Champoux, M. (1999). Play and stress: cortisol as a negative correlate of play in *Saimiri*. In: S. Reifel (ed.), *Play and Cultures Study*, vol. 2, 91–208.

Blaustein, J. D. (2010). Feminine reproductive behavior and physiology in rodents: integration of hormonal, behavioral, and environmental influences. In: D. W. Pfaff, *et al.* (eds), *Hormones, Brain and Behavior*, 2nd ed. New York: Academic Press, pp. 67–107.

Boccia, M. L., Reite, M. & Laudenslager, M. (1989). On the physiology of grooming in a pigtail macaque. *Physiology and Behavior*, 45, 667–670.

Bourin, M., Petit-Demoulière B., Nic Dhonnchadha, B. & Hascöet, M. (2007). Animal models of anxiety in mice. *Fundamental & Clinical Pharmacology*, 21, 567–574.

Breen, K. M. & Karsch, F. J. (2004). Does cortisol inhibit pulsatile luteinizing hormone secretion at the hypothalamic or pituitary level? *Endocrinology*, 145, 692–698.

Brockman, D. K. & Whitten, P. L. (1996). Reproduction in free-ranging *Propithecus verreauxi*: estrus and the relationship between multiple partner matings and fertilization. *American Journal of Physical Anthropology*, 100, 57–96.

Brockman, D. K., Whitten, P. L., Richard, A. F. & Schneider, A. (1998). Reproduction in free-ranging male *Propithecus verreauxi*: The hormonal correlates of mating and aggression. *American Journal of Physical Anthropology*, 105, 137–151.

Brockman, D. K., Whitten, P. L., Richard, A. F. & Benander, B. (2001). Birth season testosterone levels in male Verreaux's sifaka, *Propithecus verreauxi*: insights into socio-demographic factors mediating seasonal testicular function. *Behavioral Ecology and Sociobiology*, 49, 117–127.

Brydges, N. M., Hall, L., Nicolson, R., Holmes, M. C. & Hall, J. (2012). The effects of juvenile stress on anxiety, cognitive bias and decision making in adulthood: A rat model. *PLoS ONE*, 7, e48143. http://dx.doi.org/10.1371/journal.pone.0048143.

Campbell, M. W. & de Waal, F. B. M. (2011). Ingroup-outgroup bias in contagious yawning by chimpanzees supports link to empathy. *PLoS ONE*, 6(4), e18283. http://dx.doi.org/10.1371/journal.pone.0018283.

Campbell, M. W., Devyn Carter, J., Proctor, D., Eisenberg, M. L. & de Waal, F. B. M. (2009). Computer animation stimulates contagious yawning in chimpanzees. *Proceedings of the Royal Society B: Biological Sciences*, 276, 4255–4259.

Campennì, M. & Schino, G. (2014). Partner choice promotes cooperation: The two faces of testing with agent-based models. *Journal of Theoretical Biology*, 344, 49–55.

Castles, D. L. & Whiten, A. (1998). Post-conflict behaviour of wild olive baboons. II. Stress and self-directed behaviour. *Ethology*, 104, 148–160.

Castles, D. L., Whiten, A. & Aureli, F. (1999). Social anxiety, relationships and self-directed behaviour among wild female olive baboons. *Animal Behaviour*, 58, 1207–1215.

Cavigelli, S. A. (1999). Behavioral patterns associated with faecal cortisol levels in free-ranging female ring-tailed lemurs, *Lemur catta*. *Animal Behaviour*, 57, 935–944.

Cavigelli, S. A., & Pereira, M. E. (2000). Mating season aggression and fecal testosterone levels in male ring-tailed lemurs (*Lemur catta*). *Hormones and Behavior*, 37, 246–255.

Cherkin, D. C., Sherman, K. J., Kahn, J., *et al.* (2009). Effectiveness of focused structural massage and relaxation massage for chronic low back pain: protocol for a randomized controlled trial. *Trials*, 10, 96.

Cilia, J. & Piper, D. C. (1997). Marmoset conspecific confrontation: an ethologically-based model of anxiety. *Pharmacology Biochemistry and Behavior*, 58, 85–91.

Coe, C. L., Mendoza, S. P., Smotherman, W. P. & Levine, S. (1978). Mother–infant attachment in the squirrel monkey: adrenal response to separation. *Behavioral Biology*, 22, 256–263.

Coles, M. E., Turk, C. L., Heimberg, R. G. & Fresco, D. M. (2001). Effects of varying levels of anxiety within social situations: Relationship to memory perspective and attributions in social phobia. *Behaviour Research and Therapy*, 39, 651–665.

Cordoni, G. & Palagi, E. (2007). Response of captive lowland gorillas (*Gorilla gorilla gorilla*) to different housing conditions: testing the aggression-density and coping models. *Journal of Comparative Psychology*, 121, 171–180.

Craig, K. J., Brown, K. J. & Baum, A. (1995). Environmental factors in the etiology of anxiety. In: F. E. Bloom & D. J. Kupfer (eds), *Psychopharmacology: The fourth generation of progress*. New York: Raven, pp. 1325–1339.

Daniel, J. R., Santos, A. J. & Vicente, L. (2008). Correlates of self-directed behaviours in captive *Cercopithecus aethiops*. *International Journal of Primatology*, 29, 1219–1226.

Darwin, C. (1872). *The Expression of the Emotions in Man Animals*. London: John Murray.

Das, M., Penke, Z., & van Hooff, J. A. R. A. M. (1998). Postconflict affiliation and stress-related behavior of long-tailed macaque aggressors. *International Journal of Primatology*, 19, 53–71.

Davitz, J. & Mason, D. J. (1955). Socially facilitated reduction of fear response in rats. *Journal of Comparative and Physiological Psychology*, 47, 941–947.

de Waal, F. B. M. (1987). Tension regulation and nonreproductive functions of sex in captive bonobos (*Pan paniscus*). *National Geographic Research*, 3, 318–335.

de Waal, F. B. M. & van Roosmaleen, A. (1979). Reconciliation and consolation among chimpanzees. *Behavioral Ecology and Sociobiology*, 5, 55–66.

Demuru, E. & Palagi, E. (2012). In Bonobos yawn contagion is higher among kin and friends. *PLoS ONE*, 7(11), e49613. http://dx.doi.org/10.1371/journal.pone.0049613.

Deputte, B. L. (1994). Ethological study of yawning in primates. 1. Quantitative analysis and study of causation in two species of old world monkeys (*Cercocebus albigena* and *Macaca fascicularis*). *Ethology*, 98, 221–245.

Dunbar, R. I. M. (1991). Functional significance of social grooming in primates. *Folia Primatologica*, 57, 121–131.

Dunbar, R. I. M. & Dunbar, E. P. (1975). *Social Dynamics of Gelada Baboons*. Basel: Karger.

Evans, C. S. & Goy, R. W. (1968). Social behaviour and reproductive cycles in captive ring-tailed lemurs (*Lemur catta*). *Journal of Zoology*, 156, 181–197.

Evers, E., de Vries, H., Spruijt, B. M. & Sterck, E. H. M. (2014). The EMO-model: An agent-based model of primate social behavior regulated by two emotional dimensions, anxiety-FEAR and satisfaction-LIKE. *PLoS ONE*, 9, 2 e87955.

Fagen, R. & Fagen, J. (2004). Juvenile survival and benefits of play behaviour in brown bears, *Ursus arctos*. *Evolutionary Ecology Research*, 6, 89–102.

Fraser, O. N., Stahl, D., & Aureli, F. (2008). Stress reduction through consolation in chimpanzees. *PNAS*, 105, 8557–8562.

Fraser, O. N., Stahl, D. & Aureli, F. (2010). The function and determinants of reconciliation in *Pan troglodytes*. *International Journal of Primatology*, 31, 39–57.

Freud, S. (1969). *A General Introduction to Psychoanalysis*, revised edition. New York: Touchstone.

Fried, R. (1994). Evaluation and treatment of "psychogenic" pruritus and self-excoriation. *Journal of the American Academy of Dermatology*, 30, 993–999.

Gallese, V., Keysers, C. & Rizzolatti, G. (2004). A unifying view of the basis of social cognition. *Trends in Cognitive Sciences*, 8, 396–403.

Galloway, A. T., Addessi, E., Fragaszy, D. & Visalberghi, E. (2005). Social facilitation of eating familiar food in tufted capuchin monkeys (*Cebus apella*): Does it involve behavioral coordination? *International Journal of Primatology*, 26, 181–189.

Gallup, A. C. & Eldakar, O. T. (2013). The thermoregulatory theory of yawning: what we know from over 5 years of research. *Frontiers in Neuroscience*, 6, 188. http://dx.doi.org/10.3389/fnins.2012.00188.

Gallup, A. C., Miller, M. L. & Clark, A. B. (2009). Yawning and thermoregulation in budgerigars (*Melopsittacus undulatus*). *Animal Behaviour*, 77, 109–113.

Geissmann, T. & Mutschler, T. (2006). Diurnal distribution of loud calls in sympatric wild indri (*Indri indri*) and ruffed lemurs (*Varecia variegata*): implications for call functions. *Primates*, 47, 393–396.

Genuth, S. M. (1993). The endocrine system. In: Berne, R. M. & Levy, M. N. (eds), *Physiology*. Vol. III. St. Louis: Mosby Yearbook, pp. 813–1024.

Giganti, F. & Zilli, I. (2011). The daily time course of contagious and spontaneous yawning among humans. *Journal of Ethology*, 29, 215–219. http://dx.doi.org/10.1007/s10164-010-0242-0.

Goodall, J. (1968). A preliminary report on the expressive movements and communication in the Gombe stream chimpanzees. In: P. C. Jay (ed.), *Primates: studies in adaptation and variability*. New York: Holt, Rinehart and Winston, pp. 313–374.

Gould, L. (1994). Patterns of affiliative behavior in adult male ringtailed lemurs (*Lemur catta*) at the Beza-Mahafaly Reserve, Madagascar. Ph.D. Dissertation, Washington University.

Gould, L. (1997). Intermale affiliative relationships in ringtailed lemurs (*Lemur catta*) at the Beza-Mahafaly Reserve, Madagascar. *Primates*, 38, 15–30.

Gould, L. (2006). Male sociality and integration during the dispersal process in *Lemur catta*: a case study. In: Jolly, A., Sussman, R. W., Koyama, N. & Rasamimanana, H. (eds), *Ringtailed Lemur Biology: Lemur catta in Madagascar*. New York: Springer, pp. 296–310.

Gould, L, & Ziegler, T. (2007). Variation in fecal testosterone levels, inter-male aggression, dominance rank and age during mating and post-mating periods in wild adult male ring-tailed lemurs (*Lemur catta*). *American Journal of Primatology*, 69, 1325–1339.

Gould, L. Ziegler, T. & Wittwer, D. (2005). Effects of reproductive and social variables on fecal glucocorticoid levels in a sample of adult male ring-tailed lemurs (*Lemur catta*) at the Beza Mahafaly Reserve, Madagascar. *American Journal of Primatology*, 67, 5–23.

Gross, C. & Hen, R. (2004). The developmental origins of anxiety. *Nature Reviews Neuroscience*, 5, 545–552.
Guggisberg, A. G., Mathis, J. H., Uli, S. & Hess, C. W. (2007). The functional relationship between yawning and vigilance. *Behavioural Brain Research*, 179, 159–166.
Guggisberg, A. G., Mathis, J., Schnider, A. and Hess, C. W. (2010). Why do we yawn? *Neuroscience & Biobehavioral Reviews*, 34, 1267–1276.
Hadidian, J. (1980).Yawning in an old world monkey (*Macaca nigra*). *Behaviour*, 75, 133–147.
Hayes, J. & Cox, C. (1999). Immediate effects of a five-minute foot massage on patients in critical care. *Intensive and Critical Care Nursing*, 15, 77–82.
Hecht, E. E., Gutman, D. A., Preuss, T. M. *et al.* (2012). Process versus product in social learning: comparative diffusion tensor imaging of neural systems for action execution-observation matching in macaques, chimpanzees, and humans. *Cerebral Cortex*, 23, 1014–1024.
Hellhammer, K. H., Buchtal, J., Gutberlet, I. & Kirschbaum, C. (1997). Social hierarchy and adrenocortical stress reactivity in men. *Psychoneuroendocrinology*, 22, 643–650.
Hennessy, M. B., Maken, D. S. & Graves, F. C. (2000). Consequences of the presence of the mother or unfamiliar adult female on cortisol, ACTH, testosterone and behavioral responses of periadolescent guinea pigs during exposure to novelty. *Psychoneuroendocrinology*, 25, 619–632.
Higham, J. P., MacLarnon, A. M., Heistermann, M., Ross, C. & Semple, S. (2009). Rates of self-directed behaviour and faecal glucocorticoid levels are not correlated in female wild olive baboons (*Papio hamadryas anubis*). *Stress: the International Journal on the Biology of Stress*, 12, 526–532.
Jolly, A. (1966). *Lemur Behavior: a Madagascar field study*. Chicago: University of Chicago Press.
Jolly, A., Rasamimanana, H. R., Kinnaird, M. F., *et al.* (1993). Territoriality in *Lemur catta* groups during the birth season at Berenty, Madagascar. In: Kappeler, P. M. & Ganzhorn, J. U. (eds), *Lemur Social Systems and their Ecological Basis*. New York: Plenum Press, pp. 85–109.
Judge, P. G. & Bachmann, K. A. (2013). Witnessing reconciliation reduces arousal of bystanders in a baboon group (*Papio hamadryas hamadryas*). *Animal Behaviour*, 85, 881–889.
Kaada, B. & Torsteinbø, O. (1989). Increase of plasma beta-endorphins in connective tissue massage. *General Pharmacology*, 20, 487–489.
Kappeler, P. M. (1990). Female dominance in *Lemur catta*: more than just female feeding priority? *Folia Primatologica*, 55, 92–95.
Keverne, E. B., Martensz, N. D. & Tuite, B. (1989). Beta-endorphin concentrations in cerebrospinal fluid of monkeys are influenced by grooming relationships. *Psychoneuroendocrinology*, 14, 155–161.
Kikusui, T., Winslow, J. T. & Mori, Y. (2006). Social buffering: relief from stress and anxiety. *Philosophical Transactions of the Royal Society B: Biological Sciences*, 361, 2215–2228.
Klein, Z. A., Padow, V. A. & Romeo, R. D. (2010). The effects of stress on play and home cage behaviors in adolescent male rats. *Developmental Psychobiology*, 52, 62–70.
Koski, S. E., Koops, K. & Sterck, E. H. M. (2007). Reconciliation, relationship quality and post-conflict anxiety: Testing the integrated hypothesis in captive chimpanzees. *American Journal of Primatology*, 69, 158–172.
Koyama, N. (1988). Mating behavior of ring-tailed lemurs (*Lemur catta*) at Berenty, Madagascar. *Primates*, 29, 163–174.
Koyama, N., Nakamichi, M., Ichino, S. & Takahata, Y. (2002). Population and social dynamics changes in ring-tailed lemur troops at Berenty, Madagascar between 1989–1999. *Primates*, 43, 291–314.
Krause, L. & Shuster, S. (1983). Mechanism of action of antipruritic drugs. *British Medical Journal*, 287, 1199–2000.
Kutsukake, N. (2003). Assessing relationship quality and social anxiety among wild chimpanzees using self-directed behaviour. *Behaviour*, 140, 1153–1171.
Kutsukake, N., & Castles, D. L. (2001). Reconciliation and variation in post-conflict stress in Japanese macaques (*Macaca fuscata fuscata*): testing the integrated hypothesis. *Animal Cognition*, 4, 259–268.

Leavens, D. A., Aureli, F., Hopkins, W. D. & Hyatt, C. W. (2001). Effects of cognitive challenge on self-directed behaviors by chimpanzees (*Pan troglodytes*). *American Journal of Primatology*, 55, 1–14.

Lehmann, H. E. (1979). Yawning. A homeostatic reflex and its psychological significance. *Bulletin of the Menninger Clinic*, 43, 123–126.

Leone, A., Ferrari, P. F. & Palagi, E. (2014). Different yawns, different functions? Testing social hypotheses on spontaneous yawning in *Theropithecus gelada*. *Scientific Reports*, 4. http://dx.doi.org/10.1038/srep04010.

Levine, S., Coe, C. L. & Smotherman, W. P. (1978). Prolonged cortisol elevation in the infant squirrel monkey after reunion with mother. *Physiology and Behavior*, 20, 7–10.

Lord, L. D., Bond, J. & Thompson, R. R. (2009). Rapid steroid influences on visually guided sexual behavior in male goldfish. *Hormones and Behavior*, 56, 519–526.

MacLean, P. D. (1990). *The Triune Brain in Evolution: Role in Paleocerebral Functions.* New York: Springer.

Maestripieri, D., Schino, G., Aureli, F. & Troisi, A. (1992). A modest proposal: displacement activities as an indicator of emotions in primates. *Animal Behaviour*, 44, 967–979.

Manogue, K. R., Leshner, A. I. & Candland, D. K. (1975). Dominance status and adrenocortical reactivity to stress in squirrel monkeys (*Saimiri sciureus*). *Primates*, 16, 457–463.

Maretti, G., Sorrentino, V., Finomana, A., Gamba, M. & Giacoma, C. (2010). Not just a pretty song: an overview of the vocal repertoire of *Indri indri*. *Journal of Anthropological Sciences*, 88, 151–165.

Massen, J. J. M., Vermunt, D. A. & Sterck, E. H. M. (2012). Male yawning is more contagious than female yawning among chimpanzees (*Pan troglodytes*). *PLoS ONE*, 7, (7) e40697. http://dx.doi.org/10.1371/journal.pone.0040697.

McFarland, R. & Majolo, B. (2011). Reconciliation and the costs of aggression in wild Barbary macaques (*Macaca sylvanus*): a test of the integrated hypothesis. *Ethology*, 117, 928–937.

Mendl, M., Brooks, J., Basse, C., *et al.* (2010). Dogs showing separation-related behaviour exhibit a 'pessimistic' cognitive bias. *Current Biology*, 20, R839–R840.

Mendoza, S. P., Coe, C. L., Lowe, E. L. & Levine, S. (1978). The physiological response to group formation in adult male squirrel monkeys. *Psychoneuroendocrinology*, 3, 221–229.

Molesti, S. & Majolo, B. (2013). Grooming increases self-directed behaviour in wild Barbary macaques, *Macaca sylvanus*. *Animal Behaviour*, 86, 169–175.

Morgan, K. N. (2006). Is autism a stress disorder? What studies of non-autistic populations can tell us. In: Baron, M. G., Groden, J., Groden, G. & Lipsitt, L. P. (eds), *Stress and Coping in Autism*. Oxford: Oxford University Press.

Morris, D. (1977). *Manwatching. A field guide to human behavior.* New York: Harry N Abrams.

Mulder, E. J. H., Robles de Medina, P. G., Huizink, A. C., *et al.* (2002). Prenatal maternal stress: effects on pregnancy and the (unborn) child. *Early Human Development*, 70, 3–14.

Norscia, I. & Borgognini-Tarli, S. M. (2008). Ranging behavior and possible correlates of pair-living in southeastern avahis. *International Journal of Primatology*, 29, 153–171.

Norscia, I. & Palagi, E. (2011). Yawn contagion and empathy in *Homo sapiens*. *PLoS ONE*, 6, (12), e28472.

Norscia, I. & Palagi, E. (2016). *She* more than *he*: gender bias supports the empathic nature of yawn contagion in *Homo sapiens*. *Royal Society Open Science*, 3, 150459. http://dx.doi.org/10.1098/rsos.150459.

Norscia, I., Carrai, V. & Borgognini-Tarli, S. M. (2006). Influence of dry season, food quality and quantity on behavior and feeding strategy of *Propithecus verreauxi* in Kirindy, Madagascar. *International Journal of Primatology*, 27, 1001–1022.

Palagi, E. & Norscia, I. (2011). Scratching around stress: hierarchy and reconciliation make the difference in wild brown lemurs (*Eulemur fulvus*). *Stress*, 14, 93–97.

Palagi, E., Cordoni, G. & Borgognini-Tarli, S. M. (2004). Immediate and delayed benefits of play behaviour: new evidence from chimpanzees (*Pan troglodytes*). *Ethology*, 110, 949–962.

Palagi, E., Paoli, T. & Tarli, S. B. (2006). Short-term benefits of play behavior and conflict prevention in *Pan paniscus*. *International Journal of Primatology*, 27, 1257–1270.

Palagi, E., Leone, A., Mancini, G. & Ferrari, P. F. (2009). Contagious yawning in gelada baboons as a possible expression of empathy. *PNAS*, 106, 19262–19267.

Palagi, E., Dall'Olio, S., Demuru, E. & Stanyon, R. (2014). Exploring the evolutionary foundations of empathy: consolation in monkeys. *Evolution and Human Behavior*, 35, 341–349.

Paukner, A. & Anderson, J. R. (2006). Video-induced yawning in stumptail macaques (*Macaca arctoides*). *Biology Letters*, 2, 36–38.

Pavani, S., Maestripieri, D., Schino, G., Turillazzi, P. G. & Scucchi, S. (1991). Factors influencing scratching behaviour in long-tailed macaques. *Folia Primatologica*, 57, 34–38.

Pellis, S. M. & Pellis, V. C. (2009). *The Playful Brain: Venturing to the Limits of Neuroscience*. Oxford: Oneworld Publications.

Pereira, M. E. & Kappeler, P. M. (1997). Divergent systems of agonistic behaviour in lemurid primates. *Behaviour*, 134, 225–274.

Platek, S. M., Critton, S. R., Myers, T. E. & Gallup, G. G. (2003). Contagious yawning: the role of self-awareness and mental state attribution. *Cognitive Brain Research*, 17, 223–227.

Preston, S. D. & de Waal, F. B. M. (2002). Empathy: Its ultimate and proximate bases. *Behavioral and Brain Sciences*, 25, 1–71.

Price, J. S. (2003). Evolutionary aspects of anxiety disorders. *Dialogues in Clinical Neurosciences*, 5, 223–236.

Provine, R. R. (1986). Yawning as a stereotyped action pattern and releasing stimulus. *Ethology*, 72, 448–455.

Provine, R. R., Hamernik, H. B. & Curchack, B. C. (1987). Yawning: relation to sleeping and stretching in humans. *Ethology*, 76, 152–160.

Pyritz, L., Kappeler, P. M. & Fichtel, C. (2011). Coordination of group movements in wild red-fronted lemurs (*Eulemur rufifrons*): Processes and influence of ecological and reproductive seasonality. *International Journal of Primatology*, 32, 1325–1347.

Pyritz, L., Fichtel, C., Huchard, E., & Kappeler, P. M. (2013). Determinants and outcomes of decision-making, group coordination and social interactions during a foraging experiment in a wild primate. *PLoS ONE*, 8, e53144. http://dx.doi.org/10.1371/journal.pone.0053144.

Richard, A. F. (1978). *Behavioral Variation: Case Study of a Malagasy Lemur*. Lewisburg: Bucknell University Press.

Romero, T., Castellanos, M. A. & de Waal, F. B. (2010). Consolation as possible expression of sympathetic concern among chimpanzees. *PNAS*, 107, 12110–12115.

Rothman, S. (1941). Physiology of itching. *Physiological Reviews*, 21, 357–381.

Rubin, K. H. & Burgess, K. (2001). Social withdrawal. In: M. W. Vasey & M. R. Dadds (eds), *The Developmental Psychopathology of Anxiety*. Oxford: Oxford University Press, pp. 407–434.

Saketos, M., Sharma, N. & Santoro, N. F. (1993). Suppression of the hypothalamic-pituitary-ovarian axis in normal women by glucocorticoids. *Biology of Reproduction*, 49, 1270–1276.

Sapolsky, R. M. (1982). The endocrine stress-response and social status in the wild baboon. *Hormones and Behavior*, 16, 279–292.

Sapolsky, R. M. (1983). Endocrine aspects of social instability in the olive baboon (*Papio anubis*). *American Journal of Primatology*, 5, 365–379.

Sapolsky, R. M. (1987). Stress, social status, and reproductive physiology in free-living baboons. In: D. Crews (ed.), *Psychobiology of Reproductive Behavior: An Evolutionary Perspective*. Englewood Cliffs: Prentice-Hall, pp. 291–322.

Sapolsky, R. M. (1990). Stress in the wild. *Scientific American*, 262, 116–123.

Sapolsky, R. M. (1992). Cortisol concentrations and the social significance of rank instability among wild baboons. *Psychoneuroendocrinology*, 17, 701–709.

Sapolsky, R. M. (2005). The influence of social hierarchy on primate health. *Science*, 308, 648–652.

Sapolsky, R. M. & Ray, J. C. (1989). Styles of dominance and their endocrine correlates among wild olive baboons (*Papio anubis*). *American Journal of Primatology*, 18, 1–13.

Sauther, M. L. (1991). Reproductive behavior of free-ranging *Lemur catta* at Beza Mahafaly Special Reserve, Madagascar. *American Journal of Physical Anthropology*, 84, 463–477.

Sauther, M. L. (1993). Resource competition in wild populations of ringtailed lemurs (*Lemur catta*): implications for female dominance. In: Kappeler, P. M. & Ganzhorn, J. U. (eds), *Lemur Social Systems and their Ecological Basis*. New York: Plenum Press, pp. 135–152.

Sauther, M. L., & Sussman, R. W. (1993). A new interpretation of the social organization and mating system of the ringtailed lemur (*Lemur catta*). In: Kappeler, P. M. & Ganzhorn, J. U. (eds), *Lemur Social Systems and their Ecological Basis*. New York: Plenum Press, pp. 111–121.

Sauther, M. L., Sussman, R. W. & Gould, L. (1999). The socioecology of the ring-tailed lemur: Thirty-five years of research. *Evolutionary Anthropology*, 8, 120–132.

Schino, G. & Aureli, F. (2009). Reciprocal altruism in primates: partner choice, cognition, and emotions. *Advances in the Study of Behavior*, 39, 45–69.

Schino, G. & Aureli, F. (2010). Primate reciprocity and its cognitive requirements. *Evolutionary Anthropology: Issues, News, and Reviews*, 19, 130–135.

Schino, G., Scucchi, S., Maestripieri, D. & Turillazzi, P. G. (1988). Allogrooming as a tension-reduction mechanism: a behavioral approach. *American Journal of Primatology*, 16, 43–50.

Schino, G., Troisi, A., Perretta, G. & Monaco, V. (1991). Measuring anxiety in nonhuman primates: effect of lorazepam on macaque scratching. *Pharmacology Biochemistry and Behavior*, 38, 391–889.

Schino, G., Perretta, G., Taglioni, A. M., Monaco V. & Troisi, A. (1996). Primate displacement activities as an ethopharmacological model of anxiety. *Anxiety*, 2, 186–191.

Schwitzer, C., Mittermeier, R. A., Johnson, S. E., *et al.* (2014). Averting lemur extinctions amid Madagascar's political crisis. *Science*, 343, 842–843.

Sclafani, V., Norscia, I., Antonacci, D. & Palagi, E. (2012). Scratching around mating: factors affecting anxiety in wild *Lemur catta*. *Primates*, 53, 247–254.

Shankly, K. (1988). Pathology of pruritus. *Veterinary Clinics of North America*, 18, 971–981.

Sheriff, M. J., Dantzer, B., Delehanty, B., Palme, R. & Boonstra, R. (2011). Measuring stress in wildlife: techniques for quantifying glucocorticoids. *Oecologia*, 166, 869–87.

Shulman, K. R. & Jones, G. E. (1996). The effectiveness of massage therapy intervention on reducing anxiety in the workplace. *Journal of Applied Behavioral Science*, 32, 160–173.

Smith, E. O. (1999). Yawning: an evolutionary perspective. *Journal of Human Evolution*, 14, 191–198.

Smith, O. A., Astley, C. A., Spelman, F. A., *et al.* (1993). Integrating behavior and cardiovascular responses: posture and locomotion. I. Static analysis. *American Journal of Physiology – Regulatory, Integrative and Comparative Physiology*, 265, R1458–R1468.

Smith, T. & O'Driscoll, M. L. (2003). Can massage induce relaxation? A review of the evidence. *International Journal of Therapy and Rehabilitation*, 10, 491–496.

Stangier, U., Heidenreich, T., Peitz, M., Lauterbach, W. & Clark, D. M. (2003). Cognitive therapy for social phobia: individual versus group treatment. *Behaviour Research and Therapy*, 41, 991–1007.

Sussman, R. W. (1977). Socialization, social structure, and ecology of two sympatric species of lemur. In: Chevalier-Skolnikoff, S. & Poirier, F. E. (eds), *Primate BioSocial Development: Biological, Social and Ecological Determinants*. New York: Garland, pp. 515–528.

Sussman, R. W. (1992). Male life histories and inter-group mobility among ring-tailed lemurs (*Lemur catta*). *International Journal of Primatology*, 13, 395–413.

Sussman, R. W. & Richard, A. F. (1974). The role of aggression among diurnal prosimians. In: Holloway, R. L. (ed.), *Primate Aggression, Territoriality, and Xenophobia*. San Francisco: Academic Press, pp. 50–76.

Tan, S. M., Jong, S. C., Chan, L. F., et al. (2013). Physician, heal thyself: The paradox of anxiety amongst house officers and work in a teaching hospital. *Asia-Pacific Psychiatry*, 5 (Suppl 1), 74–81.

Terry, R. L. (1970) Primate grooming as a tension reduction mechanism. *The Journal of Psychology*, 76, 129–136.

Thierry, B. (2004). Social epigenesis. In: B. Thierry, M. Singh & W. Kaumanns (eds), *Macaque Societies: A Model for the Study of Social Organization*, Cambridge University Press, pp. 267–290.

Thorsteinsson, E. B., James, J. E. & Gregg, M. E. (1998). Effects of video-relayed social support on hemodynamic reactivity and salivary cortisol during laboratory-based behavioral challenge. *Health Psychology*, 17, 436–444.

Tilbrook, A. J., Turner, A. I. & Clark, I. J. (2002). Effects of stress on reproduction in non-rodent mammals: the role of glucocorticoids and sex differences. *Reviews of Reproduction*, 5, 105–113.

Tinbergen, N. (1951). *The Study of Instinct*. New York: Oxford University Press.

Tran, B. W., Papoiu, A. D., Russoniello, C. V., et al. (2010). Effect of itch, scratching and mental stress on autonomic nervous system function in atopic dermatitis. *Acta Dermato-Venereologica*, 90, 354–361.

Troisi, A. (2002). Displacement activities as a behavioral measure of stress in nonhuman primates and human subjects. *Stress*, 5, 47–54.

Troisi, A., Aureli, F., Schino, G., Rinaldi, F. & de Angelis, N. (1990). The influence of age, sex, and rank on yawning behavior in two species of macaques (*Macaca fascicularis* and *Macaca fuscata*). *Ethology*, 86, 303–310.

Troisi, A., Belsanti, S., Bucci, A. R., et al. (2000). Affect regulation in alexithymia – an ethological study of displacement behavior during psychiatric interviews. *The Journal of Nervous and Mental Disease*, 188, 13–18.

Tuinier, S., Verhoeven, W. M. A. & Van Praag, H. M. (1996). Serotonin and disruptive behaviour; a critical evaluation of the clinical data. *Human Psychopharmacology: Clinical and Experimental*, 11, 469–482.

Valenstein, E. S. & Young, W. C. (1955). An experimental factor influencing the effectiveness of testosterone proprionate in eliciting sexual behavior in male guinea pigs. *Endocrinology*, 56, 173–185.

van Goozen, S. H. M. (2005). Hormones and the developmental origins of aggression. In: Tremblay, R. E., Hartup, W. W. & Archer, J. (eds), *Developmental Origins of Aggression*. New York: The Guilford Press.

van Moffaert, M. (2003). The spectrum of dermatological self-mutilation and self-destruction: common issues. In: Koo, J. Y. M. & Lee, C. S. (eds), *Psychocutaneous Medicine*. New York: Marcel Dekker, pp. 139–155.

van Riezen, H., & Segal, M. (1988). *Introduction to the Evaluation of Anxiety and Related Disorders. Comparative Evaluation of Rating Scales for Clinical Psychopharmacology*. New York: Elsevier, pp. 225–228.

van Wolkenten, M. L., Davis, J. M., Gong, M. L. & de Waal, F. B. (2006). Coping with acute crowding by *Cebus apella*. *International Journal of Primatology*, 27, 1241–1256.

Vick, L. G. & Pereira, M. E. (1989). Episodic targeting aggression and the histories of *Lemur* social groups. *Behavioral Ecology and Sociobiology*, 25, 3–12.

Vick, S. & Paukner, A. (2009). Variation and context of yawns in captive chimpanzees (*Pan troglodytes*). *American Journal of Primatology*, 71, 1–8.

Virgin, C. E. & Sapolsky, R. M. (1997). Styles of male social behavior and their endocrine correlates among low-ranking baboons. *American Journal of Primatology*, 42(1), 25–39.

Weaver, A. & de Waal, F. (2003). The mother-offspring relationship as a template in social development: reconciliation in captive brown capuchins (*Cebus apella*). *Journal of Comparative Psychology*, 117, 101–110.

Weissman, C. (1990). The metabolic response to stress: an overview and update. *Anesthesiology*, 73, 308–327.

Welch, J. M., Lu, J., Rodriguiz, R. M., *et al.* (2007). Cortico-striatal synaptic defects and OCD-like behaviours in Sapap3-mutant mice. *Nature*, 448, 894–900.

Whitten, P. L., Brockman, D. K. & Stavisky, R. C. (1998). Recent advances in noninvasive techniques to monitor hormone-behavior interactions. *Yearbook of Physical Anthropology*, 41, 1–23.

Wilson, J. H. (2001). Prolactin in rats is attenuated by conspecific touch in a novel environment. *Cognitive, Affective, & Behavioral Neuroscience*, 1, 199–205.

Wingfield, J. C., Hegner, R. E., Dufty, A. M. & Ball, G. F. (1990). The challenge hypothesis: theoretical implications for patterns of testosterone secretion, mating systems, and breeding strategies. *American Naturalist*, 136, 829–846.

Wingfield, J. C. & Ramenofsky, M. (1999). Hormones and the behavioral ecology of stress. In: Balm, P. H. M. (ed.), *Stress Physiology in Animals*. Sheffield: Sheffield Academic Press, pp. 2–51.

Wright, A. J., Soto, N. A., Baldwin, A. L., *et al.* (2007). Anthropogenic noise as a stressor in animals: a multidisciplinary perspective. *International Journal of Comparative Psychology*, 20, 250–273.

Zannella, A., Norscia, I., Stanyon, R. & Palagi, E. (2015). Testing yawning hypotheses in wild populations of two strepsirrhine species: *Propithecus verreauxi* and *Lemur catta*. *American Journal of Primatology*, 77, 1207–1215.

Zilli, I., Giganti, F. & Uga, V. (2008). Yawning and subjective sleepiness in the elderly. *Journal of Sleep Research*, 17, 303–308.

Part III

Why lemurs keep in touch

7 Playing lemurs: why primates have been playing for a long time

We do not stop playing because we get old. We get old because we stop playing

Stanley G. Hall, 1904

7.1 Playing, an everyday life activity: easy to spot, hard to define

After focusing for a long time on animal grief, trauma, violence, pain and suffering, comparative psychologists have been lately putting their efforts into dealing with happiness, laughter, joy and affection both in humans and other animals (the so-called 'positive psychology'). Most of the 'positive emotions' which are under the psychologists' magnifying glass are also indissolubly linked to play. Every day of our lives is punctuated by play and humour. We play during conversation with friends on the phone or on Skype, or when we post messages on Facebook – frequently inserting a smiling face like this :-) – etc. We play with our conspecifics and also with members of other species. Dogs, cats and even rats can be wonderful playmates. We play because it is pleasurable and rewarding. By social play and humour we experience emotions that can pass individuals' boundaries and create a network of social and affective bonds (Power, 2000). Despite its pervasive nature, play is one of the most mysterious behaviours an ethologist can come across.

The difficulty with studying play behaviour starts with its theoretical definition. In his *Confessions* (Book 11, Chapter 14), St Augustine pondered the meaning of time, '*Quid ergo est tempus? Si nemo ex me quaerit, scio: si quaerenti explicare velim, nescio*' [What is time? If no one asks me, I know. If I want to explain it, I no longer know]. Maybe the frustration experienced by St Augustine is the same as that experienced by most ethologists when dealing with the definition of play (Palagi and Paoli, 2007, 2008). We have no problems in recognising a playful interaction of our dogs, cats and children; however, as we go through the extensive literature dedicated to play we soon realise that there are as many definitions of play as there are the authors who studied it (for an extensive review see Burghardt, 2005). Play has been often defined via *litotes*,[1] by not specifying what it is but by specifying what it is *not*. As play lacks certain characteristics that are typical of the *serious* and *functional* behaviours, play has been defined as a *non-functional* behaviour with no obvious immediate benefits (Martin and Caro, 1985; Bekoff and

---

[1] From the Greek word λιτότης – a figure of speech achieved by using negation with a term in place of using an antonym of that term.

Allen, 1998). Although many, if not all, authors underlined the absence of obvious advantages, we have to stress, however, that, due to its phylogenetic depth, play must have both delayed and immediate benefits, which represent the winning cards that made it favoured by natural selection.

As a matter of fact, compared to other behaviours, whose modalities and functions are easier to define and detect, play remains an intriguing challenge for a number of researchers interested in the study of this ephemeral, versatile and controversial phenomenon. However, investigators need a range of objective criteria with which to positively define play for what it is (and not just for what it is *not*) for two main reasons. Firstly, data coming from non-human animals can be brought into closer correspondence with those coming from humans and, secondly, data can be compared across diverse animal species. In his book, *The Genesis of Animal Play*, Burghardt (2005) suggested five criteria to define play. The first criterion assumes that play is an activity that is not completely functional in the structure or circumstance in which it is performed because it seems not to contribute to immediate survival of individuals. The second criterion encompasses a series of characteristics by defining play as spontaneous, voluntary, intentional, pleasurable, rewarding, reinforcing or autotelic ('done for its own sake'). The third criterion establishes that, compared to other behaviours, play is incomplete, exaggerated or precocious. Moreover, differently from other functional behaviours, play involves patterns modified in their form, sequencing or targeting. According to the fourth criterion, during a playful session the patterns are often repeated in a highly variable combination. Finally, the fifth criterion indicates that play occurs when animals are relatively free from environmental and social stressors.

Play joins and cuts across a variety of disciplines leading directly to inquiries relating to individual developmental changes and species adaptation. Psychologists are interested in discovering the consequences of play on behavioural development, anthropologists are focused on identifying its role in the evolution of social and cognitive skills, and evolutionary biologists are concentrated on searching for the functions of an apparently non-functional behaviour (Burghardt, 2005).

We have not yet reached a sufficient theoretical knowledge of play; however, the last two decades saw many efforts in understanding and clarifying some aspects of this phenomenon, both in humans and other animals. The emergent scenario tells us that play is indeed functional and its function can vary according to the play modality performed. Additionally, the function of similar play modalities can be diversified according to a number of variables, such as species, age, sex, context and relationship quality of the players.

Play can simply involve a physical activity (locomotor-rotational and acrobatic body movements) or it can also entail the use of objects (Figure 7.1). Both these types of play can be carried out in solitary or social contexts. Play, both solitary and social, is widespread in mammals (Fagen, 1981). It is not always easy to disentangle each type of play from the other. Although the presence of a possible phylogenetic diversity (ground squirrels, Pasztor *et al.*, 2001; grasshopper mice, Pellis *et al.*, 2000) in some species different types of play can be mixed together in extremely sophisticated

7.1 Playing, an everyday life activity: easy to spot, hard to define 187

Figure 7.1 Object, social play in *Macaca sylvanus*: individuals engaging in a session of social play involving the use of a rag. Parco Natura Viva (Bussolengo, Italy). Photo: Elisabetta Palagi.

sequences of actions that can flow into a single natural category (Candland *et al.*, 1978; Thompson, 1998). For this reason, no single, simple definition of play category can be satisfactorily applied (Power, 2000). Nonetheless, categorisation and tentative definitions help ethologists to identify, measure, describe and quantify, in a more or less standardised way, the different playful activities of animals.

Here we focus on social play and, in particular, on the most common and at the same time challenging form of social play, rough and tumble (Aldis, 1975; Fagen, 1981; Jolly, 1985; Burghardt, 2005). Rough and tumble, or play fighting, is a physically vigorous interaction comprising a particular set of social behavioural patterns, such as chase, jump, bite, slap, stamp on, coming along with a positive affect between playmates (Burghardt, 2005; Palagi *et al.*, 2015). Rough and tumble was first labelled as such by Karl Groos (1896) and later picked up by Harlow and Harlow (1965) to describe play wrestling in rhesus macaques. Although play fighting resembles serious fighting, a number of features distinguish play fighting from serious fighting (Smith, 1982). In play fighting, (1) a resource is not gained or protected, (2) the contact is restrained, or, at least, combat-induced injuries do not typically result, (3) the sequence of attack, defence and counterattack can be repeated many times, with the partners reversing roles of attacking and being attacked, (4) such contact can lead to further affiliation between the participants, and, (5) for many species, including humans, special facial and bodily gestures are used to signal that it is play (Palagi, 2008).

Play fighting in strepsirrhines has been reported since the 1960s in various species as a common (if not the commonest) form of social play. Doyle (1974), in a review on prosimian behaviour, lingers on the patterns used during play wrestling by *Propithecus verreauxi*, *Lemur catta*, *Phaner furcifer*, *Microcebus murinus*, and other non-lemur strepsirrhines. Although reporting mainly anecdotal play incidents, the review indicates that the behavioural patterns used for play fighting can be quite elaborate. According to Doyle (1974), '*Propithecus verreauxi* play wrestle with one another largely with hands...Animals sit facing one another sparring with hands and feet...They grab wildly at one another's wrists, ankles and noses attempting to penetrate each other's guard...Both may drop simultaneously to hang from a branch by their feet and spar with their hands or hang by their hands and kick... *Lemur catta* play involves many varied postures. They play with each other, leaping about from tree to tree, their most common game being to chase one another in circles, one attempting to jump on the other and wrestle...*Microcebus murinus* play among themselves, jumping from place to place following one another in a sort of follow-the-leader game...*Phaner furcifer* are reported to hang by their feet and spar like *Propithecus verreauxi*...'. Social play has been observerd also in young *Hapalemur simus* (reclassified as *Prolemur simus*; Wilson and Reeder, 2005) at around two years of age (Mutschler and Tan, 2004) and in infant aye-aye during mother–offspring interactions (Feistner and Ashbourne, 1994). Despite these promising preliminary observations, social play in lemurs has been excluded from quantitative studies until recently (Figure 7.2).

Whatever the patterns and animal species considered, play fighting can be fair or unfair. Like other interactions, play fighting can go either way (see Bekoff's box (Box 7.1)). The first step is to discriminate between potential 'friends' and 'enemies' and the way play is conducted is crucial to reach this goal. When playmates are able to balance competition and cooperation, thus playing fairly, they are likely to reiterate their interactions in the future and show higher levels of social contacts. When one of the playmates starts cheating and does not cooperate any more (e.g., a play bite becomes aggressive), the play session is likely to escalate into a real aggression. In this case, the interaction is interrupted and the 'aggressee' will be unlikely to play again with the unfair opponent (Bekoff, 1977a). Accordingly, Dugatkin and Bekoff (2003) proposed a mathematical model based on game theory, in which they showed that failure to negotiate and cooperate prevents animals from continuing to play together and, successively, leads to higher levels of social isolation. In this perspective, animals that fail to engage fully in play pay a high social cost: they have difficulty in creating a web of social relationships.

The studies conducted so far on play fighting (see also Pellis' box (Box 7.2)) show that this behaviour in the early phases of life is fundamental to obtain socially competent and skilled adults able to deal with complex and unexpected situations (Fagen 1981, 1993; Špinka *et al.*, 2001; Bell *et al.*, 2010; Palagi, 2011). For example, play fighting experience can increase behavioural flexibility, and help animals to better regulate emotional responses and to react appropriately to environmental and social challenges (Pellis and Pellis, 2007).

Probably owing to its relevance to the management of social relationships, play fighting can be found in many eutherian and marsupial mammals as well as in

Figure 7.2 Sequence of play fighting (or rough and tumble) in *Propithecus verreauxi*. Berenty forest, Madagascar. Photo: Ivan Norscia.

some other vertebrates, including birds (Burghardt, 2005). It is also a consistent behaviour showing strong structural similarities across many different human cultures and across many different human and non-human species (Fry, 2005). This suggests that play fighting has biological roots, although a social influence is present. In fact, a cross-cultural examination at a finer scale reveals that in humans some stylistic variations are present and clearly learned in different social environments (Fry, 1992, 2005). In other animals, social and physical environmental factors influence the frequency of play fighting and its design (Pellis and Pellis, 2009). In this view, play fighting provides an interesting example of how evolutionary factors and elements of the social and physical environment interact.

Box 7.1 | by Marc Bekoff

Speaking of which: why studying play is so fascinating

Early in my career, various researchers told me that it was a waste of time to study play behaviour. Some people also told me that 'real ethologists' do not study dogs because they are artefacts – merely 'creations of man' – and we cannot really learn much about the behaviour of wild animals by studying them. In the 1970s, it seemed that only veterinarians and those people interested in practical applications of behavioural data studied dogs. I disagreed and at two meetings held in 2013, this historical mistake was revisited and soundly rejected, and now we know just how important studies of dogs truly are.

Years ago, some people thought of play as a wastebasket into which behaviour patterns that were difficult to understand should be tossed or that understanding play was not important for researchers interested in the evolution of behavioural development (e.g., Lazar and Beckhorn, 1974). Others, including my PhD mentor, Michael W. Fox, realised that play was essential to normal social, cognitive and physical development and that people just had not taken the time to study it in detail (for discussions about possible functions of play see Bekoff and Byers, 1981; Fagen, 1981; Špinka *et al.*, 2001; Burghardt, 2005; Pellis and Pellis, 2009; Palagi, 2011). One reason why studying play has been so difficult is because it is a hodge-podge or kaleidoscope (Bekoff, 1972, 1974a, b) of lots of different activities from various social contexts including predatory, mating and agonistic behaviour, and it takes a lot of time to learn about the details of this fascinating behaviour. For example, it can take many hours to conduct frame-by-frame analyses of as little as ten minutes of play captured on video, but these sorts of analyses are essential to gaining an understanding of this behavioural phenotype.

In 1973, I organised a symposium that centred on this fascinating behaviour (Bekoff, 1974a). I also began detailed studies focusing on what animals *do* when they play that have lasted more than four decades. My studies of play are based on careful observation and analyses of videotapes. My students and I watch tapes of play one frame at a time to see what the animals are doing and how they exchange information about their motivations, intentions and desires to play. Following ethological traditions, my first step was to develop a lengthy and detailed ethogram (a list of actions; Bekoff, 1972, 1974b). In my studies I have always taken a strongly evolutionary and ecological approach using Niko Tinbergen's (1951, 1963) integrative ideas about the questions with which ethological studies should be concerned: namely, evolution, adaptation, causation and ontogeny (development and the emergence of individual differences). I always wanted to know more about what animals feel when they are playing (Bekoff, 2007). They clearly enjoy it and are having fun. My interest in play continues today because it is a fascinating behavioural phenotype, extremely important for animals.

I began my research by collecting detailed data on young dogs, coyotes and wolves in captivity using direct observation and filming, some of which had no obvious connection to the then scanty extant theory about the evolution and development of play. However, over time, the *zeitgeist* changed and many data found homes as new hypotheses, and theories materialised. I am thrilled that I and others did not give in because now it is clear that accurate studies of play can inform the development of 'big theories' concerning the evolution of social behaviour, fairness, cooperation, moral behaviour, cognitive capacities including whether animals have a 'theory of mind', and individual survival and reproductive fitness (Bekoff, 2004, 2013; Bekoff and Pierce, 2009; Pierce and Bekoff, 2012).

One example of how play is related to larger issues in behavioural evolution and development is the obvious fact that play is a voluntary and cooperative activity (linked to the development of social skills), through which animals come to know 'right' from 'wrong'. This process is important in the development of fairness and moral sentiments (Bekoff, 2004; Bekoff and Pierce, 2009; Cordoni and Palagi, 2011; Pierce and Bekoff, 2012), including social justice (Pierce and Bekoff, 2012).

My early research focused on the importance of 'bows' in the initiation of play (conveying the message 'I want to play with you'; Bekoff 1974b, 1977b), but I did not then see how they were related to how canids punctuate play sequences (Bekoff, 1995) and tell others, in essence, 'I am going to bite you hard but it is still play' or 'I am sorry I just bit you so hard, please forgive me'. Dogs, coyotes and wolves, for example, ask another individual to play by bowing: crouching on their forelimbs, raising their hind end in the air, and often barking and wagging their tail as they bow. Bows rarely occur outside of the context of social play. Bows occur throughout play sequences, but most commonly are performed at the beginning or towards the middle of playful

Box 7.1 (continued)

Box 7.1 (*cont.*)

encounters. We learned that individuals should not bow if they do not want to play and across species the general rules of play are *ask first, be honest, follow the rules and admit when you're wrong*. In addition, in wild coyotes, individuals who do not play fair are avoided or their invitations to play are ignored and they tend to wander off from their group and suffer higher mortality than those youngsters who remain in their pack. There seems to be a close relationship between fairness and fitness, but we need more data to assess how robust this relationship is.

Of course, bows have to be seen by other dogs. In her research on play in dogs, Horowitz (2008) discovered that play signals were 'sent nearly exclusively to forward-facing conspecifics while attention-getting behaviours were used most often when a playmate was facing away, and before signalling an interest to play.' There also are auditory (play sounds such as play panting), olfactory (play odours) and tactile (touch) play signals (Fagen, 1981; Bekoff and Byers, 1981; Burghardt, 2005; Pellis and Pellis, 2009).

Our research also showed that the bow is a highly ritualised and stereotyped movement resembling Modal Action Patterns (MAP; Barlow, 1977). Bows are distinctive, and recognisable, but they are not always performed precisely the same as are Fixed Action Patterns (Bekoff, 1977b).

I also did not see at the time how individual patterns of play could be related to the development of social bonds and individual dispersal patterns (Bekoff, 1977a) or to evolutionary questions about individual reproductive fitness (Bekoff, 2004; Bekoff and Pierce, 2009). And, while I focused on visual signals I did not pay attention to how dogs sought attention from others using vocal signals. In the early 2000s I had the pleasure of helping to train Alexandra Horowitz as she began her work on visual attention and play (Horowitz, 2008).

Clearly, play behaviour is an essential part of the life of individuals of numerous different species. We need more comparative research to learn how it is related to various life-history strategies and also more research on how different species negotiate fair play 'on the run' – the subtle signals that are used to maintain the play mood – and why play can be considered a foundation of fairness. We also need to know more about how play research can inform ideas about animals having a theory of mind. These data are essential for coming to a further understanding of the evolution of play across diverse species, how ecological variables influence the development of play in individuals of the same species, and how an individual's playful experiences or the lack thereof influence his or her future behaviour. Furthermore, the use of neuroimaging (e.g. Berns, 2013) will be very helpful to learn about what is happening in the brains of animals as they play.

> Social play clearly is fun for the players. Thus, future research should also consider whether it is possible for animals to have 'too much fun' or behave 'too fairly', following up on research on humans in the field called 'diagonal psychology' (Nesse, 2004). Research shows that there can be negative consequences associated with being too full of oneself or for being 'too happy' (see for example, June et al., 2014). Thus, one can ask if there is stabilising selection for fun and fairness that reins in the possible costs of behaviour that can be too risky or too costly, including social play. While there are bits of data that inform this area of inquiry, this field of investigation is wide open for further detailed study.
>
> I look forward to seeing more and more comparative research that centres on the evolution and development of play in a wide variety of species. Play clearly is adaptive but there still is much to learn. After more than forty years I am still amazed at how exciting it is to study play behaviour. And, I am thrilled that I didn't stop 'way back when'.
>
> Acknowledgements: Parts of this paper have appeared in some of my earlier essays as noted and also in an essay titled 'The Significance of Ethological Studies: Playing and Peeing' published in *Dog Behavior and Cognition*, edited by Alexandra Horowitz (Springer, New York, 2014).

7.2 Playful signals: a theoretical framework

Play fighting, or rough and tumble, is closely linked to communication. Rough and tumble involves competition for some advantage (e.g. contacting a particular body target) and its patterns can be derived from other functional contexts like aggression, sex and predation. The competition, unlike that seen in serious fighting, is tempered by some degree of cooperation, thus reducing the risk of escalation (Pellis et al., 2010b). However, given that the balance between competition and cooperation can at times be ambiguous, playful intention also needs to be well communicated to avoid accidental transgressions that can lead to rough and tumble escalating to serious fighting. Nonetheless, most of the time animals are perfectly able to manage play sessions in order to avoid any kind of escalation. This means that, time after time, a sophisticated communication tactic is put in place to avoid any situation of ambiguity between the animal sending the playful signal and the animal that is meant to receive it (Palagi et al., 2015). The signals used during playful sessions can rely on different sensory modalities. Hence, body postures, vocalisations, facial expressions can communicate the potential intention of a play mate (Fagen, 1981; Bekoff, 2001; Palagi, 2006; Mancini et al., 2013a, b). As much as punctuation can change the meaning of a sentence, signals can determine whether a session can continue in the form of a playful interaction or not (see also Box 7.1). Playing and, most of all, playing in a correct way by using appropriate signals can help the players to establish durable cooperative relationships, well beyond the playful context itself (Palagi and Cordoni, 2012). In a recent review, Palagi *et al.* (2015) organised signals along two axes or dimensions: 'borrowed' to play-exclusive signals and emotion- to intention-driven signals. The first dimension highlights the fact that some signals are unique to playful interactions whereas others are drawn from other contexts, decontextualised and used in a different way. At one extreme we can find species using many signals that are exclusive of play fighting. For example, in Hanuman langurs (*Semnopithecus entellus*), a third of the play repertoire consists of patterns that are unique to rough and tumble, such as eyes closing, play gallop, head rotation, and play face (Petrů *et al.*, 2009). Other primate species, instead, possess few play specific signals and borrow the majority of playful signals from other contexts. For example, ring-tailed lemurs mostly recycle motor patterns of non-playful contexts for play invitation (tail wave, run, bite, jump, etc.). The open mouth display – produced by uncovering lower and upper teeth – is the only play-exclusive signal known so far (Palagi *et al.*, 2014). Ring-tailed lemurs' tail use to inform partners of their playful mood is an example of species-specific motor patterns borrowed from other functional contexts (Jolly, 1966; Palagi, 2009). In the so-called 'serious' contexts, adult males anoint and wave their tails towards other males to signal competition (agonistic context) and towards females as signals of appeasement or submission during courtship (sexual context; Jolly, 1966; Pereira and Kappeler, 1997; see also Chapter 2 for a detail description of tail use in *Lemur catta*). In the playful version of the communication pattern involving tail use (Jolly, 1966) lemurs anoint their tails neither facing nor gazing at the playmate (tail play). Infant lemurs begin to perform tail play during the weaning period (about 6 months; Palagi *et al.*, 2002). Tail use by ring-tailed infants and juveniles during play is often entirely anointing, rarely followed by waving (Figure 7.3). The way the tail is used in *Lemur catta* is species-specific but ring-tailed lemurs are not the only species that use the tail for communicating play. Dogs use tail wagging, combined with other body movements, for play invitation (Bekoff and Pierce, 2009; Bekoff, Box 7.1 in this chapter). Tail wagging is not exclusively used during play whereas the open mouth, also in dogs, is play specific.

The second dimension described by Palagi *et al.* (2015), is related to the proximate mechanisms that produce the signal. Some signals, linked to emotions, are relatively involuntary and automatic. Other signals, driven by cognition, are intentional and strategically used because they are directed to a specific receiver when the receiver is able to detect them. Intentional signals, according to the definition by Call and Tomasello (2007), are those that are directed to a specific target (audience dependent; Palagi *et al.*, 2015).

The dualism between emotional and intentional signals, while nice in textbooks, is not easy to apply in practice due to the complex relationships between the emotional and cognitive spheres in animals. Complex cognitive–emotional behaviours have their basis in dynamic coalitions of networks of brain areas, none of which should be defined as either affective or cognitive (Pessoa, 2009; Cattaneo and Pavesi, 2014). As a consequence, the degree to which the intentional

7.2 Playful signals: a theoretical framework

Figure 7.3 Tail play in *Lemur catta*. Drawing: Carmelo Gómez González.

and emotional communication systems intermingle for the emission of a given signal cannot be known with certainty (Demuru *et al.*, 2015). It is, however, possible to attempt a categorisation of the playful signals according to whether they are mainly emotion- or cognition-driven. This is possible for example by verifying if the target audience is present or not when the signal is emitted and by analysing the correlates that may indicate that the potential receiver's attention is wanted by the emitter when they release the signal. If the play message is sent in the absence of an audience, this suggests that the proximate factor eliciting the signal has more of an emotional than an intentional basis. On the contrary, if the sender performs some behaviour to get the attention of the possible playmate before emitting the playful signal this may indicate the intentional nature of the signal. Palagi (unpublished data) observed juvenile gorillas clapping their hands to catch the attention of another member of their group and displaying an open mouth at their face once the group-mate had turned to watch them. Tanner and Byrne (1993) observed a female gorilla repeatedly hiding her play face with the hand so that group members could not perceive it. In both cases the hand gesture (clapping or mouth covering) associated with the play face clarifies the nature of the signal itself. Yet, in the first example the gesture unveils the intentional basis of the signal whereas in the second the gesture clarifies that the signal was being emitted out of a positive emotion. In general, playful signals are important for expressing positive emotions, making the session enjoyable and rewarding for the play mates (Kuczaj and Horback, 2013).

The adoption of two dimensions to categorise play signals, despite its limits, is useful because it accommodates many different and differing kinds of signals that communicate play. Some playful signals, such as the relaxed open mouth, have ancient evolutionary roots and are, therefore, shared among many species. This play-specific signal generally conveys a message related to the positive emotions experienced by the emitter while playing. Other signals, such as play invitations, are more variable across species because they are used also in non-playful contexts and their use can be mediated by individual life history and experience. The ontogeny of a signal also informs its possible nature and adds to signal categorisation. Emotion driven signals normally emerge early in life and occurs also in solitary contexts (and in fact they are not audience dependent) (Pellis and Pellis, 2011; Ross et al., 2014). Intentional signals emerge at a later developmental stage thus suggesting the important role of social experience in shaping their modulation, adaptation and optimisation.

7.3 I see it in your face: the role of facial expression in signalling the playful mood

In humans, the association of play face (in this case smile or laughter) with a motor pattern can charge that pattern with a new meaning. For example, if a person shows a play face while jumping, balancing, swinging, sliding, running in an exaggerated manner or running with a variable sequence (e.g. zig-zagging) it is clear that that person is in a playful mood. In two- and four-year-old children, there is a strong association between a laughing play face and play fighting (Blurton Jones, 1972).

Darwin (1872) noted that human facial expressions strongly resemble those of other animals. The origin of certain human facial expressions, such as smiling, dates back to an ancestral non-human primate (de Waal, 2003; van Hooff and Preuschoft, 2003). Analogously (or homologously), apes' play vocalisations associated with the play face are similar in many features to the laughing associated with smiling in humans during play fighting (Vettin and Todt, 2005). Smiling and laughing are widespread among humans and regularly punctuate play fighting sessions. Smiling occurs mostly in social contexts where and when the signal can be perceived (Bainum et al., 1984; Provine and Fischer, 1989). Smiling can be enhanced by social interactions since early infancy (starting around the end of the first month), thus being one of the first signals of positive emotions (Lewis, 2000; Messinger et al., 2012; Ross et al., 2014). Laughter comes soon after (around the third to fourth month of life) in response to social stimulation and tickling (Srofe and Waters, 1976; Field, 1982). In humans, it is possible to distinguish between 'Duchenne (stimulus-driven and emotionally valenced) and non-Duchenne (self-generated and emotionless) laughter' (Gervais and Wilson, 2005, p. 396), with the latter not being spontaneous but more controlled and detached from emotion (Davila-Ross et al., 2011). The Duchenne laughter (or true laughter) has been recognised also in orangutans (*Pongo pygmaeus*), gorillas (*Gorilla gorilla*), chimpanzees (*Pan troglodytes*), and bonobos (*P. paniscus*) and is associated with play (Davila-Ross et al., 2009). At present, the most likely hypothesis is that Duchenne laughter arose in pre-hominid species. On the bright side, the use of laughter has been extended to non-playful

Figure 7.4 Play faces in *Theropithecus gelada* (a; Wilhelma zoo, Stuttgart, Germany), *Gorilla gorilla* (b; ZooParc de Beauval, France), and *Pan troglodytes* (c; ZooParc de Beauval, France). Photos: Elisabetta Palagi.

contexts such as courtship and satisfaction. During the biological and cultural evolution of humans, laughter has acquired a 'dark side' (non-Duchenne laughter) being used as a strategic form of social manipulation (Gervais and Wilson, 2005). The non-Duchenne laughter has not been found (yet!) in non-human apes (Davila-Ross *et al.*, 2011) but behavioural evidence suggests that playful facial expression can have a manipulatory function in great apes as well. Adolescent chimpanzees perform playful facial expressions more frequently when the mothers of their younger playmates are present. It seems that the adolescent chimpanzees are able to adjust their facial displays not only to manage the play session but also to manipulate the audience not directly involved in the session (Flack *et al.*, 2004).

The play face produced by opening the mouth is widespread among primates and is probably an ancestral signal. Of course, variations on the theme are possible (Figure 7.4).

In humans, there are three main types of smile observed among children: the upper smile (exposing the upper teeth only), closed smile (no teeth visible) and broad smile (both upper and lower teeth are exposed). In children, the upper smile is most common during friendly interactions, the closed smile is commonly observed in solitary play, and the broad smile characterises social play (Blurton Jones, 1971; McGrew, 1972; Cheyne, 1976). In different monkey and ape species, the open mouth display can be performed by exposing both lower and upper teeth (full play face, corresponding to the broad smile in children) or by uncovering the lower teeth only (play face) (Palagi and Mancini, 2011). While the play face can be consistently found among primates, the full play face does not follow phylogeny (Preuschoft and van Hooff, 1997). Within hominoids, the full play face is regularly observed in humans (*Homo sapiens*), bonobos (*Pan paniscus*) and gorillas (*Gorilla gorilla*) (Figure 7.5), whereas the play face is more common in chimpanzees (*Pan troglodytes*) (Palagi, 2006; Palagi *et al.*, 2007; Cordoni and Palagi, 2011; Palagi and Cordoni, 2012).

In some cercopithecoids, the distribution of the play face versus the full play face seems to be dependent on the tolerance levels of each species (Thierry *et al.*, 1989; Petit *et al.*, 2008). For example, in crested macaques (*Macaca nigra*), mandrills (*Mandrillus sphinx*) and geladas (*Theropithecus gelada*), the full play face is a combination of

Figure 7.5 Social play session between immatures of *Gorilla gorilla* (ZooParc de Beauval, France). The individual on the left is play biting the arm of the individual on the right, who is emitting a full play face. Photo: Elisabetta Palagi.

play face and the silent bared-teeth display, another facial expression used to affiliate with others (van Hooff and Preuschoft, 2003; Bout and Thierry, 2005). Macaques are a good example to investigate which social factors influence the distribution of the two variants of playful facial displays. In fact they are all organised in multimale, multifemale groups but at the same time vary in their tolerance levels, ranging from intolerant (despotic) to tolerant (egalitarian) species (Thierry, 2000; Chapter 4). Taking the two extremes of the tolerance gradient, it has been observed that the full play face is much more frequent in the more tolerant Tonkean macaques (with about 90% of all open mouths corresponding to the full play face version; Figure 7.6) than in the highly despotic Japanese macaques (Pellis *et al.*, 2011).

Based on the available literature, no final conclusion can be made on why the full play face is more frequent than play face in tolerant species, and especially between adults. A possible explanation may lay in the way individuals of tolerant groups play. Play sessions are characterised by a higher number of degrees of freedom, less prefixed motor schemes and more uninhibited motor patterns. The playful sessions are less regulated by intrinsic factors (ranking position, size disparity, etc.) and can variably include a high number of interruptions and role reversals, exaggerated patterns and unfinished sequences. In tolerant species, the players can afford 'daring' more during a playful session, making it rougher and riskier but also more rewarding. Consequently, there is a higher necessity for the play mates to declare their intention through manifest signals able to convey with no ambiguity the motivation to play. This hypothesis is supported by the fact that adults – more than juveniles – make use of the full play face, which can be directed to either another adult playmate or the mother of an immature playmate to state with no

7.3 I see it in your face: the role of facial expression in signalling the playful mood

Figure 7.6 Play sequence with full play faces in *Macaca tonkeana* (Zoo et parc de Thoiry, France). A case of rapid facial mimicry is evident (Mancini *et al.*, 2013a, b): one individual mirrors the facial expression of the other within one second (photo on the right). Photo: Chiara Scopa.

doubt that the interaction is not serious (Palagi, 2008; Palagi and Mancini, 2011; Palagi and Scopa, unpublished data). The almost total absence of data on the play specific facial expressions in New World monkeys makes the comparison between catarrhines and platyrrhines not possible at the present stage of knowledge. The use of the relaxed open mouth during play has been reported in some platyrrhines, such as *Cebus apella* and *Callithrix jacchus* (De Marco *et al.*, 2008; Voland, 1977) but modalities, distribution and functions are largely unknown.

Pellis and Pellis (1997) found that in *Lemur catta* the motor patterns used during play fighting strongly resemble those of real fighting, which is in line with the competitive nature of this species. Yet, the authors observed that only 1.8% of the playful sessions ended up with an aggression, thus suggesting that the animals are able to cope with possible ambiguous situations, using play signals.

In a recent study, Palagi *et al.* (2014) demonstrated that lemurs signal their playful motivation by performing a relaxed open mouth display, besides by manipulating their tail (tail play, Jolly, 1966; Palagi, 2009). Researchers have long wondered whether lemurs would use a relaxed open mouth display as a proper play signal or, more simply, to start a play bite (Jolly, 1966; Pellis and Pellis, 1997). In the former case, the mouth opening would be separated from biting via ritualisation, the process through which expressive displays become disconnected from their original function to serve a new function (Tinbergen, 1952). The ritualisation hypothesis has been largely supported in haplorrhines. In this primate taxon the relaxed open mouth is normally considered a ritualised signal that simulates the intention to bite for play, without actually doing it because the biting pattern is interrupted before its completion (Palagi, 2006; van Hooff and Preuschoft, 2003). Palagi *et al.* (2014) showed that the relaxed open mouth is a communicative signal also in lemurs by demonstrating that its presence is not strictly linked to a subsequent play bite. *Lemur catta* were found to rely more on the relaxed open mouth signals in the mating period compared to the pre-mating period (cf. Pellis and Pellis, 1997; Palagi *et al.*, 2014). During mating, the relationships in social groups are very tense. Competition and aggressiveness increase in both females and males, with agonistic encounters being especially severe between males (Jolly, 1966; Palagi *et al.*, 2003; 2004a; Sclafani *et al.*, 2012; Gabriel *et al.*, 2014). Some forms of conflict, such as jump fights, exclusive of the mating season can escalate, causing serious injury (e.g. wounds 4–5 cm long on thighs, arms or sides; Jolly, 1966; Sussman and Richard, 1974). In parallel with the increase in the aggression levels there is also a significant decrease of affiliation between males. This phenomenon was observed in four sites in Madagascar (Beza Mahafaly Special Reserve in southwestern Madagascar; spiny bush at Cap Sainte-Marie in southern Madagascar; rocky-outcrop forest fragments at Anja Reserve and Tsaranoro Valley in Madagascar's south-central highlands; Gabriel *et al.*, 2014). Consistently, Palagi and Norscia (2015) found that reconciliation levels in the social groups plummets during the mating period both in captivity and in the wild (forest of Berenty, south Madagascar) (see also Chapter 5). In this highly aggressive and competitive context, playing is probably much more risky than in other phases of the year and a redundancy of clear playful signals is required to limit the possibility that play sessions turn into serious and dangerous fights. Indeed, the overabundance of play signals, both in terms of exaggeration and occurrence, somehow reflects the high levels of aggressiveness of a species (bears, Henry and Herrero, 1974; coyotes, Bekoff, 1974b; hyenas, Drea *et al.*, 1996) and the degree of asymmetry of the play sessions. Palagi *et al.* (2014) observed that the relaxed open mouth in ring-tailed lemurs was more frequent when the players engaged in asymmetric sessions, with one player largely prevailing over the other. This further indicates that signalling benign intent is particularly necessary when the session involves a greater amount of risk (Silk, 1997; Bekoff, 2001).

The investigation of tail play, another signal frequently used by ring-tailed lemurs, provides further support for the importance of signalling during particularly hazardous playful interactions (Palagi, 2009). When adults play together, not just the relaxed open mouth but also tail play is used to fine-tune the session. Males direct tail play more to females probably because females are dominant and potentially dangerous playmates, even though they play less with females than with males. Another ambiguous situation is represented by polyadic play sessions, during which a player must face more than one partner at the same time. The danger born by unclear signalling is even higher when more than one receiver is involved because misinterpretation can cause the sender to be attacked by the other players. A further parameter that influences the distribution of tail play is the familiarity between players. Players that affiliate more through grooming in baseline conditions use tail signals less than other dyads. Play fighting between two individuals that socially interact at a very low frequency can be particularly unsafe because of the limited information they have on each other, for example regarding physical strength, movement rapidity, and propensity to play fairly (Palagi, 2009).

7.3 I see it in your face: the role of facial expression in signalling the playful mood

Contrary to tail play, relaxed open mouth is almost exclusively directed to a specific playmate during one to one interactions in support of the face-to-face interaction hypothesis. There is evidence that several mammal species, other than great apes and humans, can choose appropriate forms of communication depending on another animal's attentional state. It has been observed, for example, that only when a dog has got the attention of another dog, she/he directs visual play signals towards it (Horowitz, 2009; Bekoff's box, Box 7.1, in this chapter). Further support for the face-to-face interaction hypothesis comes from the response of receivers. In ring-tailed lemurs, Palagi *et al.* (2014) found that relaxed open mouth (Figure 7.7) was more frequently replicated within 5 seconds by the playmate compared to tail play. The replication of the open mouth and its almost exclusive use in dyadic sessions strongly indicates the high degree of directionality of the signal. If an animal is able to respond in a congruent way to the facial expression of another it is also supposedly able to recognise the attention level of the receiver. In an adaptive view, facial replication allows animals to appropriately respond to the stimuli of the social environment and synchronise their activity (Provine, 1996; 2004; Palagi and Mancini, 2011). Therefore, the relaxed open mouth in lemurs is detached from play bite (ritualised), more redundant when the session is imbalanced, preferably used during dyadic interactions, and characterised by a high number of congruent replications (open mouth for open mouth). These different lines of evidence converge on the conclusion that the relaxed open mouth in lemurs is homologous and analogous to the play face of haplorrhines.

Figure 7.7 Play face in *Lemur catta*. Drawing: Carmelo Gómez González.

> **Box 7.2** | **by Sergio M. Pellis**
>
> **Speaking of which: the neurobiology of play**
>
> In my senior undergraduate years, I developed an abiding passion for ethology and noticed that the study of development was the least mature of the topics covered by the discipline. Moreover, in the mid-1970s, when my interests were coming together, the discipline went through one of those interminable swings of the pendulum in the debate between the relative importance of nature and nurture. It seemed obvious to me that the eclectic biological perspective offered by classical ethology was the way forward in resolving this issue. Therefore, in casting about for a graduate thesis project, I was influenced by Marler's classic study of the development of begging behaviour in gull chicks. My initial foray was a comparative study of begging in sea birds. However, practical concerns, such as the lack of money, quickly put an end to that project.
>
> My prospective supervisor showed me a 16 mm movie he had made of two juvenile Australian magpies engaged in rough and tumble play. Two things intrigued me about what I saw. First, what I expected to see only in mammals, like kittens and puppies, was present in a whole different lineage, birds. The behaviour clearly needed to be made sense of in a broader comparative framework than simply viewing play as some quirk of mammals. Second, the behaviour as performed was not utilitarian – it did not buy the animals anything seemingly tangible. In the heyday of the outgoing behaviourism and the burgeoning influence of behavioural ecology, animals should not be doing anything in the absence of an immediate reward or a clear functional enhancement of fitness. The scant literature on play then available convinced me that a detailed developmental study was in order. Most importantly, Australian magpies were common in the parks in Melbourne, and so cheap to study (no long trips to breeding islands of sea birds required). I was hooked, and with the arrogance of youth, thought that solving the problem of play was a suitable, and attainable, goal for a doctoral thesis.
>
> While studying the play of Australian magpies, I also came across the play of another species, the oriental small-clawed otter. While there was much beauty in the movements performed, decoding the structure of the routines amid the variation seen across age and between species was a challenge and much of my graduate work was devoted to making sense of those movements. Even once the structure was decoded, further problems remained, that of unravelling the mechanisms that regulate that structure and deciphering the functions of such performances. Needless to say, I did not solve the problem of play in a four year PhD thesis, but I am happy to say that nearly 40 years on, more about the behaviour makes sense to me than it did at the beginning. At the least, I learned enough to know how to ask better questions, and after

all, that is how science progresses, by asking ever more precise questions (Pellis and Pellis, 2009).

The earliest one, dating to my PhD thesis, arose from my training in classical ethology, which imbued me with the view that before attempting to explain why a particular behaviour exists, one must first describe what it is that needs explaining. Then, as now, there are more theories about play than there are reliable descriptions about what animals do when they play. Using detailed analyses of filmed sequences of play, I found that, contrary to prevailing opinion, sequences of play are not infinitely variable, but built around an underlying routine (Pellis, 1981). That insight had two important repercussions later in my career. First, while rough and tumble play, or play fighting, superficially resembles serious fighting, not all species play by attacking and defending species-typical body targets that are competed over during serious fighting. Many species compete over body targets typical of adult sexual behaviour, or other amicable behaviour, such as body areas groomed, or attacked during predation (Pellis, 1988). Second, despite species differences in the targets competed over, when play fighting there is a common rule structure that distinguishes play from its serious counterparts. The rule is that the animals follow patterns of interaction that ensure that play fighting is reciprocal, with neither animal having or maintaining the upper hand. Indeed, building on this insight we have recently begun to show how different species, or lineages of species, have evolved different solutions to this fairness rule (Pellis et al., 2010b).

Attention to the theme underlying sequences of play fighting revealed that – again, contrary to prevailing opinion – in rats, juvenile play does not morph into adult aggression or sex, but rather, it remains a distinct way of interacting well into adulthood (Pellis and Pellis, 1987; Pellis et al., 1993). In the juvenile period, the function of play can be a delayed one – that is, play now, benefit later. In contrast, in adulthood, the functions of play are immediate – the benefits arise at the time the play is performed. My detailed studies of adult play in rats had shown that play fighting is used for social assessment and manipulation, especially in dominance relationships. Such play provides a safer way to jockey for improved status while limiting the risk of injury (Pellis, 2002). That is, my work showed that play, albeit not as frequent as it is in the juvenile period, is also a feature of adulthood.

My zoological background also meant that whatever aspect of play that I studied, I put whatever I found into a comparative context. I found that among a wide range of rodents, only a few, such as rats, retain play into adulthood. Exploration of the literature available on another lineage, primates, showed that 50%, maybe more, of species retain play fighting to some level of complexity into adulthood. This suggests that play is retained in

Box 7.2 (continued)

Box 7.2 (*cont.*)

adulthood when some particular functional contexts promote such retention. The likely enabling factor was suggested by the work I did on rats – that is, that play seems to be used in situations in which there is some uncertainty as to the relationship between the participants. Therefore, I predicted that, in primates, play among adults is more likely to occur in species in which there is unpredictability in the relationships among the members of potential social groups. Indeed, we showed that species that had mating systems and social systems that involved reduced degrees of social contact or unpredictable contact, were statistically more likely to use play in adulthood than ones with more stable systems (Pellis and Iwaniuk, 1999, 2000). These findings provided comparative support for the findings from rats, that adult–adult play is used functionally for social assessment and manipulation (Pellis, 2002).

Back to juvenile play, work from many decades, across many different laboratories, particularly on rats, has revealed that at least in part, some of the social, emotional and cognitive deficits that emerge when juveniles are reared in social isolation are due to their lack of social play experience (see Pellis and Pellis, 2006, for a review). We then showed that damage to particular areas of the prefrontal cortex (PFC), the executive part of the brain, in adult rats that were reared with other juveniles, mimic some of these deficits (Bell *et al.*, 2009; Pellis *et al.*, 2006). Moreover, by denying juveniles the experience of play fighting, but not other social experiences, the development of the architecture of the neurons of the medial prefrontal cortex (mPFC) is altered, whereas, by providing juveniles with the experience of multiple social partners, whether those partners play or not, alters the neurons of the orbital frontal cortex (OFC) (Bell *et al.*, 2010). Damage to the mPFC reduces the rat's ability to coordinate its movements effectively with those of a social partner (Bell *et al.*, 2009), and damage to the OFC reduces the rat's ability to modify its behaviour with the identity of its partner (e.g., dominant versus subordinate) (Pellis *et al.*, 2006). That is, my laboratory demonstrated a causal link between juvenile play experience, adult social skills and the brain mechanisms that underlie those skills (Pellis *et al.*, 2010a).

There are, in my opinion, two big questions to answer, why is play a relatively rare phenomenon, arising sporadically in the animal kingdom, and, once present, what factors promote its transformation as it is co-opted for different functions (Burghardt, 2005; Pellis and Pellis, 2009)? Within a broad phylogenetic framework, several, more specific, questions can be asked. What are the social, ecological and developmental conditions that make play possible in particular lineages? Are different forms of play, social versus non-social, alternative expressions from a common playful wellspring, or is each type of play uniquely evolved? That is, is there a playful brain circuit that can be co-opted in multiple contexts, or are there unique neural circuits for each type of play? In the domain of social play, what is the minimum set of neurobehavioural mechanisms that make negotiating fairness possible, and how can that minimum be modulated by mechanisms that may only be present in some lineages?

In mammals, for example, the ability to play fairly seems to involve neural mechanisms that do not involve the cortex, as decorticated animals are able to sustain reciprocity in their play fighting (Pellis *et al.*, 1992). However, cortical mechanisms are able to modulate that reciprocity to modify the rules of fairness to navigate social relationships, depending on context (Pellis *et al.*, 2006). That is, play fighting involves the interaction of many neural systems and levels of control (Graham, 2011). What are the neural mechanisms that can be modified and what socioecological factors promote their evolution? A tantalising possibility is provided by studies on an interesting branch of Old World monkeys. Macaques comprise a suite of 20 species in the genus *Macaca* that, while all having the same basic social organisation, vary in the degree of tolerance in how rigidly to which dominance relationships are adhered (Thierry, 2007). More tolerant species have a form of play that is more cooperative, exaggerating the incidence of polyadic play that is hyper-cooperative (Petit *et al.*, 2008; Reinhart *et al.*, 2010) and, as befits their more ambivalent social relationships, have more prevalent adult–adult play (Ciani *et al.*, 2012). In part, intra-species and inter-species differences in tolerance are related to the incidence of particular receptor types for serotonin (Wendland *et al.*, 2006). Serotonin is widespread in the brain, where it helps set the tone of brain activity, to low, to laid back (or, indeed depressed) and to high, to manic. One way such regulation takes place is by affecting impulse control, which involves the prefrontal cortex attenuating the activity of other neural systems. Thus, in the macaque system, there is variation in impulsivity, and depending on the social system high impulsivity or low impulsivity is useful. Selection for the alleles that control the expression of the different serotonin receptors provide the evolutionary mechanism, and how the changes in receptors affects the brain mechanisms that regulate impulsivity provide the immediate route by which play is affected. In turn, the play permissible by these changes in impulsivity can be co-opted in the juvenile period to train social skills and in adulthood for social assessment and manipulation. While this schema is speculative, it highlights what I think are the most exciting prospects for the future study of play, an approach that integrates molecular and neural mechanisms with behavioural studies over development and across species.

7.4 Beyond signals: the role of play in increasing social tolerance and fairness in cooperation

After analysing play in *Callithrix jacchus*, Voland (1977) embraced Beach's (1945) idea that 'play is an end in itself' and concluded that 'in the short run, play appears to have no actual functional necessity, since no rank order is formed through play'. However, the absence of short-term benefits would have left adult play – costly but maintained by evolution – unexplained (Bekoff and Byers, 1998; Fagen and Fagen, 2004; Palagi *et al.*, 2004b; Burghardt, 2005). In the very species used by Voland to suggest that play provided no immediate benefit (*Callithrix jacchus*), we found that play is used by adults as an anxiety reliever in the short run (Norscia and Palagi, 2011). The role of play in buffering anxiety has been reported also in other primate species, including lemurs (*Propithecus verreauxi*). Because this issue has been extensively discussed in Chapter 6, here we focus on other possible functions of adult play in primates and to what extent such functions apply to lemurs. Although play behaviour is typical of immature individuals (Fagen, 1993; Burghardt, 2005), in a number of species play is maintained into adulthood. They are not very numerous but all of them are extremely social and possess a high neuronal complexity (Pellis and Pellis, 2009). The fact that adult play is not as pervasive as juvenile play suggests that this behaviour may have specific functions and for this reason is employed on purpose (Palagi, 2011). Play in adults predominantly has a social function. It has been demonstrated that adult play has a role in social assessment and in the strategic manipulation of particular situations that may give rise to conflicts of interest (Palagi *et al.*, 2004b, 2006). Play in adults seems to covary with the degree of tolerance that characterises different species and populations (Palagi *et al.*, 2004b; Ciani *et al.*, 2012). In despotic species organised in strongly hierarchically arranged societies where relationships are bridled and crystallised, there is little space for adult play. Here, social relationships are codified and mostly unidirectional (Flack and de Waal, 2004; Ciani *et al.*, 2012; Thierry's box, Box 4.2, in Chapter 4). The relationships featured in these types of societies limit the potential expression of play because play requires the maximum degree of plasticity, versatility and bidirectionality. In order for play to occur, the two players must be equally important both in performing motor patterns and in the level of reward derived from the playful session. From both the biological and the cultural point of view play, especially between adults, suffers from the lack of freedom but on the other hand tolerance is promoted by play, thus triggering a positive feedback.

Antonacci *et al.* (2010) pointed out that the power of play in promoting tolerance in humans has been known since ancient times. According to ancient Greek mythology, Apollo, queried through Delphi's oracle, told Iphitos (the King of Elis) that the wars devastating Peloponnese would be ended by staging a game competition at Olympus. After the re-establishment of the Olympic Games (and this is where the myth ends and history begins), the longest-standing peace agreement of the history (the Olympic Truce) was signed between the areas of Peloponnese (Swaddling, 2002).

7.4 Beyond signals: the role of play in increasing social tolerance and fairness in cooperation

The role of play in limiting aggression and favouring positive social interactions starts with free play fighting and reaching its climax with the production of ruled games (Pellegrini and Smith, 1998). In human populations that are more bridled into social, legal, religious, military rules (etc.) play between adults is unfavoured (Gray, 2009; Fouts *et al.*, 2013). In hunter-gatherer societies, characterised by informal and noncompetitive social relationships (Marshall, 1976; Sutton-Smith and Roberts, 1970) the playful attitude persists in all spheres of social life such as hunting and gathering activities, religious beliefs and practices, meat sharing, and even punishment methods, which are performed through 'humour and ridicule' of others (Norbeck, 1974; Gray, 2009, 2012). These societies are also more open to new comers (Gray, 2009). Two Bofi populations of Central Africa, one of foragers and the other of farmers, show profound cultural differences in managing social relationships probably due to their diverse ecological systems of subsistence. The Bofi foragers are egalitarian and do not typically base their status or power on age or gender. In their society, playful interactions mainly realised via rough joking and storytelling, are used to mediate cooperation and sharing. Instead, in Bofi farmers social roles are more crystallised as a function of status and gender. Consistent with the framework presented above, in this population physical and verbal play are inhibited (Fouts *et al.*, 2013). It has been hypothesised that in highly cooperative groups of early hominids, play probably involved adults that not only played with infants but also between each other (Narvaez and Gleason, 2012).

In non-human primates, as in humans, sufficient levels of freedom allow using play in a strategic way during adulthood. Play, and in particular play fighting, seems to have an important role between individuals that do not have regular social interactions either because they are mostly solitary or because they live in fission–fusion social groups (Pellis and Iwaniuk, 1999, 2000). For example, play fighting has been more or less anecdotally reported during courtship in solitary species, including strepsirrhines, where adult males and females rarely meet (e.g. the genera *Mirza, Daubentonia, Galago, Perodicticus,* and *Pongo;* MacKinnon, 1974; Charles-Dominique, 1977; Pages, 1980) and during social encounters between subgroups in species characterised by a fluid and loose composition of groups (e.g. *Ateles, Cacajao,* and *Pan*; Eisenberg, 1976; Pellis and Iwaniuk, 1999, 2000). However, the key social feature that favours the playful attitude in adults is the propensity for tolerance. *Macaca tonkeana* and *M. fuscata*, two species at the opposite extremes in the social tolerance gradient of the genus (Thierry, 2000), strongly differ in the distribution of social play according to the gender and age of the players. Different from the Japanese macaque, in the tolerant Tonkean macaque play is well represented among adults with no preference with respect to age or sex of the play mate. In Tonkean macaques the social play of juveniles is affected by the degree of mothers' permissiveness because the more permissive they are the more the juveniles have the opportunity to interact with other group members and increase their propensity for tolerance (Ciani *et al.*, 2012). Conversely, the strict control on infants limits their relational sphere, thus creating the conditions for a tightened social canalisation (Berman, 1982). The limitation of social play interactions during the immature phase leads to a reduced propensity to play later in life, during adulthood.

Another good example is represented by chimpanzees and bonobos, which, despite possessing the same social structure (fission–fusion society) and being sister species due to their extreme phylogenetic proximity, are characterised by dramatic differences in adult social play. Palagi (2006, 2007) demonstrated that adult bonobos play at significantly higher frequencies than chimpanzees, a finding confirmed by long-term studies carried out on several chimpanzee and bonobo colonies (Palagi and Demuru, unpublished data). Moreover, in bonobos playful sessions can also involve more than two individuals (polyadic sessions). According to the Social Bridge Hypothesis (Palagi, 2011), two adult animals that rarely play together in dyadic interactions may find a contact point in polyadic play through the other play mates involved in the session. These individuals would represent a sort of bridge between two socially unconnected individuals, thus enlarging the individuals' social network. In the long run this mechanism may enhance the formation of large parties, a peculiarity of bonobo social organisation (Kano, 1984). Bonobo society is characterised by a wide array of cooperative activities in which adults establish and re-establish their relationships through alliances and affiliation (Palagi, 2006; Furuichi, 2011; Demuru and Palagi, 2012; Clay and de Waal, 2013; Palagi and Norscia, 2013).

Contrary to chimpanzee communities which are highly xenophobic and aggressive towards the individuals residing in other communities (Goodall, 1986; Wrangham, 1999; Wrangham and Glowacki, 2012), bonobos of different groups have been observed affiliating with each other (Furuichi, 2011; Tan and Hare, 2013) and to do so through play (Behncke, 2015).

The higher frequencies of adult play in bonobos and its variety of forms (intergroup and polyadic) strongly supports the Social Tolerance Hypothesis presented above, which foresees that the more a society is tolerant, the more play is freely expressed. When this occurs, play becomes a strategic tool to manage social situations and multiply the social relationships of an individual with others, thus favouring social integration.

Although more quantitative studies on strepsirrhine play are necessary, the Social Tolerance Hypothesis seems to explain the distribution of play in lemurs. If we look at the only two lemur species, *Propithecus verreauxi* and *Lemur catta*, in which play has been quantitatively studied, we can draw a parallel between the use of play and the tolerance levels characterising each species. The despotic system of *Lemur catta* strongly limits the expression of adult–adult play in this species. Play sessions involving exclusively adults have been observed in captivity during the pre-reproductive period between males and females, thus probably related to sexual assessment (Palagi, 2009). Otherwise, play sessions with adults normally involve at least one juvenile subject (Palagi *et al.*, 2014). Ring-tailed lemur troops rarely allow intrusion by strangers and intergroup interactions involve either avoidance or (more frequently) aggression (Jolly, 1966, 2003; Nakamichi and Koyama, 1997; Soma and Koyama, 2013). The proximity of individuals of neighbouring groups

7.4 Beyond signals: the role of play in increasing social tolerance and fairness in cooperation

and intergroup aggression elicits play between group members (Palagi *et al.*, unpublished data), following the xenophobia principle, which suggests that a peak of cohesive behaviours among insiders is evoked by newcomers (Wilson, 2002). *Propithecus verreauxi*, which is characterised by more relaxed relationships and mild levels of aggressiveness (Norscia *et al.*, 2009) use play as an access key to enter new groups. In the pre-mating period, males can start roaming and visiting other groups in search of receptive females, which experience a single oestrus period per year (up to 72 h). Subjects of both sexes can mate with multiple partners in their own and neighbouring groups (Brockman, 1999; Lawler, 2007) but owing to female dominance, females are the choosing sex (Richard, 1992; Brockman, 1999; Norscia *et al.*, 2009). Antonacci *et al.* (2010) observed that after the first attempts to keep strange males away via mild fighting and chasing, adult males began to play with them, and aggression plunged after the first playful contact. The frequency of play between in-group and out-group members was higher than that between residents, thus indicating that play is used in a strategic way for a particular purpose. Hence, adult play in sifakas appears to have a role in managing new social situations more than in maintaining 'old' relationships. In particular, play appears to be the interface between strangers with the specific function of reducing xenophobia, normally expressed by this species via aggressive chases. Some data on lorises and galagos suggest a dissociation of the function of play, similar to that observed in *L. catta* and *P. verreauxi*. Literature reports that when the lorises were confronted with strangers, it led to avoidance or serious fighting. In galagos, in the same circumstances, non-aggressive social contact was frequent, with much of that contact involving play fighting. Therefore, play fighting seems to be used to strengthen bonds with familiar individuals in lorises and (also) to test relationships with strangers in galagos (Newell, 1971; Ehrlich and Musicant, 1975; Ehrlich, 1977; Pellis and Pellis, 2010).

Antonacci *et al.* (2010) also found that in-group males initiated play sessions as much as out-group males, thus indicating that ice-breaking via play is worthwhile and beneficial for both parties. The authors pointed out that this strategy is clearly advantageous because it promotes good relations between unfamiliar individuals thus reducing to a minimum the costs that xenophobia would bring, in terms of aggression and group stability. Interestingly, grooming seems to have a different function from play. Indeed, while play was mostly exchanged between residents and out-group members, grooming significantly increased between residents after the arrival of the intruders. Therefore, grooming is used to maintain pre-existing social relations and not to establish new ones. In this respect, there is probably a functional dichotomy between social play and grooming, with social play used as an ice-breaker to avoid overt aggression and grooming possibly used to reinforce the relationships within group mates, thus reassessing group membership. On the other hand, engaging in contest competition would probably be useless to unbeneficial to males because mate choice is *not* up to them.

The specific function of play increasing tolerance over same sex competitors among males is confirmed by the different situation found in females, which played with resident males as much as they played with intruders (Antonacci *et al.*, 2010). This result can be explained in the light of the criteria adopted by females to choose mating partners. As extensively explained in Chapter 8, to be selected by females, males have to be good scent releasers and groomers more than good players. In fact, females concede mating priority to those males that are most active in scent marking and copulated repeatedly with those males providing more grooming to them. Therefore, sifaka seem to differ from other strepsirrhine species that were reported to use courtship play as a social tool for overcoming female reticence when male–female association is loose (*Galago demidovii, Perodicticus potto*; *Mirza coquereli*; *Ateles* spp.; Charles-Dominique, 1977; Pages, 1980).

In summary, play in lemurs involves the same criteria demonstrated in other primates and it provides an exceptional background to study complex forms of communication or meta-communication, thus being a window on the cognitive abilities of the players. The distribution of play reflects the degree of tolerance expressed by the different species. Additionally, it can be used in a strategic way to handle challenging situations and reduce the probability of potentially dangerous aggression. Play can be used as an ice-breaker, which enhances friendly interactions in the critical process that upgrades a stranger to a familiar individual. All in all, the data accumulated so far on social play, although scarce, provides support to the idea that 'fun', even in adult lemurs, has an evolutionary meaning.

References

Aldis, O. (1975). *Play Fighting*. New York: Academic Press.
Antonacci, D., Norscia, I. & Palagi, E. (2010). Stranger to familiar: wild strepsirrhines manage xenophobia by playing. *PLos ONE*, 5(10), e13218.
Bainum, C. K., Lounsbury, K. R. & Pollio, H. R. (1984). The development of laughing and smiling in nursery school children. *Child Development*, 55, 1946–1957.
Barlow, G. (1977). Modal action patterns. In: S. Thomas (ed.), *How Animals Communicate*. Bloomington: Indiana University Press, pp. 98–134.
Beach, F. A. (1945). Current concepts of play in animals. *American Naturalist*, 79, 523–541.
Behncke, I. (2015). Play in the Peter Pan ape. *Current Biology*, 25, R24–R27.
Bekoff, M. (1972). The development of social interaction, play, and metacommunication in mammals: An ethological perspective. *Quarterly Review of Biology*, 47, 412–434.
Bekoff, M. (editor) (1974a). Social play in mammals. *American Zoologist*, 14, 265–436.
Bekoff, M. (1974b). Social play and play-soliciting by infant canids. *American Zoologist*, 14, 323–340.
Bekoff, M. (1977a). Mammalian dispersal and the ontogeny of individual behavioral phenotypes. *American Naturalist*, 111, 715–732.
Bekoff, M. (1977b). Social communication in canids: Evidence for the evolution of a stereotyped mammalian display. *Science*, 197, 1097–1099.
Bekoff, M. (1995). Play signals as punctuation: The structure of social play in canids. *Behaviour*, 132, 419–429.
Bekoff, M. (2001). Social play behaviour: cooperation, fairness, trust, and the evolution of morality. *Journal of Consciousness Studies*, 8, 81–90.
Bekoff, M. (2004). Wild justice and fair play: cooperation, forgiveness, and morality in animals. *Biology & Philosophy*, 19, 489–520.

Bekoff, M. (2007). *The Emotional Lives of Animals.* Novato, CA: New World Library.

Bekoff, M. (2013). *Why Dogs Hump and Bees Get Depressed. The Fascinating Science of Animal Intelligence, Emotions, Friendship, and Conservation.* Novato, CA: New World Library.

Bekoff, M. & Allen, C. (1998). Intentional communication and social play: how and why animals negotiate and agree to play. In: Bekoff, M. & Byers, J. A. (eds), *Animal Play: Evolutionary, Comparative, and Ecological Perspectives.* Cambridge University Press, pp. 97–114.

Bekoff, M. & Byers, J. A. (1981). A critical reanalysis of the ontogeny and phylogeny of mammalian social and locomotor play: An ethological hornet's nest. In: K. Immelmann, G. Barlow, M. Main & L. Petrinovich (eds), *Behavioral Development: The Bielefeld Interdisciplinary Project.* Cambridge University Press, pp. 296–337.

Bekoff, M. & Byers, J. A. (1998). *Animal Play: Evolutionary, Comparative and Ecological Approaches.* New York: Cambridge University Press.

Bekoff, M. & Pierce, J. (2009). *Wild Justice: The Moral Lives of Animals.* Chicago: University of Chicago Press.

Bell, H. C., McCaffrey, D., Forgie, M. L., Kolb, B. & Pellis, S. M. (2009). The role of the medial prefrontal cortex in the play fighting of rats. *Behavioral Neuroscience*, 123, 1158–1168.

Bell, H. C., Pellis, S. M. & Kolb, B. (2010). Juvenile peer play experience and development of the orbitofrontal and medial prefrontal cortices. *Behavioural Brain Research*, 207, 7–13.

Berman, C. M. (1982). The ontogeny of social relationships with group companions among free-ranging infant rhesus monkeys: II. Differentiation and attractiveness. *Animal Behaviour*, 30, 163–170.

Berns, G. (2013). *How Dogs Love Us: A Neuroscientist and his Adopted Dog Decode the Canine Brain.* New York: New Harvest/Amazon.

Blurton Jones, N. (1971). Criteria for use in describing facial expressions of children. *Human Biology*, 43, 365–413.

Blurton Jones, N. (1972). Categories of child-child interactions. In: N. Blurton Jones (ed.), *Ethological Studies of Child Behaviour.* London: Cambridge University Press, pp. 97–127.

Bout, N. & Thierry, B. (2005). Peaceful meaning for the silent bared-teeth displays of mandrills. *International Journal of Primatology*, 26(6), 1215–1228.

Brockman, D. K. (1999). Reproductive behavior of female *Propithecus verreauxi* at Beza Mahafaly, Madagascar. *International Journal of Primatology*, 20, 375–398.

Burghardt, G M. (2005). *The Genesis of Animal Play: Testing the Limits.* Cambridge, MA: MIT Press.

Call, J. & Tomasello, M. (2007). *The Gestural Communication of Apes and Monkeys.* Mahwah: Lawrence Erlbaum Associates.

Candland, D. G., French, D. K. & Johnson, C. N. (1978). Object play: test of a categorized model by the genesis of object play in *Macaca fuscata*. In: E. O. Smith (ed.), *Social Play in Primates.* New York: Academic Press, pp. 259–296.

Cattaneo, L. & Pavesi, G. (2014). The facial motor system. *Neuroscience & Biobehavioral Reviews*, 38, 135–159.

Charles-Dominique, P. (1977). *Ecology and Behaviour of Nocturnal Primates. Prosimians of Equatorial West Africa.* New York: Columbia University Press.

Cheyne, J. A. (1976). Development of forms and functions of smiling in preschoolers. *Child Development*, 47, 820–823.

Ciani, F., Dall'Olio, S., Stanyon, R. & Palagi, E. (2012). Social tolerance in macaque societies: a comparison with different human cultures. *Animal Behaviour*, 84, 1313–1322.

Clay, Z. & de Waal, F. B. (2013). Bonobos respond to distress in others: consolation across the age spectrum. *PLos ONE*, 8(1), e55206.

Cordoni, G. & Palagi, E. (2011). Ontogenetic trajectories of chimpanzee social play: similarities with humans. *PLoS ONE*, 6(11), e27344.

Darwin, C. (1872). *The Expression of the Emotions in Man and Animals.* Chicago, IL: University of Chicago Press.

Davila-Ross, M., Owren, M. J., Zimmermann, E. (2009). Reconstructing the evolution of laughter in great apes and humans. *Current Biology*, 19, 1106–1111.

Davila-Ross, M., Allcock, B., Thomas, C. & Bard, K. A. (2011). Aping expressions? Chimpanzees produce distinct laugh types when responding to laughter of others. *Emotion*, 11, 1013–1020.

De Marco, A., Petit, O. & Visalberghi, E. (2008). The repertoire and social function of facial displays in *Cebus capucinus*. *International Journal of Primatology*, 29, 469–486.

de Waal, F. B. M. (2003). Darwin's legacy and the study of primate visual communication. *Annals of the New York Academy of Sciences*, 1000, 7–31.

Demuru, E. & Palagi, E. (2012). In bonobos yawn contagion is higher among kin and friends. *PLoS ONE*, 7(11), e49613. http://dx.doi.org/10.1371/journal.pone.0049613.

Demuru, E., Ferrari, P. F. & Palagi, E. (2015). Emotionality and intentionality in bonobo playful communication. *Animal Cognition*, 18, 333–344.

Doyle, G. A. (1974). *Behavior of Prosimians*. In: A. M. Schrier & F. Stollnitz (eds), *Behavior of Nonhuman Primates: Modern Research Trends*, vol. 5, pp. 155–353, New York: Academic Press, pp. 155–353.

Drea, C. M., Hawk, J. E. & Glickman, S. E. (1996). Aggression decreases as play emerges in infant spotted hyaenas: preparation for joining the clan. *Animal Behaviour*, 51, 1323–1336.

Dugatkin, L. A. & Bekoff, M. (2003). Play and the evolution of fairness: a game theory model. *Behavioural Processes*, 60, 209–214.

Eisenberg, J. F. (1976). Communication mechanisms and social integration in black spider monkey, *Ateles fusciceps robustus*, and related species. *Smithsonian Contributions to Zoology*, 213, 1–108.

Erhlich, A. (1977). Social and individual behaviors in captive greater galago. *Behaviour*, 63, 192–214.

Erhlich, A. & Musicant, A. (1975). Social and individual behaviors in captive slow lorises. *Behaviour*, 60, 195–200.

Fagen, R. (1981). *Animal Play Behavior*. New York: Oxford University Press, p. 684.

Fagen, R. (1993). Primate juveniles and primate play. In: M. E. Pereira & L. A. Fairbanks (eds), *Juvenile Primates*. Chicago: University of Chicago Press, pp. 182–196.

Fagen, R. & Fagen, F. (2004). Juvenile survival and benefits of play behaviour in brown bears, *Ursus arctos*. *Evolutionary Ecology Research*, 6, 89–102.

Feistner, A. T. & Ashbourne, C. J. (1994). Infant development in a captive-bred aye-aye (*Daubentonia madagascariensis*) over the first year of life. *Folia Primatologica*, 62, 74–92.

Field, T. M. (1982). Affective displays of high-risk infants during early interactions. *Emotion and Early Interaction*, 101–125.

Flack J. & de Waal, F. B. M. (2004). Dominance style, social power, and conflict. In: Thierry, B., Singh, M. & Kaumanns, W. (eds), *Macaque Societies: A Model for the Study of Social Organization*. Cambridge University Press, pp. 157–185.

Flack, J. C., Jeannotte, L. A. & de Waal, F. (2004). Play signaling and the perception of social rules by juvenile chimpanzees (*Pan troglodytes*). *Journal of Comparative Psychology*, 118, 149–159.

Fouts, H. N., Hallam, R. A. & Purandare, S. (2013). Gender segregation in early-childhood social play among the Bofi foragers and Bofi farmers in Central Africa. *American Journal of Play*, 5, 333–356.

Fry, D. P. (1992). 'Respect for the rights of others is peace': Learning aggression versus nonaggression among the Zapotec. *American Anthropologist*, 94, 621–639.

Fry, D. P. (2005). Rough-and-tumble social play in humans. In: A. D. Pellegrini & P. K. Smith (eds), *The Nature of Play: Great Apes and Humans*. New York: The Guilford Press, pp. 54–85.

Furuichi, T. (2011). Female contributions to the peaceful nature of bonobo society. *Evolutionary Anthropology: Issues, News, and Reviews*, 20, 131–142.

Gabriel, D. N., Gould, L. & Kelley, E. A. (2014). Seasonal patterns of male affiliation in ring-tailed lemurs (*Lemur catta*) in diverse habitats across southern Madagascar. *Behaviour*, 151(7), 935-961.

Gervais, M. & Wilson, D. S. (2005). The evolution and functions of laughter and humor: A synthetic approach. *The Quarterly Review of Biology*, 80(4), 395-430.

Goodall, J. (1986). *The Chimpanzees of Gombe: Patterns of Behavior*. Cambridge, MA: Belknap Press of Harvard University Press.

Graham, K. L. (2011). Coevolutionary relationship between striatum size and social play. *American Journal of Primatology*, 71, 1-9.

Gray, P. (2009). Play as a foundation for hunter-gatherer social existence. *American Journal of Play*, 1, 476-522.

Gray, P. (2012). The value of a play-filled childhood in development of the hunter-gatherer individual. In: D. Narvaez, J. Panksepp, A. Shore & T. Gleason (eds), *Evolution, Early Experience and Human Development: From Research to Practice and Policy*. New York: Oxford University Press.

Groos, K. (1896). *Die Spiele der Tiere*. Jena ed.

Hall, S. G. (1904). *Adolescence: Its Psychology and Its Relations to Physiology, Anthropology, Sociology, Sex, Crime, Religion and Education*. New York: Appleton.

Harlow, H. F. & Harlow, M. K. (1965). The affectional systems. In: A. M. Schrier, H. F. Harlow, & F. Stollnitz (eds), *Behavior of Nonhuman Primates* (vol. 2). New York: Academic Press, pp. 287-334.

Henry, J. D. & Herrero, S. M. (1974). Social play in the American black bear: its similarity to canid social play and an examination of its identifying characteristics. *American Zoologist*, 14, 371-389.

Horowitz, A. (2008). Attention to attention in domestic dog (*Canis familiaris*) dyadic play. *Animal Cognition*, 12, 107-118.

Horowitz, A. (2009). Disambiguating the 'guilty look': salient prompts to familiar dog behavior. *Behavioural Processes*, 81, 447-452.

Jolly, A. (1966). *Lemur Behavior: a Madagascar field study*. Chicago: University of Chicago Press.

Jolly, A. (1985). *The Evolution of Primate Behavior*, 2nd ed. New York: Macmillan.

Jolly, A. (2003). *Lemur catta*, ring-tailed lemur, Maky. In: Goodman, S. M. & Benstead, J. P. (eds), *The Natural History of Madagascar*. Chicago, MA: The University of Chicago Press: pp. 1329-1331.

June, G., Devlin, H. C. & Moskowitz, J. (2014). Seeing both sides: An introduction to the light and dark sides of positive emotion. In: Gruber, J. & Moskowitz, J. (eds), *Positive Emotion: The Light Sides and Dark Sides*. New York, NY: Oxford University Press.

Kano, T. (1984). Distribution of pygmy chimpanzees (*Pan paniscus*) in the central Zaire basin. *Folia Primatologica*, 43(1), 36-52.

Kuczaj, S. A. & Horback, K. M. (2013). Play and Emotion. In: S. Watanabe & S. Kuczaj (eds), *Emotions of Animals and Humans: Comparative Perspectives*. New York: Springer, pp. 87-111.

Lawler, R. R. (2007). Fitness and extra-group reproduction in male Verreaux's sifaka: an analysis of reproductive success from 1989-1999. *American Journal of Physical Anthropology*, 132, 267-277.

Lazar, J. W. & Beckhorn, G. D. (1974). Social play or the development of social behavior in ferrets (*Mustela putorius*)? *American Zoologist*, 14(1), 405-414.

Lewis, M. (2000). The emergence of human emotions. *Handbook of Emotions*, 2, 265-280.

MacKinnon, J. (1974). The behaviour and ecology of wild orangutans (*Pongo pygmaeus*). *Animal Behaviour*, 22, 3-74.

Mancini, G., Ferrari, P. F. & Palagi, E. (2013a). In play we trust. Rapid facial mimicry predicts the duration of playful interactions in geladas. *PLoS ONE*, 8(6) e66481. http://dx.doi.org/10.1371/journal.pone.0066481.

Mancini, G., Ferrari, P. F. & Palagi, E. (2013b). Rapid facial mimicry in geladas. *Scientific Reports*, 3, 1527-1533.

Marshall, L. J. (1976). *The !Kung of Nyae Nyae*. Cambridge, MA: Harvard University Press.

Martin, P. & Caro, T. M. (1985). On the functions of play and its role in behavioral development. *Advances in the Study of Behavior*, 15, 59–103.

McGrew, W. C. (1972). Interpersonal spacing of preschool children. *The Development of Competence in Early Chilhood*, 72-96.

Messinger, D. S., Mattson, W. I., Mahoor, M. H. & Cohn, J. F. (2012). The eyes have it: Making positive expressions more positive and negative expressions more negative. *Emotion*, 12, 430–436.

Mutschler, T. & Tan, C. L. (2003). *Hapalemur*, bamboo or gentle lemurs. In: Goodman, S. M. & Benstead, J. (eds), *The Natural History of Madagascar*. Chicago: The University of Chicago Press, pp. 1324–1329.

Nakamichi, M. & Koyama, N. (1997). Social relationships among ring-tailed lemurs (*Lemur catta*) in two free-ranging troops at Berenty Reserve, Madagascar. *International Journal of Primatology*, 18, 73-93.

Narvaez, D. & Gleason, T. (2012). Developmental optimization. In: D. Narvaez, J. Panksepp, A. Shore & T. Gleason (eds), *Evolution, Early Experience and Human Development: From Research to Practice and Policy*. New York: Oxford University Press.

Nesse, R. M. (2004). Natural selection and the elusiveness of happiness. *Philosophical Transactions of the Royal Society, B.*, 359, 1333–1347.

Newell, T. G. (1971). Social encounters in two prosimian species: *Galago crassicaudatus* and *Nycticebus coucang*. *Psychon Soc*, 2, 128–130.

Norbeck, E. (1974). Anthropological views of play. *American Zoologist*, 14, 267–273.

Norscia, I. & Palagi, E. (2011). When play is a family business: adult play, hierarchy, and possible stress reduction in common marmosets. *Primates*, 52(2), 101–104.

Norscia, I., Antonacci, D. & Palagi, E. (2009). Mating first, mating more: biological market fluctuation in a wild prosimian. *PLoS ONE*, 4(3), e4679.

Pages, E. (1980). Ethoecology of *Microcebus coquereli* during the dry season. In: P. Charles-Dominique, H. M. Cooper, A. Hladik, *et al.* (eds), *Nocturnal Malagasy Primates: Ecology, Physiology, and Behaviour*. New York: Academic Press, pp. 97–116.

Palagi, E. (2006). Social play in bonobos (*Pan paniscus*) and chimpanzees (*Pan troglodytes*): implications for natural social systems and interindividual relationships. *American Journal of Physical Anthropology*, 129, 418–426.

Palagi, E. (2007). Play at work: revisiting data focusing on chimpanzees (*Pan troglodytes*). *Journal of Anthropological Sciences*, 85, 63–81.

Palagi, E. (2008). Sharing the motivation to play: the use of signals in adult bonobos. *Animal Behaviour*, 75, 887–896.

Palagi, E. (2009). Adult play fighting and potential role of tail signals in ringtailed lemurs (*Lemur catta*). *Journal of Comparative Psychology*, 123, 1–9.

Palagi, E. (2011). Playing at every age: modalities and potential functions in non-human primates. In: Pellegrini, A. D. (ed.), *The Oxford Handbook of the Development of Play*. Oxford: Oxford University Press, pp. 70–82.

Palagi, E. & Cordoni, G. (2012). The right time to happen: play developmental divergence in the two *Pan* species. *PLoS ONE*, 7(12), e52767.

Palagi, E. & Mancini, G. (2011). Playing with the face: Playful facial 'chattering' and signal modulation in a monkey species (*Theropithecus gelada*). *Journal of Comparative Psychology*, 125(1), 11–21.

Palagi, E. & Norscia, I. (2013). Bonobos protect and console friends and kin. *PLoS ONE*, 8(11), e79290. http://dx.doi.org/10.1371/journal.pone.

Palagi, E. & Norscia, I. (2015). The season for peace: reconciliation in a despotic species (*Lemur catta*). *PLoS ONE* 10(11), e0142150. http://dx.doi.org/10.1371/journal.pone.0142150.

Palagi, E. & Paoli, T. (2007). Play in adult bonobos (*Pan paniscus*): modality and potential meaning. *American Journal of Physical Anthropology*, 134, 219–225.

Palagi, E. & Paoli, T. (2008). Social play in Bonobos: Not only an immature matter. In: Furuichi, T. & Thomson, J. (eds), *The Bonobos. Developments in Primatology: Progress and Prospects.* pp. 55–74.

Palagi, E., Gregorace, A. & Borgognini-Tarli, S. (2002). Development of olfactory behavior in captive ring-tailed lemurs (*Lemur catta*). *International Journal of Primatology*, 23, 587–599.

Palagi, E., Telara, S. & Borgognini-Tarli, S. (2003). Sniffing behaviour in *Lemur catta*: seasonality, sex, and rank. *International Journal of Primatology*, 24, 335–350.

Palagi, E., Telara, S. & Borgognini-Tarli, S. (2004a). Reproductive strategies in *Lemur catta*: balance among sending, receiving, and counter-marking scent. *International Journal of Primatology*, 25, 1019–1031.

Palagi, E., Cordoni, G. & Borgognini-Tarli, S. M. (2004b). Immediate and delayed benefits of play behaviour: new evidence from chimpanzees (*Pan troglodytes*). *Ethology*, 110, 949–962.

Palagi, E., Paoli, T. & Tarli, S. B. (2006). Short-term benefits of play behavior and conflict prevention in *Pan paniscus*. *International Journal of Primatology*, 27, 1257–1270.

Palagi, E., Antonacci, D. & Cordoni, G. (2007). Fine-tuning of social play in juvenile lowland gorillas (*Gorilla gorilla gorilla*). *Developmental Psychobiology*, 49, 433–445.

Palagi, E., Norscia, I. & Spada, G. (2014). Relaxed open mouth as a playful signal in wild ring-tailed lemurs. *American Journal of Primatology*, 76, 1074–1083.

Palagi, E., Burghardt, G. M., Smuts, B., *et al.* (2015). Rough-and-tumble play as a window on animal communication. *Biological Reviews*, http://dx.doi.org/10.1111/brv.12172.

Pasztor, T. J., Smith, L. K., MacDonald, N. K., Michener, G. R. & Pellis, S. M. (2001). Sexual and aggressive play fighting of sibling Richardson's ground squirrels. *Aggressive Behavior*, 27(4), 323–337.

Pellegrini, A. D. & Smith, P. K. (1998). Physical activity play. *Child Development*, 69, 577–598.

Pellis, S. M. (1981). A description of social play by the Australian magpie *Gymnorhina tibicen* based on Eshkol-Wachman notation. *Bird Behaviour*, 3, 61–79.

Pellis, S. M. (1988). Agonistic versus amicable targets of attack and defense: Consequences for the origin, function and descriptive classification of play-fighting. *Aggressive Behavior*, 14, 85–104.

Pellis, S. M. (2002). Keeping in touch: Play fighting and social knowledge. In: M. Bekoff, C. Allen & G. M. Burghardt (eds), *The Cognitive Animal: Empirical and Theoretical Perspectives on Animal Cognition*. Cambridge, MA: MIT Press, pp. 421–427.

Pellis, S. M. & Iwaniuk, A. N. (1999). The problem of adult play: A comparative analysis of play and courtship in primates. *Ethology*, 105, 783–806.

Pellis, S. M. & Iwaniuk, A. N. (2000). Adult-adult play in primates: Comparative analyses of its origin, distribution and evolution. *Ethology*, 106, 1083–1104.

Pellis, S. M. & Pellis, V. C. (1987). Play-fighting differs from serious fighting in both target of attack and tactics of fighting in the laboratory rat *Rattus norvegicus*. *Aggressive Behavior*, 13, 227–242.

Pellis, S. M. & Pellis, V. C. (1997). Targets, tactics, and the openmouth face during play fighting in three species of primates. *Aggressive Behavior*, 23, 41–57.

Pellis, S. M. & Pellis, V. C. (2006). Play and the development of social engagement: A comparative perspective. In: P. J. Marshall & N. A. Fox (eds), *The Development of Social Engagement: Neurobiological Perspectives*. Oxford: Oxford University Press, pp. 247–274.

Pellis, S. M. & Pellis, V. C. (2007). Rough-and-tumble play and the development of the social brain. *Current Directions in Psychological Science*, 16, 95–98.

Pellis, S. M. & Pellis, V. C. (2009). *The Playful Brain: Venturing to the Limits of Neuroscience*. Oxford: Oneworld Publications.

Pellis, S. M. & Pellis, V. (2010). Social play, social grooming, and the regulation of social relationships. In: A. V. Kalueff, J. L. La Porte & C. L. Bergner (eds), *Neurobiology of Grooming Behaviour*. Cambridge University Press, pp. 66–87.

Pellis, S. M. & Pellis, V. C. (2011). To whom the play signal is directed: a study of headshaking in black-handed spider monkeys (*Ateles geoffroyi*). *Journal of Comparative Psychology*, 125, 1–10.

Pellis, S. M., Pellis, V. C. & Whishaw, I. Q. (1992). The role of the cortex in play fighting by rats: Developmental and evolutionary implications. *Brain, Behavior & Evolution*, 39, 270–284.

Pellis, S. M., Pellis, V. C. & McKenna, M. M. (1993). Some subordinates are more equal than others: Play fighting amongst adult subordinate male rats. *Aggressive Behavior*, 19, 385–393.

Pellis, S. M., Pasztor, T. J., Pellis, V. C. & Dewsbury, D. A. (2000). The organization of play fighting in the grasshopper mouse (*Onychomys leucogaster*): Mixing predatory and sociosexual targets and tactics. *Aggressive Behavior*, 26, 319–334.

Pellis, S. M., Hastings, E., Shimizu, T., *et al.* (2006). The effects of orbital frontal cortex damage on the modulation of defensive responses by rats in playful and non-playful social contexts. *Behavioral Neuroscience*, 120, 72–84.

Pellis, S. M., Pellis, V. C. & Bell, H. C. (2010a). The function of play in the development of the social brain. *American Journal of Play*, 2, 278–296.

Pellis, S. M., Pellis, V. C. & Reinhart, C. J. (2010b). The evolution of social play. In: C. Worthman, P. Plotsky, D. Schechter & C. Cummings (eds), *Formative Experiences: The Interaction of Caregiving, Culture, and Developmental Psychobiology*. Cambridge University Press, pp. 404–431.

Pellis, S. M., Pellis, V. C., Reinhart, C. J. & Thierry, B. (2011). The use of the bared-teeth display during play fighting in Tonkean macaques (*Macaca tonkeana*): Sometimes it is all about oneself. *Journal of Comparative Psychology*, 125(4), 393–403.

Pereira, M. E. & Kappeler, P. M. (1997). Divergent systems of agonistic behaviour in lemurid primates. *Behaviour*, 134(3), 225–274.

Pessoa, L. (2009). *Nature Reviews Neuroscience*, 9, 148–158.

Petit, O., Bertrand, F. & Thierry, B. (2008). Social play in crested and Japanese macaques: Testing the covariation hypothesis. *Developmental Psychobiology*, 50, 399–407.

Petrů, M., Špinka, M., Charvátová, V. & Lhota, S. (2009). Revisiting play elements and self-handicapping in play: a comparative ethogram of five Old World monkey species. *Journal of Comparative Psychology*, 123, 250–263.

Pierce, J. & Bekoff, M. (2012). Wild justice redux: What we know about social justice in animals and why it matters. *Social Justice Research* (special issue edited by Sarah Brosnan), 25, 122–139.

Power, T. G. (2000). *Play and Exploration in Children and Humans*. Mahwah, NJ: Lawrence Erlbaum Associates.

Preuschoft, S. & van Hooff, J. A. (1997). The social function of 'smile' and 'laughter': Variations across primate species and societies. In: U. Segerstrale & P. Mobias (eds), *Nonverbal Communication: Where Nature Meets Culture*. New Jersey: Erlbaum, pp. 252–281.

Provine, R. R. (1996). Laughter. *American Scientist*, 84, 38–45.

Provine, R. R. (2004). Laughing, tickling, and the evolution of speech and self. *Current Directions in Psycological Sciences*, 13, 215–218.

Provine, R. P. & Fischer, K. R. (1989). Laughing, smiling, and talking: Relation to sleeping and social context in humans. *Ethology*, 83, 295–305.

Reinhart, C. J., Pellis, V. C., Thierry, B., *et al.* (2010). Targets and tactics of play fighting: Competitive *versus* cooperative styles of play in Japanese and Tonkean macaques. *Journal of Comparative Psychology*, 23, 166–200.

Richard, A. F. (1992). Aggressive competition between males, female-controlled polygyny and sexual monomorphism in a Malagasy primate, *Propithecus verreauxi*. *Journal of Human Evolution*, 22, 395–406.

Ross, K. M., Bard, K. A. & Matsuzawa, T. (2014). Playful expressions of one-year-old chimpanzee infants in social and solitary play contexts. *Frontiers in Psychology*, 5, 741. http//dx.doi.org/10.3389/fpsyg.2014.00741.

Sclafani, V., Norscia, I., Antonacci, D. & Palagi, E. (2012). Scratching around mating: factors affecting anxiety in wild *Lemur catta*. *Primates*, 53, 247–254.

Silk, J. B. (1997). The function of peaceful postconflict contacts among primates. *Primates*, 38, 265–279.

Smith, P. K. (1982). Does play matter? Functional and evolutionary aspects of animal and human play. *The Behavioural and Brain Sciences*, 5, 139–184.

Soma, T. & Koyama, N. (2013). Eviction and troop reconstruction in a single matriline of ring-tailed lemurs (*Lemur catta*): what happened when 'grandmother' died? In: Masters, J., Gamba, M. & Génin, F. (eds), *Leaping Ahead – Advances in Prosimian Biology*. New York: Springer, pp. 137–146.

Špinka, M., Newberry, R. C. & Bekoff, M. (2001). Mammalian play: training for the unexpected. *Quarterly Review of Biology*, 76, 141–168.

Srofe, L. A. & Waters, E. (1976). The ontogenesis of smiling and laughter: A perspective on the organization of development in infancy. *Psychological Review*, 83(3), 173–189.

Sussman, R. W. & Richard, A. F. (1974). The role of aggression among diurnal prosimians. In: Holloway, R. L. (ed.), *Primate Aggression, Territoriality, and Xenophobia*. San Francisco: Academic Press, pp. 50–76.

Sutton-Smith, B. & Roberts, J. M. (1970). Cross-cultural and psychological study of games. In: G. Lüschen (ed.), *Cross-Cultural Analysis of Sport and Games*. Champaign, Illinois: Stipes, pp. 100–108.

Swaddling, J. (2002). *The Ancient Olympic Games*. Austin: Texas University Press.

Tan, J. & Hare, B. (2013). Bonobos share with strangers. *PLoS ONE*, 8, e51922. http://dx.doi.org/10.1371/journal.pone.0051922.

Tanner, J. E. & Byrne, R. W. (1993). Concealing facial evidence of mood: evidence for perspective-taking in a captive gorilla? *Primates*, 34, 451–456.

Thierry, B. (2000). Covariation of conflict management patterns across macaque species. In: Aureli, F. & de Waal, F. B. M. (eds), *Natural Conflict Resolution*, Berkeley, CA: University of California Press, pp. 106–128.

Thierry, B. (2007). Unity in diversity: Lessons from macaque societies. *Evolutionary Anthropology*, 16, 224–238.

Thierry, B., Demaria, C., Preuschoft, S. & Desportes, C. (1989). Structural convergence between silent bared-teeth display and relaxed open-mouth display in the Tonkean macaque (*Macaca tonkeana*). *Folia Primatologica*, 52, 178–184.

Thompson, K. V. (1998). Self assessment in juvenile play. In: Bekoff, M. & Byers, J. A. (eds), *Animal Play – Evolutionary, Comparative and Ecological Perspectives*. Cambridge University Press, pp. 183–204.

Tinbergen, N. (1951). *The Study of Instinct*. New York: Oxford University Press.

Tinbergen, N. (1952). 'Derived' activities, their causation, biological significance and emancipation during evolution. *The Quarterly Review of Biology*, 27, 1–32.

Tinbergen, N. (1963). On aims and methods of ethology. *Zeitschrift für Tierpsychologie*, 20, 410–433.

van Hooff, J. A. R. A. M. & Preuschoft, S. (2003). Laughter and smiling: the intertwining of nature and culture. In: de Waal, F. B. M. & Tyack, P. L. (eds), *Animal Social Complexity*. Cambridge, MA: Harvard University Press, pp. 260–287.

Vettin, J. & Todt, D. (2005). Human laughter, social play, and play vocalizations of non-human primates: An evolutionary approach. *Behaviour*, 142(2), 217–240.

Voland, E. (1977). Social play behavior of the common marmoset (*Callithrix jacchus* Erxl., 1777) in captivity. *Primates*, 18, 883–901.

Wendland, J. R., Lesch, K-P., Newman, T. K., et al. (2006). Differential variability of serotonin transporter and monoamine oxidase A genes in macaque species displaying contrasting levels of aggression-related behavior. *Behavior Genetics*, 36, 163–172.

Wilson, E. O. (2002). *Sociobiology: the new synthesis*. Cambridge, MA: Harvard University Press.

Wilson, D. E. & Reeder, D. M. (eds), (2005). *Mammal Species of the World: a taxonomic and geographic reference* (vol. 12). JHU Press.

Wrangham, R. W. (1999). Evolution of coalitionary killing. *American Journal of Physical Anthropology*, (Suppl 29), 1–30.

Wrangham, R. W. & Glowacki, L. (2012). Intergroup aggression in chimpanzees and war in nomadic hunter-gatherers: evaluating the chimpanzee model. *Human Nature*, 23, 5–29.

8 Sex is not on discount: mating market and lemurs

If we understand particular exchanges as being the result of a particular social system...it may become clear why exchange is a powerful metaphor for social relationships and their transformation.

Sitta von Reden, 1998

8.1 Biological (or pseudo-cultural?) market

It is well known that both Darwin and Wallace were influenced by the work of the economist and demographist Thomas Malthus (*An Essay on the Principle of Population*, published in 1797) while elaborating their theories on evolution. Probably, the mechanisms underlying the economic system are the outcome of an evolutionary process, which has acted on the biological system and reflects on our behaviour. The analogy between economical and biological systems is not limited to this example. In biology, as well as in economics and politics, power is a key concept for understanding asymmetric relationships between subjects (Dahl, 1957). Distributive power, derived from Russell's definition of social power (1938), focuses upon who has got the power over whom. It can originate from coercion, when an individual can force another to do something, or from leverage, when the use of force is not possible. Specifically, coercive power depends on fear, suppression of free choice and/or use of threat, for its existence. It is normally based on an imbalance of social, psychological and physical power between individuals. Examples of this kind of power are theft with violence to obtain valuable goods (Matsueda *et al.*, 2006), bullying to aggressively impose domination over others (Rayner and Hoel, 1997) and rape (Thornhill and Thornhill, 1983; Emery-Thompson, 2009; Muller and Wrangham, 2009). Human, gorilla, chimpanzee and orangutan males perform sexual coercion over females to obtain forced copulations (Goetz and Shackelford, 2006; Muller and Wrangham, 2009). Wild orangutans provide one of the most dramatic examples of this phenomenon. It has been reported that in this species 'most copulations by subadult males and nearly half of all copulations by adult males occur after the female's fierce resistance has been overcome through violent restraint' (Smuts and Smuts, 1993, p. 6). Sexual coercion is also present in many other animals spanning invertebrates and mammals. It is worth mentioning, among others, that scorpion flies (Thornhill, 1980), guppies (Head and Brooks, 2006), salamanders (Ethan *et al.*, 2006), snakes (Shine and Mason, 2005), mallards (Barash, 1977), bottlenose dolphins (Scott *et al.*, 2005), and a huge variety of non-human primates (Smuts and Smuts, 1993) force their mates into sex.

Different from coercive power, leverage comes in place whenever an individual possesses something that the other needs but cannot acquire through coercion (Lewis, 2002). In this case, trading becomes essential for mutually beneficial interactions within social groups, in economical, political and biological markets (Noë and Hammerstein, 1994, 1995). In an illegal system, doing a favour in exchange of votes – which in a democracy depend on individual choice (expressed in the polling booth) – is an example of the leverage possessed by voters over someone who wants to gain a political position. In different countries, the exchange of academic positions often involving nepotism is another example of the use of leverage (e.g. see the situation in Italy, Allesina, 2011).

Recent studies (e.g. Clarke *et al.*, 2010) have pointed out that trading can be also present when coercion is possible, thus questioning *de facto* the classical Manichean division between coercive power and leverage. Indeed, unless power asymmetry is so strong as to ensure a totalitarian control over a resource, the ability of individuals to use coercion does not preclude their need of reducing competition or gaining partner cooperation especially in a social setting.

An important feature of market models is that the expected future gains are actively influenced by playing off potential partners against each other (Noë *et al.*, 1991; Noë, 2001). The typical game theory approach includes only two players and, although this is changing within economics as well as biology, the classical models do not take into account partner choice (Noë and Hammerstein, 1995). In contrast, the biological market theory includes multi-player models, that is theoretical games with at least three or more 'players' (traders, in the market systems) (Nunn and Lewis, 2001). Two or more classes of traders (sex classes, rank classes, etc.) exchange commodities in biological markets to their mutual benefit. As it occurs in the economic market, the presence of multiple players influences the intrinsic value of the commodity exchanged which also depends on the proportion of players wanting it (demand) and those providing it (supply) (Noë and Hammerstein, 1994). Fruteau *et al.* (2009) showed that wild vervet monkeys offered a larger amount of grooming to the individual (first provider) that had been granted access to food secured in a container and could provided it to the rest of the group. Interestingly, the amount of grooming decreased when another individual (second provider) was granted access to the secured food, thus confirming that when the supply of a commodity increases, the value of such commodity decreases and the providers are paid less for that commodity, in this case using grooming as a currency.

After all, if Leonardo da Vinci had serially painted a thousand copies of La Gioconda, its value on the market would have plummeted because its originality would have been lost. Dozens of copies of da Vinci's masterpiece were reproduced in the sixteenth and seventeenth centuries but they are what they are: replicas.

Different group members can offer different kinds of commodities in exchange for alternative ones that they do not possess (Noë and Hammerstein, 1995). Usually, competition acts as the driving force within the same trader class which includes all members offering the same kind of commodity. On the other hand, cooperation can occur between different trader classes (Noë and Hammerstein, 1995; McNamara *et al.*, 2008). For example, many candidates to the same position (same class) in a company compete for the same job, trading their expertise for a salary, which is provided by the employer (other class). A group of oestrus females all offering to males the possibility to reproduce belong to the same class; whereas, males ready to offer services to 'conquer' them (e.g. invitation out for a dinner, roses, jewels, protection, grooming) belong to another trader class. Of course, the individuals belonging to the different classes have to cooperate and reach an agreement over the service exchange. In the case of the candidates (many) for the same job position, the leverage is totally skewed towards the employer (one) who can dictate conditions.

In the mating market, the balance of power tilts in favour of females whenever males cannot force females into mating, as it happens for example in sexually monomorphic species or when females form coalitions (Lewis, 2002). Consequently, males depend on females for breeding opportunities and must compete to prove their superiority to females, thus increasing their possibility to be selected (Lewis, 2002; Wong and Candolin, 2005).

Males can engage in both contest competition via physical or ritualised fighting and outbidding competition, in which a male plays off rivals by making a better offer (Noë and Hammerstein, 1995). In the latter case, males can secure the favours of a female by being more generous than others in providing a commodity in exchange for female access (competitive altruism) or by advertising their quality (e.g. the dominance status) through visual, acoustic and olfactory displays (Blaustein, 1981; Coleman *et al.*, 2004) which can be proposed singularly or combined in a multimodal advertisement (Jones and van Cantfort, 2007). The famous courtship dances of the golden-collared manakin males are one of the most elucidating examples of the art of impressing females and not going unnoticed (Fusani and Schlinger, 2012). During the mating period, they dance and snap their wings together behind their backs to win the attention of females. A recent study showed that the females prefer the males that perform the elements of the dance faster and demonstrate better motor coordination (Barske *et al.*, 2014).

Male nursery web spiders which bring their intended mate a nutritious gift in the form of an insect wrapped in silk have more reproductive success than their counterparts which show up empty-handed. Female spiders prefer the sperm of gift-giving males, which mate for almost ten times longer than empty-handed males and a larger proportion of their sired eggs successfully hatch into spider-hatchlings (Albo *et al.*, 2013).

In short, the ingredients to cook up the demonstration of the presence of the biological market are the following: (1) recognising the presence of desired commodities and/or services, (2) identifying the different trader classes, and lastly (3) assessing that the value of commodities or services varies according to the supply/demand rule (the crucial and the most difficult point to be verified) (Werner *et al.*, 2014).

8.2 Grooming, the coin of the primate realm

Grooming is a crucial behaviour in the primate world (Figures 8.1, 8.2, 8.3, 8.4 and 8.5). In primates, as in other mammals, grooming is used for ectoparasite and debris removal (Hutchins and Barash, 1976; Reichard and Sommer, 1994; Mooring *et al.*, 1996; Hawlena *et al.*, 2007) and anxiety and tension reduction through the release of beta-endorphins (Schino *et al.*, 1988; Keverne *et al.*, 1989). The importance of grooming in the primate social network has been highlighted via a meta-analysis performed on 48 social groups belonging to 22 different species and 12 genera of primates, which showed that in females grooming is proportionally reciprocated (Schino and Aureli, 2008). Indeed, grooming is extensively exchanged in primates, which use it as a social glue to start, consolidate or repair relationships (Aureli *et al.* 1989). However, grooming comes also with some costs, such as decreased vigilance towards potential predators and competitors and the loss of time that could be dedicated to other activities, such as foraging (Maestripieri, 1993; Cords, 1995).

Due to the associated costs for the groomer and the benefits for the groomee, grooming evolution and maintenance has traditionally been explained by kin selection and reciprocal altruism theories (Kapsalis and Berman, 1996; Brosnan and de

Figure 8.1 Grooming session in *Macaca sylvanus*. NaturZoo Rheine, Germany. Photo: Elisabetta Palagi.

Figure 8.2 Grooming in *Lemur catta* at Faunia Park, Madrid, Spain. Photo: Ivan Norscia.

Waal, 2002). In particular, the kin selection theory has been used to explain the disproportionate amount of grooming among related compared to unrelated individuals (Hamilton, 1964; Seyfarth and Cheney, 1984). However, kin selection does not adequately explain grooming in non-kin dyads. Trivers (1971) proposed that grooming among non-kin individuals represents a form of reciprocal altruism. This hypothesis assumes that altruistic behaviour is favoured if individuals benefit from the reciprocal interaction, and is employed for the maintenance of social bonds and coalitionary support (Seyfarth, 1977; Hemelrijk, 1994). The biological market framework expands this view and emphasises that the balance between providing and receiving grooming fluctuates following the ever changing economic forces, such as the demand/supply ratio (Henzi and Barrett, 2002; Tiddi *et al.*, 2010). In this perspective, grooming is viewed as a service to be exchanged for itself or to be interchanged with other commodities or services, such as tolerance over food, access to infants, agonistic support or breeding opportunities (Noë *et al.* 1991; Noë and Hammerstein 1994, 1995; Barrett and Henzi 2001; Noë, 2001, 2006; Ventura *et al.*, 2006; Balasubramaniam *et al.*, 2011).

For example, grooming is a service that is traded among females within groups of grey-cheeked mangabeys (*Lophocebus albigena*). In this species, Chancellor and Isbell (2009) observed that within reciprocated bouts, the duration of time spent grooming by the recipient was positively and significantly correlated with the duration of time spent grooming by the initiator (time matching). Besides, the highest percentages of grooming reciprocity were found to be present in groups characterised by the lowest rates of female agonism, thus indicating that local markets within the same population may set different values on commodities or services such as grooming, depending on the social and ecological environment (Chancellor and Isbell, 2009).

In wild chimpanzees of the Budongo Forest, it was found that some grooming was directed by lower- to higher-ranked individuals and that, on average, higher-ranked individuals groomed each other more reciprocally. Instead, individuals close in rank did not show higher levels of reciprocity (Newton-Fisher and Lee, 2011). These mixed results, not fully confirming the presence of a grooming market, were connected by the authors to the delay observed between action (first grooming) and reciprocation. Such delay introduces inflation into a biological market through potential devaluing of the unit of 'resource' (in this case, grooming effort). In fact, if the 'investor' can accumulate proximate benefits from other social partners (providing grooming), the value of subsequent interactions diminishes.

In a wild group of chimpanzees of Côte d'Ivoire, grooming was found to be reciprocated more symmetrically when measured on a long-term, rather than on an immediate or short-term basis, with dyads grooming an average of once every seven days. These results suggested that chimpanzees, similar to humans, are able to keep track of past social interactions, at least for a one-week period, and balance services over repeated encounters (Gomes *et al.*, 2009).

In captive chimpanzees (Figure 8.3), de Waal (1997) compared the success rate of each adult, A, to obtain food from another adult, B, with grooming interactions between A and B in the two hours prior to each food trial. The tendency of B to share with A was higher if A had groomed B than if A had not done so.

Figure 8.3 Grooming between chimpanzees (*Pan troglodytes*) at Krefeld Zoo, Germany. Photo: Elisabetta Palagi.

Figure 8.4 Grooming between capuchin monkeys (*Cebus apella* or *Sapajus apella*) at Faunia Park, Madrid, Spain. Photo: Ivan Norscia.

The exchange was partner-specific, that is, the effect of previous grooming on the behaviour of food possessors was limited to the grooming partner. Interestingly, the effect of grooming was greatest for pairs of adults who rarely groomed, probably because of the higher value of the service exchanged (rarer).

Tiddi and coworkers (2012) demonstrated that interindividual variation in the distribution of grooming among wild female tufted capuchin monkeys (*Cebus apella nigritus*; Figure 8.4) is affected by both dominance rank and the potential for exchanges of grooming for rank-related benefits. In particular, capuchin females derived rank-related benefits from grooming higher-ranking females in terms of increased tolerance during feeding. Indeed, dominant females preferentially tolerated those females that groomed them most. Moreover, capuchin females preferentially directed their grooming up the hierarchy and appeared to compete for access to higher-ranking females as preferred grooming partners.

In different primate species it has been observed that the fewer infants in a group, the higher the price that has to be paid in the form of grooming to gain the mother's permission to inspect and handle it ('baby market'). This was first described for chacma baboons, but similar results have been found in other primate species, such as sooty mangabeys (Fruteau *et al.*, 2011), long-tailed macaques (Gumert, 2007), spider monkeys (Schaffner and Aureli, 2005; Slater *et al.*, 2007) and golden snub-nosed monkeys (Wei *et al.*, 2013). In this species (*Rhinopithecus roxellana*) the authors observed that grooming for infant access was more likely to be initiated by potential handlers (without infants) and less likely to be reciprocated by mothers. Moreover, frequent grooming partners were allowed to handle, and maintained access to, infants longer than infrequent groomers. In support of the supply and demand market effect, the authors found that grooming bout duration was inversely related to the number of infants per female present in each one-male unit. Besides, with increasing infant age, the duration of grooming provided by handlers was shorter, thus suggesting that the 'value' of older infants was lower.

As specified above, the demonstration of the fluctuation of the service value according to the supply and demand market effect is crucial to assess the presence of a biological market. For example, contrary to snub-nose monkeys, Tiddi *et al.* (2010) excluded a market effect in infant handling in tufted capuchin monkeys because, even if the probability of the potential handlers to gain access to infants increased when they groomed the mothers, the number of infants in the group did not affect the amount of grooming needed for access to infants.

If money can be used by humans to obtain 'copulating opportunities' other mammals (not just primates) can use grooming to increase mating opportunities. Stopka and MacDonald (1999) found, for example, that females of *Apodemus sylvaticus* (a mouse species characterised by a promiscuous mating system without any paternal investment) require grooming before allowing a male to progress towards sex. The same authors hypothesised that females could obtain grooming through a process of 'unintentional bargaining': in this species, grooming was the only commodity which males had been seen to provide in the process of mate selection.

We can find another good example of grooming exchanged for sex in a lesser ape of Thailand's forests. Here, it was observed that male gibbons (*Hylobates lar*) groom females more than vice versa and females copulate more frequently on days when they received more grooming (Barelli *et al.*, 2011). Gumert (2007), in a macaque species (*Macaca fascicularis*), found that males groomed females at higher rates when females were sexually receptive. In fact, most of the grooming performed by males was concurrent with mating, inspection or presentation of female hindquarters. Moreover, females appeared to engage in higher levels of sexual activity with the most active groomers. Low ranking males had to groom females longer than high ranking ones to obtain mating opportunities, especially if the female occupied a high-ranking position. It is clear then that dominance hierarchy can skew leverage towards certain members of the group and, therefore, the value of the grooming service changes depending on who provides and/or receives the service.

The eggs ready to be fertilised of dominant females are the most wanted by males because the offspring will probably have many advantages, such as high-quality food, food access priority, and protection. Thus, the males siring those eggs will probably gain higher fitness. In support of the biological market interpretation, Gumert (2007) also found that the amount of grooming a male performed on a female during grooming-mating interchanges was related to the current supply of females around the interaction.

Clarke *et al.* (2010) found elements of cooperation between males and females (trader classes) in chacma baboons (*Papio ursinus*), a subspecies in which males can force females into mating. They found the grooming exchanged by males and females to be associated with the possibility that females spontaneously initiate copulations. The authors interpreted this form of cooperation by suggesting that female compliance has a 'market value' because the differences in the power between males

Figure 8.5 Grooming in olive baboons (*Papio anubis*). Gombe Stream National Park, Tanzania. Photo: Ivan Norscia.

are small and do not allow anyone to have the total control over females. Owing to the small power asymmetry, males are able to out-compete each other and lose part of their coercive power. These findings introduce another variable triggering intersexual cooperation that is the possibility to suppress competition more than just the opportunity to control a resource.

Box 8.1 | **by Karen B. Strier**

Speaking of which: reflections on primate sexual behaviour with special reference to muriquis

It would be almost impossible to study the northern muriqui (*Brachyteles hypoxanthus*)[1] and not become intrigued by its sexual behaviour. For starters, males and females are sexually monomorphic in body size and canine size, so the ability of males to harass or threaten females should be limited compared to many other anthropoids where males are up to twice as large as females and

Box 8.1 (continued)

---

[1] Previously, the muriqui was considered to be a monotypic genus, *Brachyteles arachnoides*, but more recently, it has been split into two species, the southern muriqui, *B. arachnoides*, and the northern muriqui, *B. hypoxanthus*. Although I have observed the same population and many of the same individuals since 1982, they are referenced as *B. arachnoides* and then *B. hypoxanthus* in my publications to reflect contemporary taxonomic classifications.

Box 8.1 (*cont.*)

pose real threats to them and their infants. Male northern muriquis also have extremely large testes relative to their body mass, a trait that has been widely associated with sperm competition in primates and other animals. According to theories of sperm competition, if males cannot monopolise sexual access to independent females, then their best alternative means of competing might be at the level of their sperm. Males with more copious sperm, or faster or more viable sperm, should have a competitive advantage over other males that copulate with the same receptive female. As a result, sperm competition has been associated with low levels of aggressive competition among males, as well as between males and females. Indeed, the northern muriqui's uniquely egalitarian society, with its high tolerance and close affiliations among philopatric males, makes more sense in the context of sperm competition (reviewed in Rosenberger and Strier, 1989).

Yet despite all of the theory and logic, nothing could have prepared me for the first copulation I observed with the muriquis more than 30 years ago. I had no comparative precedence for the leisurely manner in which muriquis mated, often in full view of one another, or for the relaxed demeanour of males as they waited for their turn to copulate in close succession with the same receptive female. At the time, just a few months into my 14-month dissertation research, I was still just beginning to decipher the peaceful nature of northern muriqui society, and I had not yet discovered that males remained in their natal groups for life while the females typically dispersed. It took many more months to put the pieces of this part of the puzzle together, and it was decades before my colleagues and I could confirm with genetic paternity data that the level of male reproductive skew was as low as the muriquis' behaviour had implied (Strier *et al.*, 2011).

Two corollaries of the genetic paternity study continue to intrigue me because of their potential implications, not only for muriquis but for other primates as well. One was the unexpectedly high grandmaternity skew we discovered by looking at the mothers of males whose paternity we examined. Although the male with the highest paternity sired 18.2% of the offspring, only four females (of 13 females) accounted for 72.8% of the grandoffspring sired by from one to three of their respective sons. Could the slow life histories and long lifespans of northern muriquis, which are estimated to live into their late thirties if not longer, also be related to their non-aggression, with both mothers and sons benefitting from their life-long associations? How are paternity and grandmaternity related in other male philopatric species, such as chimpanzees and bonobos?

A second question that intrigues me is how much of the muriquis' tolerant behaviour and low reproductive skew will persist as demographic conditions in their population change? Since the onset of my study in 1982, the population at the Reserva Particular do Patrimônio Natural – Feliciano Miguel Abdala (RPPN – FMA; formerly, the Biological Station of Caratinga), in Caratinga, Minas Gerais, Brazil has increased nearly 7-fold, from about 50 to nearly 350 individuals (as of September 2014). What were previously two mixed-sex groups have split into four, and their once cohesive groups are now largely fission–fusion (Strier *et al.*, 1993, 2006). Habitat saturation in this small (< 1000 ha) forest fragment is most likely responsible for the muriquis expansion of their vertical niche to include the ground (Tabacow *et al.*, 2009), which in turn may be related to an unexpected increase in their fertility (Strier and Ives, 2012). Nonetheless, despite this evidence of high levels of behavioural flexibility in response to demographic changes, low rates of aggression and high tolerance among males have persisted (Strier, 2011). Would even more extreme demographic fluctuations than have occurred so far ultimately lead to increasing levels of aggression, or is the muriqui's peaceful egalitarianism – and the morphological adaptations associated with it – sufficiently resilient that it will not change?

This particular suite of morphological and behavioural traits is unique to the muriquis, but it gives us insights into a much broader range of primate behavioural diversity than was previously thought to exist. The muriquis also challenge us to rethink long-standing assumptions about the prevalence of male-dominated societies and about the ways in both males and females negotiate their relationships with one another when overt competition can be avoided (Strier, 1994).

As with nearly everything we have learned about this critically endangered species, understanding muriqui sexual behaviour is important for successful conservation management programmes. Whether these programmes involve translocating solitary females to help increase gene flow among isolated populations, or captive breeding efforts, our success will depend on how well we can incorporate naturalistic behaviour patterns into our designs. After more than 30 years of studying muriquis, it is the synergy that has emerged at this intersection between science and conservation that I consider to be the most important.

8.3 Male election by females: lemurs as a model

Here we are again. Lemurs. Little attention has been given, in literature, to grooming and biological markets of social strepsirrhines, which have been kept grooming each other for a long time, and possibly longer than other primate groups.

Port *et al.* (2009) showed that grooming in wild *Eulemur fulvus rufus* from the Kirindy forest, does represent a service exchanged in a market place. The authors started from the basic assumption that, following the market theory, grooming is predicted to be approximately reciprocated within a dyad when no other services are being exchanged, but it should be more asymmetrical if partners have different quantities of other services to offer. The authors observed that grooming was highly reciprocated (with partners alternating in the roles of groomer and groomee within the same session) but there were asymmetries in the duration of grooming given and received. In fact, the degree of reciprocity was offset by power differential between individuals. Specifically, more grooming was directed from the low-ranking individual towards the high-ranking individual than the other way around. Once more, being the leader makes the difference in the value of the service exchanges. The grooming provided by a dominant is worthier than the grooming of a subordinate, therefore, dominants can afford to provide less grooming (and receive more!) than other group members. In brown lemur males the difference in grooming duration became more pronounced as the number of subordinates increased. With more suppliers of grooming packages, the unitary cost of the service decreases and more units (grooming packages) must be offered. Dominants can pay back subordinates with tolerance. This is probably why exchanging grooming with a dominant is more valuable to the subordinates to the point that they accept grooming more to be groomed less. Indeed, it was observed that aggression occurred at high frequencies between classes of individuals that were characterised by non-reciprocal grooming.

Also in *Propithecus verreauxi*, from the same study site as *Eulemur fulvus*, some hints suggest that grooming has a market value (Figure 8.6). Lewis (2010) found that unidirectional grooming (not immediately reciprocated) was performed by low ranking individuals at higher rates than high ranking individuals. Moreover, males initiated and performed unidirectional grooming at significantly higher rates than females which, in this species, are at the top of the hierarchy (Richard and Nicoll, 1987; Palagi *et al.*, 2008; Lewis, 2010). Interestingly, an analysis of the immigrants' grooming behaviour suggested that immigrant sifaka use grooming as a means of negotiating their way into new social groups (Lewis, 2008). In fact, in the same

Figure 8.6 Grooming session in *Propithecus verreauxi*. Berenty forest, Madagascar. Photo: Ivan Norscia.

sifaka population, immigrant subadult males devoted more time towards grooming the dominant males than any other resident of the group. An immigrant subadult female entering a new small group, instead, spent more time grooming the resident female than the resident male which, instead, groomed the new female at a 20% higher rate than he groomed the resident female. Finally, resident females groomed immigrant males nearly four times more than they groomed resident males. These results show that grooming is probably used in a manipulative way to enter new groups, being this service preferentially offered by the would-be residents to 'win over' their competitors (male to male, female to female, within the same trader class) and by residents to future collaborative (mating) partners (female to male and male to female). Intriguingly, in another population of the same species (Berenty, south Madagascar), during the mating period, play instead of grooming was used to reduce male–male competition, whereas the role of grooming was marginalised to intragroup interactions (Antonacci et al., 2010), probably because extra-group males were not trying to make their way into the new group. They were trying, instead, to make their way to females by increasing resident males' tolerance. Once reaching the females, as it will be discussed later on, grooming became an essential currency to be spent to buy females' propensity to mate (Norscia et al., 2009).

By studying the same population of *Propithecus verreauxi* in the secondary forest of Ankoba at Berenty, Norscia et al. (2009) demonstrated, for the first time, the existence of a mating market, involving grooming, in strepsirrhines. The sifaka, as few other gregarious lemurs, combines a powerful olfactory system and puzzling features like group living, female priority over resources and absence of sexual dimorphism (Wright, 1999). Such combination of features makes this species a good model to understand the biological bases of the mating market in primates. In social groups we can find more than two players, with the different classes of traders being obviously represented by males and females of the group. Because females equal males in size (or are even larger), and probably because of other proximate factors (e.g. higher female aggression propensity), females cannot be accessed by force. Therefore, females are the most choosing sex and in order for males to have access to them and their eggs (valuable, inalienable commodities) trading becomes essential. Yet, males cannot use food as an exchange commodity because females have unquestioned priority access to the best feeding spots (Norscia et al., 2006). Grooming, instead, remains a service that a male can offer to win over females, once the other male competitors have been defeated or outbid.

In the Berenty forest, researchers tried to understand how males competed for females and what were the main criteria used by females to select mate candidates. Apparently, males undergo a 'two-step' procedure to obtain mating opportunities: the first serves to gain priority and the second to increase the number of copulations. In their study, Norscia et al. (2009) reported that females could mate with multiple males during the receptive days. After ranking males according to their priority access to a female, the authors found that male priority rank correlated with the frequency of male scent marks applied on the spots previously marked by females (countermarking) and did not correlate with the proportion of fights won by males when females were present. This result is not surprising. In fact, although aggressive interactions are a widespread form of competitive strategy adopted by males to gain female access (Qvarnström and Forsgren, 1998), the sifaka society is characterised by female dominance and philopatry and male fighting ability is unimportant in female mate choice (Richard, 1992). Winning a fight does not necessarily confer sexual access to males. In fact, females can base their mate choice on other features (e.g., age, time spent in the group, male physiological status, etc.) especially in those species in which females can acquire a dominant or co-dominant status, individually or by forming coalitions (East and Hofer, 2001; Paoli and Palagi, 2008; Rasmussen et al., 2008).

In scent-oriented species, such as the sifaka, male competition for females can be translated into an olfactory tournament (outbidding competition) more than into an arena of aggressive encounters (contest competition) (Blaustein, 1981; Hurst and Beynon, 2004; Wong and Candolin, 2005; Heymann, 2006; Gosling and Roberts, 2001). Scent marks provide a reliable signal of competitive ability (Hurst and Beynon, 2004; Jordan, 2007; Rasmussen et al., 2008). Consistently, sifaka males competed for females by countermarking female odour depositions: in the end, the most active males got the green light and could mate first. As a matter of fact, as extensively explained in Chapter 3, sifaka males can use scent marking as a form of self-advertisement for mating purposes because odour signals can convey information on dominance status, which is one of the main choice criteria adopted by females (Qvarnström and Forsgren, 1998; Lazaro-Perea et al., 1999; Lewis, 2005). The importance of olfactory male competition in female mate choice has been provided for non-primate species (Hurst and Beynon, 2004) as well as for primate ones, including New World monkeys (Lazaro-Perea et al., 1999) and strepsirrhines. In particular, females of *Nycticebus pygmaeus* (a nocturnal strepsirrhine) rely on olfactory deposition frequency to select mating partners (Fisher et al., 2003). Moreover, during the pre-mating period *Lemur catta* males compete for female access via ritualised stink fights and females increase their tolerance towards males based on the outcome of such fights (Jolly, 1972; Palagi et al., 2004).

Coming back to the sifaka, Norscia et al. (2009) found that the frequency of grooming directed by males to females did not correlate with mating priority. Offering grooming did not help males to mate with a 'wanted' female before others. Hence, as described above, only the olfactory tournament appeared functional in that respect.

Nonetheless, grooming becomes important when the males want to increase the number of copulations with a certain female. As a matter of fact, Norscia et al. (2009) found that the mating frequency correlated neither with the proportion of fights won by males in presence of females nor with the frequency of male countermarking on female depositions. The mating frequency correlated, instead, with the frequency of grooming directed by males to females. It is important to remark that, in the pre-mating period grooming performed by males to females positively correlated with grooming performed by females to males (grooming market), whereas in the mating period, as previously stated, the grooming performed by males to females was not reciprocated being, instead, rewarded with more copulations. These results indicate that grooming was traded for itself in the pre-mating period and for mating opportunities in the mating period. In short, males used the same commodity across the study period, whereas females switched from grooming to breeding availability during the mating period.

In summary, in the sifaka, and probably in other lemurs, the higher mating priority gained by males via scent marking activity does not necessarily match with a higher number of copulations. In fact, mating first does not mean mating more. A sifaka male, after winning female access in the 'primary elections' based on self-promotion via odour messages, had to move from theory to facts, by offering a service in exchange for increased probability of siring offspring.

As reported above, different studies have shown the effect of rank in influencing service exchange in primate biological markets (Port *et al.*, 2009; Barrett and Henzi, 2001). The fact that sifaka females can mate also with out-group individuals (Brockman, 1999; Chapter 3 in the present book), indicates that mate choice by females goes beyond the relative ranking status within males belonging to a stable foraging group. New is also good!

In a subsequent study, the same research group found that male status could indeed influence the distribution of grooming exchange (Dall'Olio *et al.*, 2012). As described in Chapter 3, sifaka males are bimorphic in that they can show a clean or a stained chest, depending on their throat-marking activity and testosterone levels (Lewis and van Schaik, 2007; Lewis, 2009). In their group of origin, clean-chested males are normally low ranking adult males or subadults and stained-chested males possess a higher dominance status (Lewis and van Schaik, 2007). It has been argued that when males start roaming around during the mating period in search of females, the chest status may indicate their quality to extra-group females (Dall'Olio *et al.*, 2012). In the mating season, females received more grooming from clean-chested males than from stained-chested ones whereas in the birth season this difference disappeared. Due to their low testosterone levels and consequent low production of secretions (Lewis, 2005), during the mating season clean-chested males invested much more in the 'grooming for sex' tactic than stained-chested males, probably trying to overcompensate with grooming the lack of the benefits associated with good quality status. During the birth season, when females' valuable commodity (the egg) is not on the market, the amount of grooming performed by clean-chested males was lower, and comparable to the service provided by their stained-chested counterparts. The 'grooming for sex' tactic adopted by clean-chested males during the mating season was not completely unsuccessful; in fact, half of the clean-chested males under study did copulate with females, even though their copulation frequency was significantly lower than that of stained-chested males. The observation that copulation frequency is higher in stained-chested males (which, as said above, are usually dominant in their social group; Lewis and van Schaik, 2007) than in clean-chested males is consistent with the paternity test results presented by Kappeler and Schäffler (2008) on the Kirindy population, showing that sifaka dominant males can sire up to 90% of infants.

The presence of the multiple mating tactics, 'marking for sex' (stained-chested males) and 'grooming for sex' (an alternative, but not completely functional, tactic used by clean-chested males) may be a means by which sifaka population buffers the inbreeding phenomenon in forest fragments (Norscia and Palagi, 2011).

Box 8.2 | **by Rebecca J. Lewis**

Speaking of which: grooming, service exchange and dominance in the Verreaux's sifaka

I have been interested in power dynamics since I was very young. Why do some people have power and others not? Why as a child was I not equal to the adults around me? Why as a female was I less powerful (I was a child in the 1970s and 1980s)? I saw no reason why I should be treated any differently than any other person. As I grew older, I became interested in the social institutions of power in addition to the power dynamics within social relationships. I decided to combine my interest in animal behaviour with my interest in power to study power in non-humans.

When I was a student, research on power in animal behaviour was limited to studies of dominance and dominance hierarchies. Moreover, researchers were not consistent in how they measured dominance: body size, winning conflicts, receiving submissive signals, leading group movement, reproducing (Lewis, 2002). I found that this understanding of power was very limited. There had been a handful of publications by researchers attempting to address problems with the dominance concept (Bernstein, 1981; de Waal, 1986; Drews, 1993; Hand, 1986), but all attempted to make changes within the current paradigm of dominance. In humans, we understand that power can take many forms and can arise for many different reasons (e.g., Dahl, 1957). It seemed to me that the solution to the dominance problem in ethology was expanding our understanding of power in animals to include all types of power (not just dominance) and then to use precise characteristics to describe the various aspects of the power being described (Lewis, 2002).

While I still find traditional dominance research interesting, the research on economic power (leverage) is what excites me the most. What resources and services give some individuals power? What conditions are necessary for these resources and services to be a source of power? For example, females are more powerful than males in many lemur species (Richard, 1987). Are all females more powerful than all males regardless of age, reproductive status or position in the female hierarchy? Does this power fluctuate over the course of the year or over longer periods of time? I study Verreaux's sifaka and male sifaka groom females much more than females groom males (Lewis, 2004; Norscia *et al.*, 2009). In fact, females spend very little time grooming males! The traditional answer that males groom females more because females are dominant is not satisfactory. Why does being dominant mean that an individual receives more of a service like grooming? One explanation based on economic theory is that grooming is a service that is exchanged for something else (Barrett et al., 1999): tolerance, coalitionary support, mating opportunities, or even grooming itself. Powerful individuals, such as female sifaka, receive grooming from less powerful individuals because these less powerful individuals hope to get something in return (e.g., Lewis, 2004; Norscia et al., 2009; Lewis, 2010).

Once I knew that I wanted to study power dynamics in primates, I decided to study lemurs because so many lemurs exhibit a phenomenon known as 'female dominance,' which has been seen as an evolutionary puzzle (Jolly, 1984; Richard, 1987). People wondered why males would give up their dominance. I wondered why the assumption was that males have the power in the first place. Most explanations have relied on the assertion that the climate of Madagascar is unusually harsh and unpredictable (e.g., Pereira et al., 1999; Wright, 1999). I have tackled this question of the evolution of 'female dominance' in two ways. First, I explored power dynamics in a single lemur species: Verreaux's sifaka. I found that female power is sometimes dominance (i.e., power that arises due to an asymmetry in fighting ability) because females are often larger than males in some sifaka populations (Lewis, 2004; Lewis and Kappeler, 2005). But female sifaka also have leverage (i.e., economic power). Females become sexually active sometime between 3 and 5 years of age, but they do not really start reproducing regularly until they are 5 or 6 years of age (Richard et al., 2002). Males do not really start being subordinate to females until they are the reproductively mature age of 5 years (Ortiz and Lewis, 2015). This finding suggests to me that female power in sifaka is based on the leverage form control over mating opportunities (Lewis, 2002; Ortiz and Lewis, 2015). I have also found that male sifaka often do not submit to females without physical coercion (Kraus et al., 1999; Lewis, 2015). A male may be feeding on a precious baobab fruit when a female approaches and tries to take the fruit from him. He often will not give up the fruit immediately. He will signal his subordinate status with a chatter vocalisation but still hold onto the fruit. The female may have to use repeated aggression to get him to relinquish his fruit. This type of behaviour suggests to me that males are not willingly giving up power, and calls into question the hypothesis that 'female dominance' arises due to male deference (Hrdy, 1981).

Second, I have taken a comparative approach to studying female power. With colleagues, I looked at the relationship between sexual dimorphism and sex-dependent 'dominance' (Lewis and Kirk, 2008; Lewis et al., unpublished data). We found that as long as males are not substantially larger than females, any power structure is possible. Male-dominant, female-dominant, and co-dominant societies occur across the primate clade in species where

Box 8.2 (continued)

Box 8.2 (*cont.*)

sexual dimorphism is not extreme. Once males are a third larger than females, however, it is very difficult for females to exhibit much power in intersexual relationships. Thus, 'female dominance' is not an evolutionary puzzle unique to primates in Madagascar.

While studying power in Verreaux's sifaka, I stumbled on something very unexpected. I was analysing male–female relationships in sifaka and found that males can be divided into two different types and that female relationships with males depend on the type of males (Lewis, 2004). This bimorphism in male sifaka is associated with different reproductive tactics (Lewis and van Schaik, 2007; Lewis, 2008, 2009; Dall'Olio *et al.*, 2012). This finding was interesting because bimorphism is rare in mammals (Oliveira *et al.*, 2008), but also because this means that exchange relationships are influenced by three categories of adults: female, male morph A and male morph B (for further explanation on the two morphs and the concept of bimorphism see Chapter 3).

I believe that the future of studies of power in non-human societies will involve three areas of research.

(1) *Economic theory*: thus far we have only just begun to understand how economic forces influence non-human social relationships. I believe that many other aspects of economic theory will lead to new insights into animal social relationships. The translation of economic theory into animal behaviour requires that the researchers understand the terminology and concepts from both fields. Some concepts will not be applicable to non-human societies, but the future of power research will involve finding the economic (and even business) concepts that can help us understand strategies that animals use in biological markets.

(2) *Institutions*: In the 1960s and 1970s, primatologists were interested in social organisation and social institutions (e.g., Kummer, 1971; Hinde and Stevenson-Hinde, 1976; Hinde, 1979). The field moved away from this topic and needs to return to these early questions. We now have a much better understanding about the diversity of social systems, social organisation and social structure in primate societies. Moreover, we now have more sophisticated methods of analysis, such as network analysis and agent-based modelling, to quantify and test the importance of various parameters. These techniques will have a major impact on the study of non-human social institutions. This return to the study of institutions will also allow for more accurate comparisons of non-human and human behaviour. We will be able to better understand human social evolution.

(3) *Methodological advances*: In addition to the continued improvement of statistical and modelling techniques, the future of studies of power in animal societies will involve refined data collection methods. In the 1970s,

data collection methods were standardised (Altmann, 1974). This advancement had a huge influence on the field. Now, however, these methods are very limiting for researchers interested in social dynamics. It is not that the old methods are bad, they just do not tell us everything we need to know. For example, when a primatologist sees two animals grooming, we record the individuals involved, who initiates the behaviour, the duration of the grooming bout, whether it is mutual or unidirectional grooming. But not all grooming is equal. I have seen a chimpanzee groom another individual for a few minutes and then get distracted. The groomer shifts from sifting through the hair of its partner to more of a light petting while it is looking elsewhere. All of this behaviour may be recorded as a single grooming bout, but behaviour actually changed over the course of the bout. In the future, researchers will use more refined methods, probably borrowed from linguistic anthropologists and others who analyse conversations, to understand the subtleties of non-human social dynamics.

References

Albo, M. J., Bilde, T. & Uhl, G. (2013). Sperm storage mediated by cryptic female choice for nuptial gifts. *Proceedings of the Royal Society of London B: Biological Sciences*, 280(1772), 20131735.

Allesina, S. (2011). Measuring nepotism through shared last names: the case of Italian academia. *PLoS ONE*, 6, e21160. http://dx.doi.org/10.1371/journal.pone.0021160.

Altmann, J. (1974). Observational study of behaviour: sampling methods. *Behaviour*, 48, 227–265.

Antonacci, D., Norscia, I. & Palagi, E. (2010). Stranger to familiar: wild strepsirrhines manage xenophobia by playing. *PLos ONE*, 5(10), e13218.

Aureli, F., van Schaik, C. P. & van Hooff, J. A. R. A. M. (1989). Functional aspects of reconciliation among captive long-tailed macaques (*Macaca fascicularis*). *American Journal of Primatology*, 19, 39–51.

Balasubramaniam, K. N., Berman, C. M., Ogawa, H. & Li, J. (2011). Using biological markets principles to examine patterns of grooming exchange in *Macaca thibetana*. *American Journal of Primatology*, 73(12), 1269–1279.

Barash, D. P. (1977). Sociobiology of rape in mallards (*Anas platyrhynchos*): response of the mated male. *Science*, 197, 788–789.

Barelli, C., Reichard, U. H. & Mundry, R. (2011). Is grooming used as a commodity in wild white-handed gibbons, *Hylobates lar*? *Animal Behaviour*, 82, 801–809.

Barrett, L. & Henzi, S. P. (2001). The utility of grooming in baboon troops. In: Noë, R., van Hooff, J. A. R. A. M. & Hammerstein, P. (eds), *Economics in Nature*. Cambridge University Press, pp. 119–145.

Barrett, L., Henzi, S. P., Weingrill, A., Lycett, J. E. & Hill, R. A. (1999). Market forces predict grooming reciprocity among female baboons. *Proceedings of the Royal Society B: Biological Sciences*, 266(1420), 665–670.

Barske, J., Fusani, L., Wikelski, M., *et al.* (2014). Energetics of the acrobatic courtship in male golden-collared manakins (*Manacus vitellinus*). *Proceedings of the Royal Society B: Biological Sciences*, 281(1776), 20132482.

Bernstein, I. S. (1981). Dominance: the baby and the bathwater. *Behavioral and Brain Sciences*, 4, 419–429.

Blaustein, A. R. (1981). Sexual selection and mammalian olfaction. *American Naturalist*, 117, 1006–1010.

Brockman, D. K. (1999). Reproductive behavior of female *Propithecus verreauxi* at Beza Mahafaly, Madagascar. *International Journal of Primatology*, 20, 375–398.

Brosnan, S. F. & de Waal, F. B. (2002). A proximate perspective on reciprocal altruism. *Human Nature*, 13(1), 129–152.

Chancellor, R. L. & Isbell, L. A. (2009). Female grooming markets in a population of gray-cheeked mangabeys (*Lophocebus albigena*). *Behavioral Ecology*, 20, 79–86.

Clarke, P. M. R., Halliday, J. E. B., Barrett, L. & Henzi, S. P. (2010). Chacma baboon mating markets: competitor suppression mediates the potential for intersexual exchange. *Behavioral Ecology*, 21(6), 1211–1220.

Coleman, S. W., Patricelli, G. L. & Borgia, G. (2004). Variable female preferences drive complex male displays. *Nature*, 428, 742–745.

Cords, M. (1995). Predator vigilance costs of allogrooming in wild blue monkeys. *Behaviour*, 132(7), 559–569.

Dahl, R. A. (1957). The concept of power. *Behavioural Science*, 2, 201–215.

Dall'Olio, S., Norscia, I., Antonacci, D. & Palagi, E. (2012). Sexual signalling in *Propithecus verreauxi*: Male 'chest badge' and female mate choice. *PLoS ONE*, 7(5), e37332.

de Waal, F. B. M. (1986). The integration of dominance and social bonding in primates. *Quarterly Review of Biology*, 61, 459–479.

de Waal, F. B. (1997). The chimpanzee's service economy: food for grooming. *Evolution and Human Behavior*, 18, 375–386.

Drews, C. (1993). The concept and definition of dominance in animal behaviour. *Behaviour*, 125, 283–313.

East, M. L. & Hofer, H. (2001). Male spotted hyenas (*Crocuta crocuta*) queue for status in social groups dominated by females. *Behavioral Ecology*, 12(5), 558–568.

Emery-Thompson, M. (2009). Human rape: revising evolutionary perspectives. In: M. N. Muller & R. W. Wrangham (eds), *Sexual Coercion in Primates and Humans: An Evolutionary Perspective on Male Aggression Against Females*. Cambridge, MA: Harvard University Press, pp. 346–374.

Fisher, H. S., Swaisgood, R. & Fitch-Snyder, H. (2003). Countermarking by male pygmy lorises (*Nycticebus pygmaeus*): do females use odor cues to select mates with high competitive abilities? *Behavioral Ecology and Sociobiology*, 53, 123–130.

Fruteau, C., Voelkl, B., van Damme, E. & Noë, R. (2009). Supply and demand determine the market value of food providers in wild vervet monkeys. *PNAS*, 106, 12007–12012.

Fruteau, C., Lemoine, S., Hellard, E., Van Damme, E. & Noë, R. (2011). When females trade grooming for grooming: testing partner control and partner choice models of cooperation in two primate species. *Animal Behaviour*, 81, 1223–1230.

Fusani, L. & Schlinger, B. A. (2012). Proximate and ultimate causes of male courtship behavior in Golden-collared Manakins. *Journal of Ornithology*, 153, 119–124.

Goetz, A. T. & Shackelford, T. K. (2006). Sexual coercion and forced in-pair copulation as sperm competition tactics in humans. *Human Nature*, 17, 265–282.

Gomes, C. M., Mundry, R. & Boesch, C. (2009). Long-term reciprocation of grooming in wild West African chimpanzees. *Proceedings of the Royal Society B: Biological Sciences*, 276, 699–706.

Gosling, L. M. & Roberts, S. C. (2001). Scent-marking by male mammals: cheat-proof signals to competitors and mates. *Advances in the Study of Behavior*, 30, 169–217.

Gumert, M. D. (2007). Payment for sex in a macaque mating market. *Animal Behaviour*, 74(6), 1655–1667.

Hamilton, W. D. (1964). The genetical evolution of social behaviour. I and II. *Journal of Theoretical Biology*, 7, 1–52. http://dx.doi.org/10.1016/0022-5193(64)90038-4.

Hand, J. L. (1986). Resolution of social conflicts: dominance, egalitarianism, spheres of dominance, and game theory. *Quarterly Review of Biology*, 61, 201–220.

Hawlena, H., Bashary, D., Abramsky, Z. & Krasnov, B. R. (2007). Benefits, costs and constraints of anti-parasitic grooming in adult and juvenile rodents. *Ethology*, 113, 394–402.

Head, M. & Brooks, R. (2006). Sexual coercion and the opportunity for sexual selection in guppies. *Animal Behaviour*, 71, 515–522.

Hemelrijk, C. K. (1994). Support for being groomed in long-tailed macaques, *Macaca fascicularis*. *Animal Behaviour*, 48, 479–481.

Henzi, S. P. & Barrett, L. (2002). Infants as a commodity in a baboon market. *Animal Behaviour*, 63, 915–921.

Heymann, E. W. (2006). Scent marking strategies of New World primates. *American Journal of Primatology*, 68, 650–661.

Hinde, R. A. & Stevenson-Hinde, J. (1976). Towards understanding relationships: dynamic stability. In: Bateson P. P. G., and Hinde, R. A. (eds), *Growing Points in Ethology*. Cambridge University Press, pp. 451–479.

Hinde, R. A. (1979). The nature of social structure. In: D. A. Hamburg & E. R. McCown (eds), *The Great Apes*. Menlo Pk, CA: Benjaming Cummings.

Hrdy, S. B. (1981). *The Woman that Never Evolved*. Cambridge, MA: Harvard University Press.

Hurst, J. L. & Beynon, R. J. (2004). Scent wars: the chemobiology of competitive signalling in mice. *Bioessays*, 26(12), 1288–1298.

Hutchins, M. & Barash, D. P. (1976). Grooming in primates: implications for its utilitarian function. *Primates*, 17(2), 145–150.

Jolly, A. (1972). Troop continuity and troop spacing in *Propithecus verreauxi* and *Lemur catta* at Berenty (Madagascar). *Folia Primatologica*, 17, 335–362.

Jolly, A. (1984). The puzzle of female feeding priority. In: M. Small (ed.), *Female Primates: Studies by Women Primatologists*. New York: Alan R. Liss, pp. 197–215.

Jones, C. B. & Van Cantfort, T. E. (2007). Multimodal communication by male mantled howler monkeys (*Alouatta palliata*) in sexual contexts: a descriptive analysis. *Folia Primatologica*, 78(3), 166–185.

Jordan, N. R. (2007). Scent-marking investment is determined by sex and breeding status in meerkats. *Animal Behaviour*, 74, 531–540.

Kappeler, P. M. & Schäffler, L. (2008). The lemur syndrome unresolved: extreme male reproductive skew in sifakas (*Propithecus verreauxi*), a sexually monomorphic primate with female dominance. *Behavioral Ecology and Sociobiology*, 62, 1007–1015.

Kapsalis, E. & Berman, C. M. (1996). Models of affiliative relationships among free-ranging rhesus monkeys (*Macaca mulatta*). *Behaviour*, 133(15), 1235–1263.

Keverne, E. B., Martensz, N. D. & Tuite, B. (1989). Beta-endorphin concentrations in cerebrospinal fluid of monkeys are influenced by grooming relationships. *Psychoneuroendocrinology*, 14(1), 155–161.

Kraus, C., Heistermann, M. & Kappeler, P. M. (1999). Physiological suppression of sexual function of subordinate males: a subtle form of intrasexual competition among male sifakas (*Propithecus verreauxi*)? *Physiology & Behavior*, 66(5), 855–861.

Kummer, H. (1971). *Primate Societies: Group Techniques of Ecological Adaptation*. Arlington Heights, IL: AHM Publishing Corporation.

Lazaro-Perea, C., Snowdon, C. & de Fatima Arruda, M. (1999). Scent-marking behavior in wild groups of common marmosets (*Callithrix jacchus*). *Behavioral Ecology and Sociobiology*, 46, 313–324.

Lewis, R. J. (2002). Beyond dominance: the importance of leverage. *Quarterly Review of Biology*, 77, 149–164.

Lewis, R. J. (2004). Male–female relationships in sifaka (*Propithecus verreauxi verreauxi*): power, conflict, and cooperation. Dissertation, Durham, NC: Duke University.

Lewis, R. J. (2005). Sex differences in scent-marking in sifaka: mating conflict or male services? *American Journal of Physical Anthropology*, 128, 389–398.

Lewis, R. J. (2008). Social influences on group membership in Verreaux's sifaka (*Propithecus verreauxi verreauxi*). *International Journal of Primatology*, 29, 1249–1270.

Lewis, R. J. (2009). Chest staining variation as a signal of testosterone levels in male Verreaux's Sifaka. *Physiology and Behavior*, 96(4), 586–592.

Lewis, R. J. (2010). Grooming pattern in Verreaux's sifaka. *American Journal of Primatology*, 72, 254–261.

Lewis, R. J. (2015). The evolution of subordination and social complexity: an analysis of power in Verreaux's sifaka. *American Journal of Physical Anthropology*, 156, 203.

Lewis, R. J. & Kappeler, P. M. (2005). Seasonality, body condition, and the timing of reproduction in *Propithecus verreauxi verreauxi* in the Kirindy Forest. *American Journal of Primatology*, 67, 347–364.

Lewis, R. J. & Kirk, E. C. (2008). Female dominance and monomorphism: Are patterns of intersexual dominance influenced by sexual dimorphism? *American Journal of Physical Anthropology*, 135, 140.

Lewis, R. J. & van Schaik, C. P. (2007). Bimorphism in male Verreaux's sifaka in the Kirindy Forest of Madagascar. *International Journal of Primatology*, 28, 159–182.

Maestripieri, D. (1993). Vigilance costs of allogrooming in macaque mothers. *American Naturalist*, 141, 744–753.

Malthus, T. R. (1797). *An Essay on the Principle of Population*. London: Printed for J. Johnson, in St. Paul's Church-Yard.

Matsueda, R. L., Kreager, D. A. & Huizinga, D. (2006). Deterring delinquents: a rational choice model of theft and violence. *American Sociological Review*, 71, 95–122.

McNamara, J. M., Barta, Z., Fromhage, L. & Houston, A. I. (2008). The coevolution of choosiness and cooperation. *Nature*, 451, 189–192.

Mooring, M. S., McKenzie, A. A. & Hart, B. L. (1996). Grooming in impala: role of oral grooming in removal of ticks and effects of ticks in increasing grooming rate. *Physiology and Behavior*, 59, 965–971.

Muller, M. N. & Wrangham, R. W. (eds). (2009). *Sexual Coercion in Primates and Humans: an evolutionary perspective on male aggression against females*. Cambridge, MA: Harvard University Press.

Newton-Fisher, N. E. & Lee, P. C. (2011). Grooming reciprocity in wild male chimpanzees. *Animal Behaviour*, 81, 439–446.

Noë, R. (2001). Biological markets: partner choice as the driving force behind the evolution of mutualism. In: Noë, R., van Hoof, J. A. R. A. M. & Hammerstein, P. (eds), *Economics in Nature*. Cambridge University Press, pp. 93–118.

Noë, R. (2006). Digging for the roots of trading. In: Kappeler, P. M. & van Schaik, C. P. (eds) *Cooperation in Primates and Humans*. Berlin, Heidelberg: Springer, pp. 233–261.

Noë, R. & Hammerstein, P. (1994). Biological markets: supply and demand determine the effect of partner choice in cooperation, mutualism and mating. *Behavioral Ecology and Sociobiology*, 35, 1–11.

Noë, R. & Hammerstein, P. (1995). Biological markets. *Trends in Ecology & Evolution*, 10, 336–340.

Noë, R., van Schaik, C. P. & van Hooff, J. A. R. A. M. (1991). The market effect: an explanation of pay-off asymmetries among collaborating animals. *Ethology*, 87, 97–118.

Norscia, I. & Palagi, E. (2011). Fragment quality and distribution of the arboreal primate *Propithecus verreauxi* in the spiny forest of South Madagascar. *Journal of Tropical Ecology*, 27, 103–106.

Norscia, I., Carrai, V. & Borgognoni-Tarli, S. M. (2006). Influence of dry season, food quality and quantity on behavior and feeding strategy of *Propithecus verrreauxi* in Kirindy, Madagascar. *International Journal of Primatology*, 27, 1001–1022.

Norscia, I., Antonacci, D. & Palagi, E. (2009). Mating first, mating more: biological market fluctuation in a wild prosimian. *PLoS ONE*, 4(3), e4679.

Nunn, C. L. & Lewis, R. J. (2001). Cooperation and collective action in animal behavior. In: Noë, R., van Hoof, J. A. R. A. M. & Hammerstein, P. (eds), *Economics in Nature*. Cambridge University Press, pp. 42–66.

Oliveira, R. F., Taborsky, M. & Brockmann, J. (2008). *Alternative Reproductive Tactics: An Integrative Approach*. Cambridge University Press.

Ortiz, K. M. & Lewis, R. J. (2015). The influence of fighting ability and reproduction in intersexual relationships in Verreaux's sifaka (*Propithecus verreauxi*). *American Journal of Physical Anthropology*, 156 (S60), 242–243.

Palagi, E., Telara, S. & Borgognini-Tarli, S. M. (2004). Reproductive strategies in *Lemur catta*: balance among sending, receiving and countermarking scent signals. *International Journal of Primatology*, 25, 1019–1031.

Palagi, E., Antonacci, D. & Norscia, I. (2008). Peacemaking on treetops: first evidence of reconciliation from a wild prosimian (*Propithecus verreauxi*). *Animal Behaviour*, 76(3), 737–747.

Paoli, T. & Palagi, E. (2008). What does agonistic dominance imply in bonobos?. In: *The Bonobos*. New York: Springer, pp. 39–54.

Pereira, M. E., Strohecker, R. A., Cavigelli, S. A., Hughes, C. L. & Pearson, D. D. (1999). Metabolic strategy and social behavior in Lemuridae. In: B. Rakotosamimanana, H. Rasamimanana, J. U. Ganzhorn & S. M. Goodman (eds), *New Directions in Lemur Studies*. New York: Plenum, pp. 93–118.

Port, M., Clough, D. & Kappeler, P. M. (2009). Market effects offset the reciprocation of grooming in free-ranging redfronted lemurs, *Eulemur fulvus rufus*. *Animal Behaviour*, 77(1), 29–36.

Prosen, E. D., Jaeger, R. G. & Hucko, J. A. (2006). Sexual coercion in the salamander *Plethodon cinereus*: is it merely a result of familiarity? *Herpetologica*, 62(1), 10–18.

Qvarnström, A. & Forsgren, E. (1998). Should females prefer dominant males? *Trends in Ecology and Evolution*, 13(12), 498–501.

Rasmussen, H. B., Okello, J. B. A., Wittemyer, G., et al. (2008). Age- and tactic-related paternity success in male African elephants. *Behavioral Ecology*, 19(1), 9–15.

Rayner, C. & Hoel, H. (1997). A summary review of literature relating to workplace bullying. *Journal of Community and Applied Social Psychology*, 7, 181–191.

Reichard, U. & Sommer, V. (1994). Grooming site preferences in wild white-handed gibbons (*Hylobates lar*). *Primates*, 35(3), 369–374.

Richard, A. F. (1987). Malagasy prosimians: female dominance. In: B. B. Smuts, D. L. Cheney, R. M. Seyfarth, R. W. Wrangham & T. T. Struhsaker (eds), *Primate Societies*. Chicago: University of Chicago Press, pp. 25–33.

Richard, A. F. (1992). Aggressive competition between males, female-controlled polygyny and sexual monomorphism in a Malagasy primate, *Propithecus verreauxi*. *Journal of Human Evolution*, 22(4), 395–406.

Richard, A. F. & Nicoll, M. E. (1987). Female social dominance and basal metabolism in a Malagasy primate, *Propithecus verreauxi*. *American Journal of Primatology*, 12(3), 309–314.

Richard, A. F., Dewar, R. E., Schwartz, M. & Ratsirarson, J. (2002). Life in the slow lane? Demography and life histories of male and female sifaka (*Propithecus verreauxi verreauxi*). *Journal of Zoology*, 256, 421–436.

Rosenberger, A. L. & Strier, K. B. (1989). Adaptive radiation of the ateline primates. *Journal of Human Evolution*, 18(7), 717–750.

Russell, B. (1938). *Power: The Role of Man's Will to Power in the World's Economic and Political Affairs*. New York: Norton.

Schaffner, C. & Aureli, F. (2005). Embraces and grooming in captive spider monkeys. *International Journal of Primatology*, 26, 1093–1106.

Schino, G. & Aureli, F. (2008). Grooming reciprocation among female primates: a meta-analysis. *Biology Letters*, 4(1), 9–11.

Schino, G., Scucchi, S., Maestripieri, D. & Turillazzi, P. G. (1988). Allogrooming as a tension-reduction mechanism: A behavioral approach. *American Journal of Primatology*, 16(1), 43–50.

Scott, E. M., Mann, J., Watson-Capps, J. J., Sargeant, B. L. & Connor, R. C. (2005). Aggression in bottlenose dolphins: evidence for sexual coercion, male-male competition, and female tolerance through analysis of tooth-rake marks and behavior. *Behaviour*, 142, 21–44.

Seyfarth, R. M. (1977). A model of social grooming among adult female monkeys. *Journal of Theoretical Biology*, 65, 671–698.

Seyfarth, R. M. & Cheney, D. L. (1984). Grooming, alliances and reciprocal altruism in vervet monkeys. *Nature*, 308, 541–543.

Shine, R. & Mason, R. T. (2005). Does large body size in males evolve to facilitate forcible insemination? A study on garter snakes. *Evolution*, 59, 2426–2432.

Sitta von Reden. (1998). The commodification of symbols. Reciprocity and its perversions in Menander. In: C. Gill, N. Postlethwaite & R. Seaford (eds), *Reciprocity in Ancient Greece*. New York: Oxford University Press.

Slater, K. Y., Schaffner, C. M. & Aureli, F. (2007). Embraces for infant handling in spider monkeys: evidence for a biological market? *Animal Behaviour*, 74, 455–461.

Smuts, B. B. & Smuts, R. W. (1993). Male aggression and sexual coercion of females in nonhuman primates and other mammals: evidence and theoretical implications. *Advances in the Study of Behavior*, 22, 1–63.

Stopka, P. & Macdonald, D. W. (1999). The market effect in the wood mouse, *Apodemus sylvaticus*: selling information on reproductive status. *Ethology*, 105, 969–982.

Strier, K. B. (1994). Myth of the typical primate. *Yearbook of Physical Anthropology*, 37, 233–271.

Strier, K. B. (2011). Social plasticity and demographic variation in primates. In: Sussman, R. W. & Cloninger, C. R. (eds), *Origins of Altruism and Cooperation*. NY: Springer, pp. 179–192.

Strier, K. B. & Ives, A. R. (2012). Unexpected demography in the recovery of an endangered primate population. *PLoS ONE*, 7(9), e44407.

Strier, K. B., Mendes, F. D. C., Rímoli, J. & Odalia Rímoli, A. (1993). Demography and social structure in one group of muriquis (*Brachyteles arachnoides*). *International Journal of Primatology*, 14, 513–526.

Strier, K. B., Boubli, J. P., Possamai, C. B. & Mendes, S. L. (2006). Population demography of northern muriquis (*Brachyteles hypoxanthus*) at the Estação Biológica de Caratinga/Reserva Particular do Patrimônio Natural-Feliciano Miguel Abdala, Minas Gerais, Brazil. *American Journal of Physical Anthropology*, 130(2), 227–237.

Strier, K. B., Chaves, P. B., Mendes, S. L., Fagundes, V. & Di Fiore, A. (2011). Low paternity skew and the influence of maternal kin in an egalitarian, patrilocal primate. *PNAS*, 108, 18915–18919.

Tabacow, F. P., Mendes, S. L. & Strier, K. B. (2009). Spread of a terrestrial tradition in an arboreal primate. *American Anthropologist*, 111, 238–249.

Thornhill, R. (1980). Rape in *Panorpa* scorpionflies and a general rape hypothesis. *Animal Behaviour*, 28, 52–59.

Thornhill, R. & Thornhill, N. W. (1983). Human rape: An evolutionary analysis. *Ethology and Sociobiology*, 4, 137–173.

Tiddi, B., Aureli, F. & Schino, G. (2010). Grooming for infant handling in tufted capuchin monkeys: a reappraisal of the primate infant market. *Animal Behaviour*, 79, 1115–1123.

Tiddi, B., Aureli, F. & Schino, G. (2012). Grooming up the hierarchy: The exchange of grooming and rank-related benefits in a New World Primate. *PLoS ONE*, 7(5), e36641. http://dx.doi.org/10.1371/journal.pone.0036641.

Trivers, R. L. (1971). The evolution of reciprocal altruism. *The Quarterly Review of Biology*, 46, 35–57.

Ventura, R., Majolo, B., Koyama, N. F., Hardie, S. & Schino, G. (2006). Reciprocation and interchange in wild Japanese macaques: grooming, cofeeding, and agonistic support. *American Journal of Primatology*, 68(12), 1138–1149.

Wei, W., Qi, X., Garber, P. A., *et al.* (2013). Supply and demand determine the market value of access to infants in the Golden Snub-Nosed monkey (*Rhinopithecus roxellana*). *PLoS ONE*, 8, e65962.

Werner, G. D., Strassmann, J. E., Ivens, A. B., *et al.* (2014). Evolution of microbial markets. *PNAS*, 111(4), 1237–1244.

Wong, B. B. M. & Candolin, U. (2005). How is female mate choice affected by male competition? *Biological Reviews*, 80, 559–571.

Wright, P. C. (1999). Lemur traits and Madagascar ecology: coping with an island environment. *American Journal of Physical Anthropology*, 110 (s29), 31–72.

Part IV

Closing remarks

9 Understanding lemurs: future directions in lemur cognition

When we started considering the idea to write this book on lemurs' behaviour in 2010 we thought of making reference to the relatively few experimental studies on lemur cognition in the introduction, mentioning them as a promising novel branch of investigation for strepsirrhines. But over the past few years the research on the subject has blossomed and many articles have been published on scientific journals, converting the lemur cognition domain in one of the most flourishing fields of investigation of primatology, anthropology and comparative psychology. We realised, then, that we could not dismiss cognitive studies with a couple of paragraphs in the introductory section and we decided to reserve the final chapter to the topic, which probably better than others points towards unexplored lemur potentials and future directions for a more comprehensive understanding of the primate world. In the following pages we try to provide the reader with the basic elements that we think are necessary to understand what is going on when lemurs are tested on cognitive tasks. We suggest that the reader uses this chapter for orientation, as a sort of compass to navigate through the different aspects of lemur cognition. If it is of interest, we invite the reader to consult the original papers to gather further details on the experimental apparatuses and procedures.

9.1 Manipulative lemurs? Maybe

In her 1966 *Science* article, Alison Jolly observed that 'some prosimians, the social lemurs, have evolved the usual primate type of society and social learning without the capacity to manipulate objects as monkeys do. It thus seems likely that the rudiments of primate society preceded the growth of primate intelligence, made it possible, and determined its nature'.

As primatological studies have advanced, some evidence of tool manipulation ability, a possible precursor of tool use, has been found in lemurs (for a review see Fichtel and Kappeler, 2010; Schilling, 2013). Only few incidents of spontaneous object manipulation have been observed in the wild. The aye-aye (*Daubentonia madagascariensis;* Figure 9.1) is known for using the probing middle finger for tapping, and extracting nectar, kernels and insects embedded in trees or branches; it can also use the fourth finger for tasks requiring strength, scooping action and deep access (Lhota et al., 2008). Black and brown lemurs sometimes manipulate insects vigorously, bouncing them from hand to hand, presumably in an effort to anoint their bodies with an insect repellent (Fichtel and Kappeler, 2010; Norscia and

Palagi, personal observation). In captivity, lemurs have shown that they can perform simple manipulation tasks. For example, brown, black, and ring-tailed lemurs (Jolly, 1964; Kappeler, 1987; Fornasieri *et al.*, 1990; Anderson *et al.*, 1992) can open simple boxes whereas aye-ayes were also found able to open complex puzzle boxes (Digby *et al.*, 2008). The ability to open boxes is crucial to test lemurs on social learning tasks, as it is reported later on in this chapter. Schilling (2013) described quite complex manipulation problems that *Microcebus murinus* learned to solve during different trials. For example, mouse lemurs learned to apply three different methods to open a box (raising the lid of the box, pulling the drawer with the cord, pushing the drawer and passing to the other side of the box). They also learned to retrieve a mealworm inside a cylinder contained in another cylinder by pulling a cord to draw the inner cylinder holding the food to the mouth of the outer cylinder and then maintaining the cord with one hand while pulling the thread with the attached worm. Mouse lemurs also proved capable of using the reversed image of a worm in a mirror as a cue to catch the worm, without seeing it (because it was place behind an opaque wall). Grey mouse lemurs are described as rapid, flexible, and persevering learners (they can stay concentrated on a task up to 10 minutes) and their ability to perform complex manipulation tasks suggest that the cognitive abilities of strepsirrhines are close to those of simian primates.

Santos and coworkers (2005a) pointed out that the reason why some primate species are less able to manipulate tools in experimental tasks can lie in anatomical constraints, other than on cognitive limitations. Lemurs could be physically unable to manipulate a stick into a tube to obtain food, as capuchins and chimpanzees do

Figure 9.1 An aye-aye (*Daubentonia madagascariensis*) at the Parc Botanique et Zoologique de Tsimbazaza (PBZT), Antananarivo (Madagascar). Photo taken by Paola Richard within the project MadAction (University of Turin).

(Visalberghi and Limongelli, 1994; Limongelli *et al.*, 1995). While testing brown and ring-tailed lemurs on object manipulation, Santos *et al.* (2005a) found that lemurs preferred larger tools and explained this finding with lemurs' hand anatomy: because the lemur thumb is more ventrally placed it is better than other primate hands at seizing larger objects. On the other hand, the variability in the manipulatory behaviour across primate species appears not to be explained by hand anatomy, but rather by habitat use. Broad-niched opportunists (versus specialists) appear to show a wider array of manipulative behaviours, possibly because they deal with a more variable habitat. Similarly highly folivorous primates such as lemurs, marmosets and leaf-eating monkeys seem to possess a less variable behaviour than frugivorous and insectivorous Old World monkeys (Fichtel and Kappeler, 2010). Lemurs' manipulative behaviour seems to be comparable with that of at least some New and Old World monkeys (Fichtel and Kappeler, 2010).

Santos *et al.* (2005a) explored if two lemur species, *Lemur catta* (three individuals) and *Eulemur fulvus* (three individuals) could comprehend the functionality (mean-end relationship) of objects for potential tool use. Two identical cane-shaped tools were placed on either side of a tray behind a mesh. In each trial one food reward was located inside the hook so that the subject could retrieve it by pulling the tool (correct choice) and the other food reward was located outside the hook so that the lemur could not retrieve it if pulling the tool (incorrect choice). If the lemur managed to manipulate the tool (e.g. twisting it to the side) to retrieve the object, the choice was also considered as correct. The lemurs succeeded in reaching 80% of correct pulls. When presented with a choice between tools with different features (e.g. differing in shape versus colour, texture versus shape, etc., Figure 9.2) lemurs

Figure 9.2 Ring-tailed lemur (*Lemur catta*) choosing between differently shaped tools presented by the experimenter. Photo: Laurie Santos.

appeared to prefer the characteristics that were more causally relevant for retrieving the food. For example, lemurs preferred size to colours but would indifferently choose between tools varying in either shape or texture. In the final experiments lemurs were sometimes able to change tool position to make its orientation effective to retrieve the food (Santos *et al.* 2005a).

All in all, it appears that although lemurs possess certain manipulation capacity, varying from species to species, their actual ability to represent tools is still a controversial issue. Indeed, the studies on object manipulation and tool representation need several replications before any conclusion can be drawn on lemurs' manipulative skills. Let us wait and see.

9.2 Lemurs that count: numerical sensitivity in strepsirrhines

The ability to comprehend that objects continue to exist even when no longer in view is called object permanence. This capacity varies depending (among other factors) on species, object location (e.g. single, multiple) and on whether the subject has the direct perception of the object or not. Deppe *et al.* (2009) tested individuals of different lemur species (*Lemur catta, Eulemur mongoz, Eulemur rufus,* and *Hapalemur griseus* × *alaotrensis*) and showed that they were able to find a food item (raisin in the case study) that was presented to the subjects by the experimenter, then visibly placed in a container (a dish, out of three possible dishes of the apparatus) and finally hidden by covering the container with cloths. Lemurs succeeded in these 'visible displacement trials', even when the difficulty increased (e.g. raisin always hidden in the same location, in different locations, with random order and/or trajectory, etc.). However, if lemurs were not allowed to search for up to 25 seconds, their performance declined with increasing time delay. Lemurs did not succeed in 'invisible trials' in which the raisin was presented to each subject in the experimenter's palm but then concealed by making a fist and subsequently placed (invisibly to the subject) in the target dish (Deppe *et al.*, 2009). Therefore, lemurs seemed unable to understand invisible displacements (hidden objects changing position). Interestingly, Mallavarapu *et al.* (2013) found no convincing evidence that black-and-white ruffed lemurs (*Varecia variegata*; Figure 9.3) could understand visible displacements either. The lemurs failed the control trials in that they preferentially selected both the containers visited by the experimenter during the trial (baited box and last box touched, in this case study; see Mallavarapu *et al.*, 2013 for details). Therefore, lemurs might be able to represent previously seen objects (at least for a short time) when they are subsequently hidden to their sight but the debate on this issue remains open due to the conflicting (but few) results accumulated so far.

Human infants acquire the ability to understand invisible displacements only later in their development (at 2 years of age approximately; Siegler, 1986). Younger human infants (at one year to one-and-a-half years), great apes, monkeys, and even dogs and various taxa of birds appear to be able to solve visible object displacements. For lemurs, Deppe *et al.* (2009) hypothesised that the natural predators

9.2 Lemurs that count: numerical sensitivity in strepsirrhines 251

Figure 9.3 A black-and-white ruffed lemur (*Variecia variegata*) at Pistoia Zoo, Italy. Photo: Elisabetta Palagi.

of lemurs and/or the type of food that they eat do not require – for survival – the ability to track moving concealed objects (invisible displacements). We that maybe the use of collective alarm calls and the arboreal locomotory skills of lemurs make fleeing the most adaptive strategy in the forest canopy.

Santos *et al.* (2005b) used a methodology originally developed for use with human infants to test the numerical abilities of lemurs. This methodology, called 'looking time paradigm', relies on the experimentally proven observation that subjects will look longer at events that they see as violations of the physical or social world. In the experiments, Santos and coworkers (2005b) tested lemurs of different species (*Eulemur fulvus*, *Eulemur mongoz*, *Lemur catta*, and *Varecia rubra*) for a 1 + 1 addition operation. The lemurs watched as two lemons were sequentially added on a stage behind an occluder preventing the subjects from seeing inside. Inside the apparatus, the experimenters affixed a shelf to the back of the occluder such that an experimenter could place a lemon behind the occluder and onto the shelf while appearing to place the lemon onto the stage. After a testing phase in which lemurs could see the objects placed onto the stage (with no occluder) and familiarise with the one lemon versus two lemons experiment, the subjects (21 lemurs) were tested to verify if they had expectations about the outcome of the 1 + 1 events. The experimenter placed a lemon visibly on the stage and then placed the occluder to block the lemon from the subject's view. The other lemon was placed behind the occluder. Then, the occluder was removed to show the lemur a 'one lemon' outcome (impossible) or a 'two lemons' outcome and the time the lemur stayed looking at the outcome was measured for the next 10 s (from the occluder removal). Lemurs significantly looked longer when they were presented with an impossible outcome.

The test was repeated in the exact same way but showing, at the end, a 'two lemons' outcome and a 'three lemons' outcome. Again, lemurs looked the impossible outcome longer. Finally, in a third experiment using the same procedure, the lemurs were presented with a 'two lemons' outcome (expected) and a 'big lemon' outcome (impossible). The size of the big lemon was more or less equivalent to the sum of the sizes of the two smaller lemons. The lemurs looked longer at the numerically inconsistent outcome of one larger object than at the numerically consistent outcome of two smaller objects (Figure 9.4). The authors, therefore, concluded that lemurs are likely to represent the outcome 1 + 1 event not in terms of 'amount of stuff present on stage' but in terms of numerosity. Like human infants less than one year of age (Wynn, 1992; Simon *et al.*, 1995; Koechlin *et al.*, 1997; Feigenson *et al.*, 2002), rhesus macaques (Hauser and Carey, 2003), and cotton-top tamarins (Uller *et al.*, 2001) adult lemurs appear to be able to successfully form expectations about the exact outcome of a 1 + 1 addition event.

After testing whether lemurs could form expectations about the exact outcome of a 1 + 1 addition event, another group of investigators checked lemurs' abilities to distinguish between different quantities (Lewis *et al.*, 2005b). According to Weber's Law as numerical magnitude increases, a large disparity is needed to obtain the same level of discrimination between quantities. This phenomenon has been observed in 10–12-month-old human infants (Feigenson *et al.*, 2002), rhesus macaques (Hauser *et al.*, 2000), cotton-top tamarins (Hauser *et al.*, 2003), which can discriminate quantities of 1 versus 2 and 2 versus 3 but not sets with 4 or more items (e.g. 4:5). Lewis and colleagues (2005) tested several individuals of *Eulemur mongoz* on their ability to discriminate between different quantities as the numerical magnitude increased. They showed lemurs with grapes being placed in a green plastic bucket. The bucket had a false bottom in which grapes could be placed through a hole 3 cm in diameter (inaccessible to the subjects), hidden by a sticky note. In different trials, after observing the experimenters placing a certain numbers of grapes into the bucket, the lemurs could either retrieve all the grapes (if they were placed on the 'real' bucket bottom) or not (if some grapes were hidden by the false bottom). For example, the experimenters could place just one grape into the bucket and allow retrieval (1–1 condition) or place two grapes and allows the retrieval of just one of them (because one was hidden by the false bottom; 2–1 condition). Different conditions were tested: 1–1 versus 2–1 condition, 2–2 versus 4–2, 2–2 versus 3–2, etc. (with the first number indicating the grapes placed into the bucket and the second indicating the grapes that could be retrieved). The experimenters, who also controlled for possible confounding variables (e.g. odours and grape size) verified whether lemurs would search for grapes significantly longer when they expected to find more than they could actually retrieve. They found that lemurs discriminated quantities that differed by a 1:2 ratio but not quantities that differed by 2:3 or 3:4 ratio (they did not search longer when more grapes than the retrievable ones were placed into the bucket). The authors hypothesised that the precision with which lemurs make numerical discriminations may be inferior to that of New and Old World monkeys, even if disentangling the actual numerical abilities of lemurs is

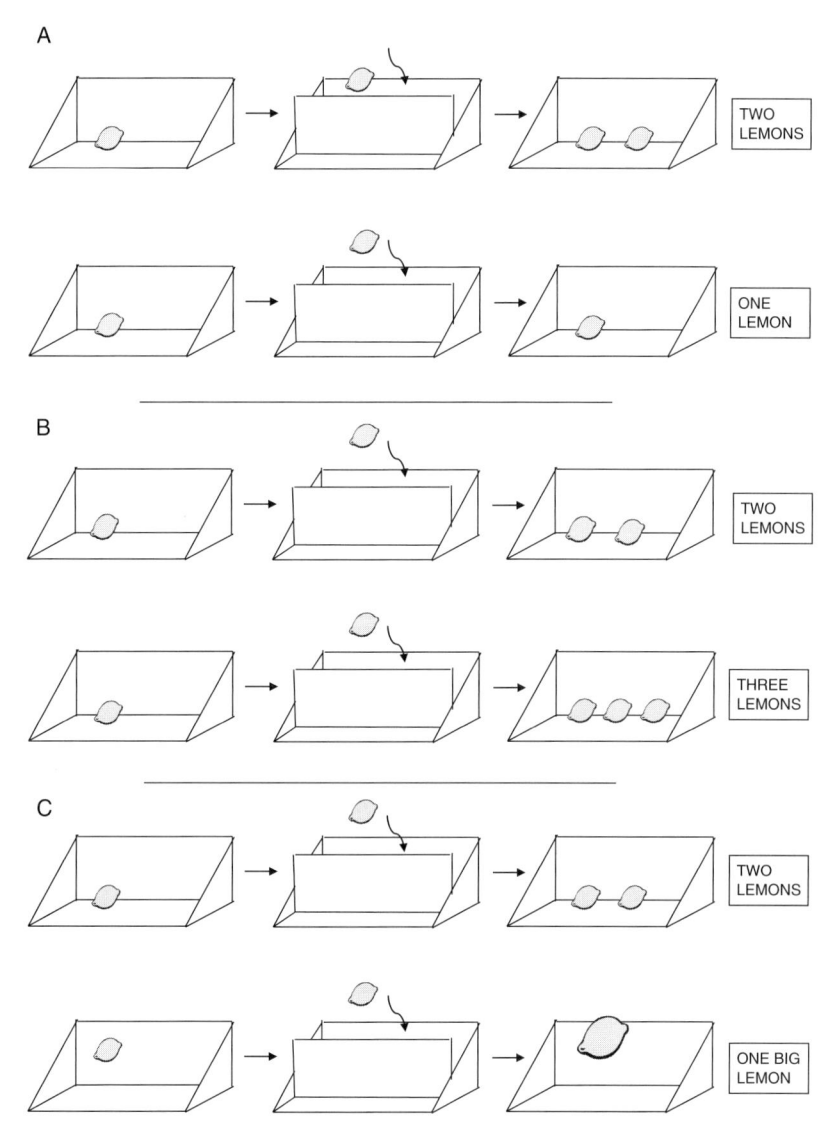

9.2 Lemurs that count: numerical sensitivity in strepsirrhines

Figure 9.4 Test trials carried out by Santos *et al.*, 2005b and experimental apparatus: (a) two lemons (correct outcome) versus one lemon (impossible outcome); (b) two lemons (correct outcome) versus three lemons (impossible outcome); (c) two smaller lemons (correct outcome) versus one bigger lemon (size ≃ size of the two smaller lemons; impossible outcome). Drawing: Carmelo Gómez González based on Santos *et al.*, 2005b.

not easy. Merritt *et al.* (2011) showed that *Lemur catta* could discriminate between numerical pairs (squares or circles presented on a touch screen) with a ratio greater than 1:2 and 2:3 even though the goodness of their performance remained ratio dependent. Jones and Brannon (2012), using food items, tested 113 subjects belonging to different species (*Eulemur* spp., *Hapalemur griseus*, *Propithecus* spp., *Varecia* spp.). They found that strepsirrhines tended to select the highest of two quantities of food differing by a 1:2 ratio but only in the condition 1 versus 2 items. They did not show preference for small quantities that differed by a 1:3 ratio. Instead, they could consistently choose the larger of two food sets when the ratio was 1:3 with no set size limitation (1 versus 3, 2 versus 6, and 4 versus 12 items). Recently, Jones *et al.* (2014) showed that despite 50-70 million years of divergence, lemurs (*Eulemur mongoz*, *Eulemur macaco* and *Lemur catta*) and macaques (*Macaca mulatta*) possess similar numerical acuity when they have to discriminate between two numerosities presented on a touch screen. In all species the accuracy decreased when the ratio between the two experimental numerosities approached one, thus confirming that the discrimination ability is ratio dependent.

Other than reassessing the ability of lemurs to numerically distinguish between different quantities, Merritt *et al.* (2011) tested if ring-tailed lemurs were able to learn numerical rules. They trained the lemurs to respond to pairwise combinations of visual 'exemplars' on a touch screen, each of which containing between one and four elements (squares and/or circles). The lemurs were rewarded (with a sucrose

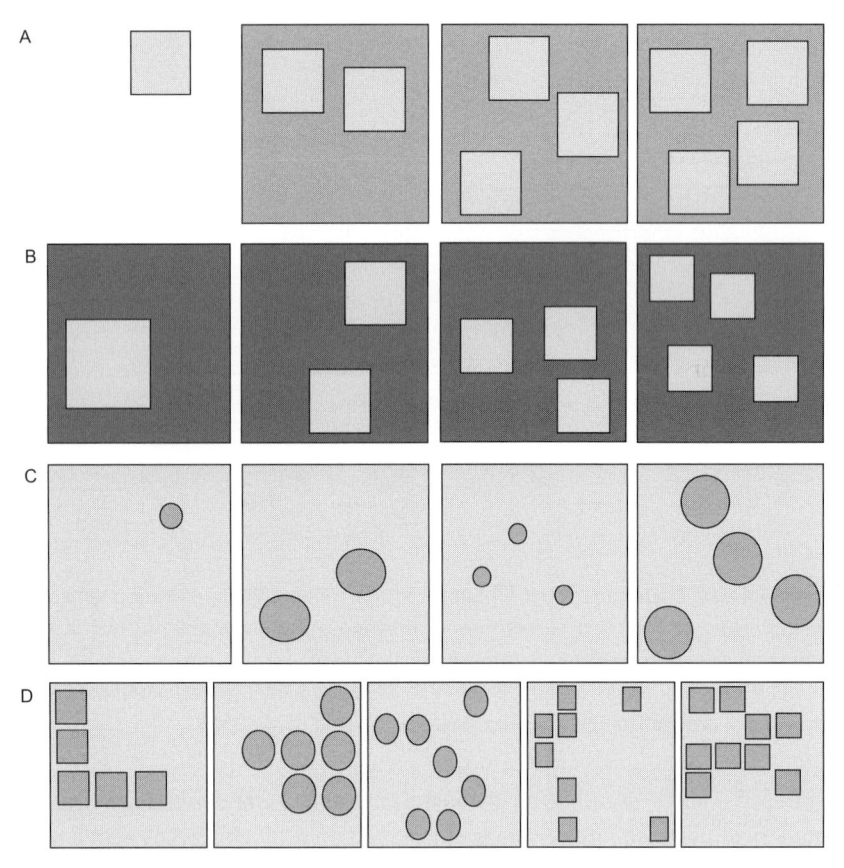

Figure 9.5 Examples of stimuli used in the experiments carried out by Merritt *et al.*, 2011: (a) Element size-constant; (b) overall surface area-constant; (c) random element size groups of stimuli; (d) computer generated stimuli for numerosities 5–9. Drawing: Carmelo Gómez González based on Merritt *et al.*, 2011.

tablet and positive visual and auditory feedback) when they responded (by touching the screen) to the numerically smaller value shown on video. The authors presented the stimuli so to control also for the surface area of the visual elements shown on video. Specifically, the surface area of the stimulus with more elements could be larger, smaller than or the same as the surface area of the stimulus with less elements (see Merritt et al., 2011 for the details of the experimental trials; Figure 9.5).

The authors found that not only could lemurs learn to order numerical values from 1 to 4 (with known stimuli presented during training and also with novel exemplars of the same numerical value) but also to transfer the numerical rules to novel values from 5 to 9. The same ability has been already found in squirrel monkeys, cebus monkeys, baboons, rhesus macaques and humans (Brannon and Terrace, 2000; Smith et al., 2002; Judge et al., 2005; Cantlon and Brennon, 2006). Therefore, Merritt et al. (2011) concluded that at a minimum, their results on lemurs suggested that within the primate lineage, the ability to form ordinal numerical rules is an evolutionary ability that emerged before the separation of strepsirrhine and haplorrhine primates (50–70 million years ago).

Box 9.1 | **by Evan L. MacLean and Brian Hare**

Speaking of which: lemur cognition and sociality

Less than two decades ago – in a comprehensive review of research on primate cognition – Tomasello and Call noted that studies of lemurs were conspicuously lacking, leaving a void in our knowledge of primate cognitive diversity (Tomasello and Call, 1997). Recently however, studies of lemur cognition have proliferated shedding light on a wide range of important evolutionary questions. First, because lemurs are believed to resemble the earliest living primates more so than monkeys or apes, studies of lemur cognition allow powerful inferences about the cognitive characteristics of early primate species (Tattersall, 1982) For example, if lemurs, monkeys and apes share common mechanisms for solving a particular problem, then this aspect of cognition was likely present in the earliest primate species. If this ability is *not* found in other mammals closely related to primates (e.g. colugos or tree shrews) it is possible that these skills reflect a key cognitive adaptation that coincided with the emergence of the primate order. However, if monkeys and apes exhibit cognitive flexibility that is not found in lemurs, these traits may have evolved after the prosimian–anthropoid divergence. Thus lemurs play a special role both in identifying the aspects of cognition that are unique to primates, and also as a phylogenetic out-group for studies with monkeys and apes.

Second, comparative studies within lemurs present unique opportunities for testing hypotheses about the socioecological factors that favour cognitive

Box 9.1 (continued)

Box 9.1 (*cont.*)

evolution. Although lemurs are a closely related clade, lemur species have radiated into remarkably divergent ecological niches, and are characterised by a variety of activity patterns, diets and social systems. Therefore lemurs present a natural experiment in which close genetic relatives, with differing behavioural ecologies can be compared with one another. Against this backdrop, scientists have powerful opportunities to test hypotheses regarding how factors such as social living and dietary strategies relate to species differences in cognition (MacLean *et al.*, 2012). Indeed, early behavioural studies with lemurs provided the inspiration for important hypotheses about primate cognitive evolution, including Jolly's initial formulation of the Social Intelligence Hypothesis (Jolly, 1966). However, until recently few ecologically motivated studies of lemur cognition had been conducted.

Much of our work has attempted to test hypotheses about the adaptive value of cognition through comparisons of lemur species that differ in key aspects of their social organisation. The first of these studies was inspired by the hypothesis that transitive inference (if $A > B$ and $B > C$, then $A > C$) is one cognitive mechanism through which individuals could rapidly infer dominance relationships between members of a social group (Cheney and Seyfarth, 1990). An earlier study with corvids had revealed an interesting species difference in which highly social pinyon jays learned to track and infer hierarchical information more readily than western scrub jays – a closely related but relatively non-social species (Bond *et al.*, 2003). Based on these findings we compared ring-tailed and mongoose lemurs – close relatives that differ markedly in their social organisation – in a similar paradigm. Consistent with the findings from corvids, highly social ring-tailed lemurs outperformed less social mongoose lemurs on tests of transitive reasoning (MacLean *et al.*, 2008). Interestingly, however, this species difference occurred only when the dyadic pairs in the hierarchy were presented in a randomly shuffled sequence prior to the test of transitive inference (e.g., $C > D$, $A > B$, $F > G$, $B > C$, $D > E$). In contrast, when the dyadic pairs comprising the hierarchy were retrained in a manner emphasising the underlying linearity of the sequence (e.g., $A > B$, $B > C$, $C > D$, etc.), both species performed similarly on subsequent tests of transitive inference. Therefore these findings suggested that the two species did not differ in their fundamental ability to make transitive inferences, but rather in their propensity to spontaneously extrapolate the hierarchy from a series of linked dyadic relationships. Interestingly, the condition in which ring-tailed lemurs outperformed mongoose lemurs is more likely to resemble the manner in which dyadic social interactions would be observed in nature. That is, an animal is likely to witness a variety of dyadic dominance interactions, occurring in a piecemeal fashion over time, and the ability to infer unobserved dominance relationships based on this information may be vital to success in large hierarchically organised societies.

Inspired by these findings we extended our focus to other lemur species, as well as additional measures of cognition. Because the Social Intelligence Hypothesis proposes that a key function of cognitive flexibility is to outwit conspecifics for access to food and mates, our next tests measured four lemur species' abilities to exploit social cues in a food competition paradigm. Similarly to previous studies with monkeys and apes, lemurs were first introduced to a human competitor who actively defended monopolisable food from the subject. After this introduction lemurs were tested in a series of trials in which they could attempt to pilfer one of two pieces of food nearby a human competitor. Critically, we varied which piece of food the human competitor could see creating the opportunity for lemurs to strategically target the unwatched item. Although all four lemur species learned to avoid the human competitor during the introductory trials, only the highly social ring-tailed lemurs systematically chose the unwatched item during test trials (Sandel et al., 2011). Thus, these data suggested that in addition to living in social groups similar to those of Old World monkeys, some aspects of ring-tailed lemur social cognition may also be convergent with anthropoid primates.

Most recently, we implemented a modified version of this feeding competition task, as well as a non-social inhibitory control task with an expanded taxonomic sample including six lemur species. Based on the studies described above, we expected that lemur species characterised by large social groups would perform better in the food competition task than lemurs characterised by smaller societies. But would the pattern of species differences be similar for a non-social cognitive task? Are the evolutionary relationships between sociality and cognition domain specific, or does group living select for cognitive flexibility more broadly? The results of this study provided strong support for a domain-specific version of the Social Intelligence Hypothesis. Across the six lemur species, performance on the food competition task was significantly correlated with species-typical group size, suggesting a close relationship between social complexity and cognitive skills for outcompeting group mates. In contrast, although lemur species also varied significantly on the non-social inhibitory control test, there was no relationship between social group size and performance on this task.

In addition to the fascinating variance within these species, lemurs have also provided a foundation for developing cognitive tasks that can be employed in broader comparative studies. For example, the study of lemur inhibitory control described above was initiated through a collaborative working group tasked with coordinating advances in experimental and phylogenetic approaches to the study of cognitive evolution. Building

Box 9.1 (*cont.*)

on our initial success with lemurs, a collaborative network of researchers collected additional data from birds, rodents, carnivores and anthropoid primates, allowing a phylogenetic comparison of 32 species tested on this task (MacLean *et al.*, 2014). Given their remarkable behavioural and morphological variance, it is likely that lemurs will continue to play an important role in the development of cognitive tasks that can be used in future broad-scale studies of comparative cognition.

The studies described above illustrate the value of lemurs in testing hypotheses about the evolution of cognition. Although we have highlighted our research exploring the links between social systems and social cognition, variance in other aspects of lemur ecology provides similarly powerful opportunities for testing a wide range of evolutionary hypotheses. For example, a recent study with four lemur species revealed that highly frugivorous ruffed lemurs exhibited more robust spatial memory in a variety of contexts than less frugivorous species (Rosati *et al.*, 2014), including ring-tailed lemurs who outperform ruffed lemurs on some measures of social cognition. Thus variance in lemur cognition has likely evolved in response to diverse selective pressures, and the magnitude and direction of species differences in cognition are also likely to be domain specific. These findings underscore the need for additional studies that probe other domains of lemur cognition, and the relationships between these skills and species differences in socioecology.

A second exciting future direction in lemur research involves the direct comparison of lemur, monkey and ape cognition. Although lemur problem solving was once believed to differ qualitatively from that of monkeys and apes, recent research challenges this notion. For example, with regard to numerical discrimination, sequence learning and causal reasoning tasks, some lemur species perform quite similarly to monkeys (Santos *et al.*, 2005b; Merritt *et al.*, 2007; Jones *et al.*, 2014). In contrast, on other social-cognitive measures, the differences between lemurs and anthropoid primates appear to be more pronounced (Fichtel and Kappeler, 2010). Therefore an important aim for future research will be to identify the extent of similarity and difference between lemurs and anthropoid primates across a wide range of cognitive measures.

In sum, recent studies on lemur cognition have led to major insights about the processes through which cognition evolves, as well as the cognitive traits that are shared across primate species, or uniquely evolved in particular primate lineages. Ultimately however, we still have much to learn about the minds of lemurs, and many of the most exciting discoveries are surely yet to come.

9.3 Should I stay or should I go? Lemurs, patience and self-control

'The black-and-white ruffed lemur dangles by her hind legs at the top of the leafy canopy, plucking figs from the branch below. At some point she has consumed many of the figs from the branch, leaving a few small fruits hidden beneath the leaves. The lemur now faces a choice: should she continue to search for the remaining figs or move on to another branch full of fruit?'. This is how Stevens and Mühlhoff (2012) depict a case of intertemporal choice (a choice between options with different time delays to reward; Read, 2004) that a lemur may face in the wild. Over the last decade, lemurs' abilities have been tested under different circumstances, using experimental procedures also applied to other primates. It has been tested whether lemurs after a training phase would learn to refrain from immediately taking few food items and wait for a larger reward (intertemporal choice; Stevens and Mühlhoff, 2012) or if they would learn to select the smallest of two rewards in order to get the largest one instead (reverse–reward contingency; Genty et al., 2004; Genty et al., 2011; Glady et al., 2012). It was also investigated if the lemurs having to decide between a safe and risky option (food reward constant in amount and reward in variable amount, respectively) would secure the certain amount of food or would 'gamble' by selecting the variable option (with the possibility of getting more but also less food) (variance sensitive choice; MacLean et al., 2012).

In risk centred trials, five lemurs (two *Lemur catta*, two *Eulemur mongoz* and one *Varecia rubra*) were faced with a couple of black and white images on a monitor as stimuli, each one associated with a particular reward amount (MacLean et al., 2012). When the subjects pressed either stimulus, the apparatus dispensed the number of pellets associated with that choice. In MacLean et al. (2012)'s first experiment an option was associated with a constant reward of 1 pellet whereas an option yielded 2 pellets half of the time and 0 pellets the other half of the time. Thus, in this case, the average reward was the same for the constant and for the variable option. Lemurs showed a strong preference for the constant reward option, even though both options yielded equal net payouts. This is in line with observations from other primate species, including humans, which violate the classical economic utility theory (von Neumann and Morgestern, 1944; Rieskamp et al., 2006). The most basic expectation of this theory is that individuals should equally choose between an option A and an option B if both options provide the same expected utility (expressed as probability × reward magnitude). For example, when individuals are presented with the choice between a safe option, allowing them to obtain 1 unit of food (expected utility = $1 \times 1 = 1$) and a risky one, allowing them to obtain either 2 units or nothing (expected utility = $0.5 \times 2 + 0.5 \times 0 = 1$), they should choose both options similarly since the average expected reward is the same. Instead, a literature review of risk sensitivity in over 25 non-primate species revealed that the majority of them were risk averse (Kacelnik and Bateson, 1996). Similarly, human and non-human primates, including lemurs, tend to avoid risky prospects (Hayden and Platt, 2007; Heilbronner et al., 2008; MacLean et al., 2012).

According to the Expected Utility Maximisation Hypothesis, when the magnitude of reward associated to the risky option becomes significantly larger than the one associated to the safe option, individuals should switch from risk-aversion to risk-proneness, in order to choose the larger prospective reward (van Neumann and Morgestern, 1944; Stevens, 2010). MacLean et al. (2012) tested this hypothesis in lemurs by gradually increasing the average reward of the variable, 'risky' option across successive conditions (each one of 10 sessions). The experiment showed that lemurs were sensitive to the increase of the risk premium but overall remained risk averse. In fact, they showed preference for the constant option when the variable option yielded 1.5-2 times the constant option's net payout, they were indifferent to the two options when the average variable reward was from 2.5-5 times higher than the constant reward and they showed a non-significant preference for the variable option when its average reward was 7.5 times more than the constant option. The main argument used to discuss this result is that risk avoidance is a useful strategy in unpredictable environments, where the behaviour that leads an animal to obtain more food today may also lead it to starvation tomorrow (MacLean et al., 2012). Madagascar is indeed characterised by an extremely unpredictable environment and it has been hypothesised that its long historic isolation, low productivity and erratic and severe climate (with cyclical but irregular droughts, storms, cyclones, etc.) have played a major role in lemur evolution, possibly including their risk-aversion propensity (Wright, 1999; MacLean et al., 2012).

The Unpredictable Environment Hypothesis appears to be supported by studies on other species.

Two human populations of Andean herders living in different environments diverge in their risk propensity (Kuznar, 2001). The population of the high sierra, which is a cold and unpredictable environment with limited resources is more risk averse than the population living in the richer and more stable Andean puna. When given a choice between a set number of livestock, or a lottery ticket with a set probability of winning two times as many livestock, people from the high-sierra community required a 90% probability of winning 100 livestock in order to decline the sure delivery of 50 animals. Instead, the people of Andean puna accepted the risky option even with a significantly lower reward (Kuznar, 2001). In apes, it was observed that when bonobo and chimpanzee individuals were tested in a task that allowed subjects to choose either a certain intermediate amount of food or a gamble with a 50% probability of receiving a larger or smaller reward, chimpanzees chose the risky option more frequently than bonobos. Also these results were interpreted in the light of the different environment where the two ape species live, with chimpanzees dealing with more variably distributed food resources than bonobos (Heilbronner et al., 2008). Lemurs' aversion towards risky options seems to be particularly high. For example, when tested on a similar task, bumblebees started to prefer the variable option when its mean reward became twice that of the constant option (Real et al., 1982), and dark-eyed juncos when the mean risky reward was 1.5 times that of the 'safe' option (Caraco and Lima, 1985). In contrast, the majority of lemurs that were tested showed significant risk-aversion to these same reward ratios. Of course, comparisons can only be qualitative and the choice can depend on a plethora of factors, including species-specific peculiarities, energy budget, problems typically faced by a species in its environment, etc. (MacLean *et al.*, 2012).

As mentioned at the beginning of section 9.3, Stevens and Mühloff (2012) tested lemurs' choice between options that differed for the delay with which the reward was yielded. Individuals can make intertemporal choices based on temporal discounting (the value given to the reward decreases as the time delay to receiving that reward increases) or by maximising their food intake rate in the short- or long-term (by considering the amount of food reward received, the delay in receiving it, the handling time to process and consume the food, and/or inter-trial interval; see Stevens and Mühloff, 2012 for a comprehensive description).

After a training phase, five lemurs out of twelve (three *Varecia variegata*, one *Varecia rubra*, and one *Eulemur macaco;* the other lemurs did not complete the experiment) were faced with the choice between two options: one giving two apple pieces immediately and the other yielding six apple pieces after a certain time (Stevens and Mühloff, 2012). The apple pieces were placed on two movable horizontal slides at either side of a box, covered with a Plexiglas barrier with two holes on the subject's side. The option could be selected by the lemurs by reaching through the holes on either side. Once the choice was made, the experimenter removed the unwanted reward and after the programmed delay pushed the slide with the chosen reward forward, so that the subject could reach the food (for further details on the procedure see Stevens and Mühloff, 2012). The smaller reward was provided immediately whereas the larger reward was given with an increasing delay over the course of the experiment. Results show that lemurs were indifferent to the two options with a delay of 9–25 seconds and that, similarly to common marmosets, they do not tend to maximise their food intake rate (Stevens *et al.*, 2005; Stevens and Mühloff, 2012). It appears that different variables such as the social structure, the body mass, and the dietary constraints influence the performance of intertemporal choice. For example, animals with larger body size (with lower metabolic rates), flexible social structure (e.g. fission–fusion, possibly favouring inhibitory abilities) or a diet relying on resources that cannot be immediately accessed (extractive foragers, sit-and-wait hunters, gummivores) may have developed the ability to wait longer for food. In this respect, the selective pressures enhancing the ability to wait may not have been so strong for lemurs (Stevens and Mühloff, 2012). As for the risky options, we cannot exclude that the unpredictability of food resource availability has instead pushed lemurs towards the development of a *hic-et-nunc* strategy aimed at quickly obtaining the amount food available at any given time, 'here and now'.

Another form of self-control is expressed in the so-called reverse–reward contingency tasks, that is the ability to inhibit a natural tendency to reach for the greater of two quantities of food (Anderson, 2001). The classical procedure used to test this ability consists in showing two food quantities to the experimental subject. When the subject reaches towards one quantity, the non-selected quantity is given as a reward (Genty *et al.*, 2004). Most of the primates in which the reverse–reward contingency has been tested have difficulties in performing the task in its original form including children over four years old (Mischel et al., 1989), great apes (*Pan troglodytes*, Vlamings et al., 2006; Uher and Call, 2008; *Pan paniscus*, Vlamings et al., 2006; *Gorilla gorilla*, Vlamings et al., 2006; Uher and Call, 2008; *Pongo pygmaeus*, Shumaker et al., 2001; Vlamings et al., 2006; Uher and Call, 2008) and monkeys (*Macaca mulatta*, Murray et al., 2005; *Cercocebus torquatus lunulatus*, Albiach-Serrano et al., 2007). Individuals of other monkey species failed to master the task (*Macaca fuscata*, Silberberg and Fujita, 1996; *Saimiri sciureus*, Anderson et al., 2000; *Saguinus oedipus*, Kralik et al., 2002).

Owing to this problem, a corrected procedure was introduced to help individuals to learn the reverse-reward contingency rule. According to this procedure, the experimental subjects receive no reward if they select the larger quantity but still receive the larger quantity if they choose the smaller one (large-or-nothing contingency; Anderson et al., 2000). The first to propose the reverse contingency task to strepsirrhine individuals were Genty et al. (2004), who tested five brown lemurs (*Eulemur fulvus*) and four black lemurs (*Eulemur macaco*) with one raisin versus four raisins presented at two different sides of a tray. The lemurs failed passing the test with the original procedures but the majority of them started showing a significant preference for the smaller quantity after 50 trials when the large-or-nothing procedure was applied (if they selected four raisins they received nothing and if they selected one raisin they received four). The lemurs that had failed also in mastering the corrected task received additional 'teaching': in the event of selection of the larger array, the experimenter left the tray in front of the cage where the subject was kept and did not withdrew it until the lemur selected the smaller quantity. After this additional training, all lemurs learned to refrain from selecting the larger quantity to chose, instead, the smaller quantity. Some lemurs maintained this ability when novel combination of quantities were presented (e.g. 2:4; 1:3; 5:6) at least with certain ratios, thus suggesting that lemurs applied a general learned rule to novel problems. After learning to select the smaller quantity using the large-or-nothing procedure, the lemurs were able to select the smaller quantity also in the classical procedure (receiving less if selecting more) that they did not master before the training with facilitated procedures (Genty et al., 2004). However, later Genty et al. (2011) found that one out of five individuals *Eulemur fulvus* succeeded mastering the reverse contingency task after 1300 trials without receiving previous training via the large-or-nothing task. The individual was also able to generalise the performance when new food arrays were proposed. A learning effect similar to that obtained via the large-or-nothing contingency procedure was observed when lemurs were trained with the qualitative version of the reverse-reward contingency task (Glady et al., 2012). After assessing animals' food preference, two different types of food were presented to the subjects and the more preferred food (e.g. fig) was yielded when the less preferred (e.g. carrot) was chosen. Two individuals of *Eulemur fulvus* out of four could master this task and proved able to apply the rule to novel food combinations (e.g. fig versus raisin, fig versus apple, raisin versus apple, etc.) and to transfer this ability to different quantities (thus selecting the smaller quantity to have the larger one, as foreseen by the classical procedure). They acquired the ability to select the smaller quantity more rapidly (600 trials) than in Genty et al. (2011)'s experiment in which no previous 'learning facilitation' was provided.

Without previous training, lemurs needed more trials to reach the same performance criterion as macaques and mangabeys (cf. Murray et al., 2005; Albiach-Serrano et al., 2007; Genty et al., 2011). It is interesting to notice that the lemur outperformed the weakest macaque (1300 trials for the lemur versus 2700 trials for the macaque; Murray et al., 2005; Genty et al., 2011) but all rhesus macaques and mangabeys eventually mastered the task (Murray et al., 2005; Albiach-Serrano et al., 2007), whereas only one lemur did so in Genty et al. (2011)'s experiment. Nevertheless, as pointed out by Genty et al. (2011), the comparison with monkeys requires caution because of differences in the experimental procedure. Similar to lemurs, a study on capuchin monkeys (*Cebus apella*) found better performance when different kinds of food were first used instead of different quantities of the same food (qualitative version of the task; Anderson et al., 2008) even if capuchin monkeys were not able transfer the ability to the quantitative task. Again, the problem here lies in the procedure. The food used for capuchin monkeys (sweet potato) was of moderate attractiveness, thus probably not eliciting enough motivation to perform the task. In different primate species the reverse-reward contingency performance is improved when food is associated with symbols and the choice is made on symbols instead of on real food. Symbols can be, for example, different colours (tamarins; Kralik et al., 2002), Arabic numerals (chimpanzees; Boysen and Berntson, 1995), tokens (in capuchin monkeys: Addessi and Rossi, 2010), images (children: Carlson et al., 2005). Human and non-human primates seem to be facilitated in mastering the quantitative reverse-contingency task through symbols probably because the symbolic stimuli serve as a representational aid for overcoming a natural perception-motor bias towards the larger quantity of food (Boysen and Berntson, 1995). We are sure that some researchers in the field are thinking about proposing (or are already proposing!) this kind of task to lemurs. It will be interesting to read in future papers if lemurs respond in a way similar to monkeys and apes when symbols are used.

In summary, lemurs can perform the reverse contingency task with different food qualities and when the choice is large or none; after being trained with these types of contingency tasks, they are also able to inhibit their choice towards the larger quantity. However, except for one case, lemurs were not able to learn – by simply increasing the number of trials – to select the smaller quantity to get the larger one, mostly because they developed a strong side preference leading them to select the food either on the left or on the right side. These results are again discussed in the light of the environmental pressures. Self-control abilities may be less advantageous in an unstable environment where food resources are sparse and temporally available and where delaying food intake would be too risky for survival (Genty et al., 2011).

> **Box 9.2** | **by Elsa Addessi and Elisabetta Visalberghi**
>
> **Speaking of which: choice-making and capuchin monkeys**
>
> We have always been interested in understanding how other individuals think and what they consider important when they choose to act in a certain way. Humans are easier to interpret than other animals since they can verbally answer to questions. Thus, we had to figure out other ways to understand the mechanisms underlying the behaviour of non-human animals. This challenge led us into animal cognition.
>
> It was by chance that we ended up studying one of the most fascinating primate species, the capuchin monkeys. Something we will never regret. These monkeys are special in many ways. They have larger brains than expected for their body size and a long life span in comparison to other primates of similar size. Capuchins are opportunistic omnivores with sophisticated foraging skills. They are strongly motivated to manipulate and combine objects and perform complex sequences of actions with objects. Moreover, they use tools both in the wild and in captivity. Capuchins are relatively tolerant of each other and interested in others' activities. Youngsters develop at a slower rate than most other monkey species and have an extended period of maternal dependency during which they learn a large variety of skills. In short, despite 35 million years of independent evolution, capuchin monkeys show many striking similarities with hominids (for an extensive review of capuchin monkeys' biology see Fragaszy et al., 2004).
>
> We had the opportunity to study a colony of tufted capuchin monkeys and spent hours and hours looking at what they did. We could ask them 'questions' by setting up problems to solve and looking at how individuals behave when alone or in the presence of group members.
>
> Humans heavily rely on cultural cumulative knowledge. Technology as well as moral norms are passed through generations thanks to a variety of processes among which imitation, active teaching, and language play very important roles. It has long been argued that human mental experiences can only be detected through the use of language, whereas cognitive ethologists consider that at least some of these processes are present also in non-human animals. Many other researchers became interested in animal cognition after the dark age of behaviourism; thus, we had the opportunity to be part of the stimulating debates on animal cognition that occurred in the last decades.
>
> In the first review, critically evaluating the evidence supporting the widespread idea that monkeys imitate, in other words, that they learn novel behaviours by watching proficient models, Visalberghi and Fragaszy (1990) negatively answered the question 'Do monkeys ape?' entitling their article. More than ten years later their answer was still the same: monkeys do

[1] 'apparent imitation resulting from directing the animal's attention to a particular object or to a particular part of the environment' (Thorpe, 1963, p. 134).

not imitate, whereas social influences, such as local enhancement,[1] stimulus enhancement,[2] and object movement re-enactment,[3] are indeed present (Visalberghi and Fragaszy, 2002).

However, the tasks considered in the studies reviewed in the above articles involved behaviours particularly difficult to acquire (such as tool use). Therefore, we started a series of experiments focused on feeding behaviour, something that should be easier to learn than using a tool. Feeding is a *condicio sine qua non* for an animal's survival, and food selection and processing are ideal behavioural domains in which to investigate social learning processes.

Visalberghi and Fragaszy (1995) showed that capuchins, as well as many other omnivorous species, are neophobic (i.e., they initially avoid novel food or eat a small amount of it) and that novel foods are eaten more when individuals are with group members than when they are alone. We further evaluated whether individuals learn about food palatability by watching what others eat (Visalberghi and Addessi, 2000, 2001). In these experiments the observer was presented with a novel food whose colour and odour could match or not the colour and odour of a different food eaten by its group members. Since observers and group members were tested in adjacent enclosures separated by a Plexiglas panel, we assume that the observer received primarily visual information on the food eaten by its group members, although the two foods had also different odours. We found that eating was socially facilitated also when the colours of the food eaten by the observer and by group members were different. Moreover, the observer did not consume more of the colour matching food than of the non-matching food, even when given a choice between two foods, only one of which matched the colour of the food eaten by group members. In other words, although an individual is more likely to eat novel foods when group members are around, the colour of the food that group members are eating is irrelevant. Therefore, the assumption that capuchins learn about food palatability from others is unwarranted. Interestingly preschool children studied with the same paradigm eat more of a novel food only when it matches in colour the food eaten by a demonstrator (Addessi *et al.*, 2005).

As several strepsirrhines and many platyrrhines, capuchin monkeys present a polymorphism in the opsin gene located on the X chromosome; thus, different alleles code for sensitivity to different wavelengths, and because the gene is X-linked, all males are dichromatic whereas about one third of females are trichromatic (Jacobs, 2008). However, it is unlikely that this factor accounted

Box 9.2 (continued)

---

[2] 'the enhancement of the particular limited aspect of the total stimulus situation to which the response is to be made' (Spence, 1937, p. 821).

[3] object-movement re-enactment occurs when watching an object or its parts moving and producing a salient outcome motivates the observer to reproduce that outcome (Huang and Charman, 2005).

Box 9.2 (*cont.*)

for the different role of social influences on food neophobia in capuchins and children. In fact, all our capuchins successfully discriminated pairs of plastic chips having the same colour of the dyed foods (Addessi, 2003).

Given the above results, how is it that capuchins end up eating the same food and making the same choices most of the times? The nutrient content of foods plays a major role in determining individual preferences about novel foods and social influences do not affect these nutrient-based preferences (Visalberghi *et al.*, 2003). Preferences for sweet foods and aversion to bitter substances (bitterness is often associated with toxic compounds), coupled with a neophobic response and food aversion learning (the capability of associating the ingestion of a food with its post-ingestive consequences), are sufficient to enable capuchins to safely choose among foods. This means that to learn what to ingest and what to avoid, an individual does not need to observe other group members, who after all may often be absent or not behaving informatively in a given situation.

Nevertheless, other social influences that do not take into account the whole set of information present in the behaviour of others are still at work. Social facilitation,[4] and local and stimulus enhancement (see Chapter 9, footnotes 1 and 2 for definitions of local and stimulus enhancement) increase the chances that a naïve individual will feed at the same time and place as its group members, with the result that its food choices are similar to theirs. In short, there is evidence that capuchin monkeys learn *with* other individuals rather than *from* them (Visalberghi and Fragaszy, 2002).

In the future, our research will be increasingly based on the integration of knowledge and techniques belonging to different disciplines. With no doubt there are issues that can be better understood by carrying out experiments in a laboratory setting (e.g., the role of social influences on food choice discussed above) and issues that only field studies can address (e.g., the adaptive value of a given behaviour). Nevertheless, as the Swiss primatologist Hans Kummer (e.g., Kummer, 1995) repeatedly stressed these two approaches should complement one another's.

An interesting example comes from neuroeconomics, in which there is a growing interest in using animals as models to understand human decision-making (Kalenscher and van Wingerden, 2011). Within a comparative framework, a combined laboratory-field approach is required to understand how the ecology of a species shapes its decision-making behaviour over time (i.e., involving outcomes that can be realised in the future) and under risk (i.e., under conditions of variability in the rate of gain) (De Petrillo *et al.*, 2015).

---

[4] 'an increase in the frequency or intensity of responses or the initiation of particular responses already in an animal's repertoire, when shown in the presence of others engaged in the same behavior at the same time' (Clayton, 1978, p. 374).

> For instance, capuchin monkeys presented with a series of choices between a certain, 'safe' option (yielding always four food items) and a variable, 'risky' option (yielding either one or seven food items, with the same probability) were risk prone, that is, they preferred to choose the risky option over the safe option, even if the two options offered on average the same amount of food. Capuchins' decision-making under risk in the laboratory mirrors their risk-prone behaviour in the wild, where they often rely on unpredictable food sources. However, only from fieldwork is possible to pinpoint which features of feeding ecology (e.g., diet breadth, percentage of fruit in the diet) are more relevant in shaping risk preferences. At the same time, only laboratory studies permit to employ the degree of experimental control required to rigorously evaluate individual decision-making behaviour.
>
> Many other ethologists are indeed integrating field and laboratory approaches. In this respect, the technologies developed in recent times (e.g., non-invasive genetic and hormonal analyses, transponders, drones) are becoming essential for gathering data whose collection was possible only in the laboratory, or just unfeasible before. The Etho*Cebus* Project (www.ethocebus.net) in which we are involved has combined experimental and observational approaches to study the ecology, behaviour and cognition of one population of wild bearded capuchin monkeys that habitually use stone tools (e.g. Visalberghi *et al.*, 2009) and will do even more along these lines in the near future.

9.4 Do lemurs ape? Lemur social intelligence, tradition and future challenges

This title is a bait; but we could not resist the temptation to start the last section of this book by aping previous titles regarding the unresolved issue of imitative social learning in non-human primates (starting: 'Do monkeys ape?', Visalberghi and Fragaszy, 1990). While social learning has been demonstrated in monkeys and apes, their actual ability to socially acquire new skills via imitation (understanding and focusing on the behaviour used by others to solve tasks and not just on the stimulus or the result of an action) is debated, with evidence pointing to the lack of actual imitative learning in monkeys (Visalberghi and Fragaszy, 1990; Tomasello, 1996; Visalberghi and Fragaszy, 2002; van Schaik and Burkart, 2011). It is not the prerogative of this chapter detailing the possible types of social learning (imitation, emulation, observational learning, etc.) for the simple reason that the quantitative investigation of social learning in lemurs has just started. And if the issue of imitative learning in haplorrhines is controversial, the possible ways through which strepsirrhines may learn from others have not even entered the debate. At this stage, experimental studies are directing their efforts in determining whether lemurs *are* able to engage in social learning at all (e.g. Kendal *et al.*, 2010; Stoinski *et al.*, 2011). Because it is premature considering the mechanisms underlying social learning in lemurs, let us take a step behind to see what is the state of the art on the social components of learning in lemurs.

A building block of social learning is probably the ability to pay attention to others. In a sight oriented world, this can be achieved in various ways, including following the gaze of group mates (Ferrari *et al.*, 2006). Rosati and Hare (2009) pointed out that whether it involves low level processes or more sophisticated social-cognitive skills, gaze following (looking in the direction that others are looking) allows individuals to learn things like food sources, predators, and conspecifics that others have detected in the environment. Gaze following has been found in all great apes and some New and Old World monkey species (e.g. see Emery *et al.*, 1997; Tomasello *et al.*, 1998; Neiworth *et al.*, 2002; Braeuer *et al.*, 2005; Burkhart and Heschl, 2006).

In lemurs, the presence of gaze following seems to be species dependent (at least based on the few available studies). Shepherd and Platt (2008) used a telemetric gaze-tracking system to record orienting behaviour of ring-tailed lemurs and found that they preferentially gazed towards other individuals and could follow an other lemur's gaze while moving and interacting in social and ecological environments. The presence of gaze following in *Lemur catta* has been confirmed by a subsequent experimental study by Sandel *et al.* (2011), who faced the lemurs with two conditions: the gaze condition, in which the experimenter fixated on the lemur and then looked directly up by moving his entire head, repeating this motion for 10 seconds; and the control condition, in which the experimenter looked directly at the lemur for 10 seconds. The authors found that *Lemur catta* looked up significantly more frequently in the gaze condition than in the control condition. However, using the same experimental procedure, they found no evidence of gaze following in other lemur species (ruffed lemurs, black lemurs and mongoose lemurs). Overall, when data were pooled, gaze following was recorded only in 10% of trials for lemurs, versus 30–85% for monkeys and apes recorded in previous studies (Sandel *et al.* 2011). It is worth remarking that in 2009, Ruiz *et al.* found evidence that two species of *Eulemur* (*E. fulvus* and *E. macaco*) could at least co-orient with conspecifics because the tested individuals were more likely to choose objects correctly after having looked in the same direction as the model. Interestingly, Ruiz *et al.* (2009) propose a mechanism for the evolutionary origins of more complex gaze following, which they call 'gaze priming'. According to this mechanism, following the gaze of others can draw an animal's attention to a place, e.g. where a food resource or a potential danger can be found, thus guiding the animal towards or away from that place. This process does not require any sophisticated cognitive ability to be enacted.

The fact that ring-tailed lemurs appear being able to follow the gaze of other conspecifics is possibly due to the complex social groups they live in. In line with this hypothesis are other results reported by Sandel *et al.* (2011) who found that ring-tailed lemurs performed better than three other closely related species in tasks aimed at assessing social-cognitive abilities. In the first experiment, a food reward was placed in proximity of an experimenter that was facing away from the subject and another (identical) food reward was placed in proximity of an experimenter looking in the direction of the subject and the food. In the second experiment, the two food rewards were placed, equidistantly, in proximity of a single experimenter oriented in profile so that one of the food trays was placed in front of his face and the other behind his head. In both experiments, *L. catta*, more than other lemur species, showed preference towards the food that was not looked at or in front of the experimenter. To explain this result, Sandel *et al.* (2011) resumed Jolly 1966's Social Intelligence Hypothesis that *L. catta* may have undergone convergent social-cognitive evolution with haplorrhines possibly because they live in the largest and most complex social groups among strepsirrhines. The support to this hypothesis also comes from a study by MacLean *et al.* (2013) who investigated possible correlates of cognitive skills in six species of lemurs. The groups were previously tested in a non-social inhibitory task to exclude that interspecific differences in social cognition may significantly bias the results (even if, of course, this possibility remains). The tests used were analogous (but not identical) to those reported above by Sandel *et al.* (2011). Lemurs, including *Lemur catta*, successfully targeted the food that a competitor (simulated by an experimenter) could not see based on his body and head orientation but they were indifferent to experimenters facing the food with either the mouth or the eyes covered. Other than the scarce importance of eyes in determining lemur response, the authors found that brain size (measured either absolutely or relatively to body mass) did not predict lemur performance whereas group size was positively correlated with social-cognitive skills relevant to assessing a competitor's awareness. MacLean *et al.* (2013) posit that larger groups should require more flexible cognitive skills for competing with conspecifics and, in fact, group size or grooming network size seem to be good indicators of the relative neocortex size, at least in anthropoid primates (Barton, 1999; Kudo and Dunbar, 2001; Dunbar and Shultz, 2007).

At this point, we gathered the first hints that lemur cognition is socially influenced and that, at a minimum, lemurs are able to orient towards the attention of other individuals, even if the presence of gaze following has not always been detected. On the other hand, lemurs are smell-oriented and can heavily rely on acoustic cues to detect environmental stimuli or for alert (see Chapters 1 and 2). Thus, visual cues are not necessarily the primary mechanism used to obtain information from or about others. Intriguingly, Fichtel and Kappeler (2011) observed that different sifaka species (*Propithecus verreauxi* and *Propithecus coquereli*, respectively) exhibit the same alarm call system but show striking differences in the usage and perception of some of the alarm calls. In their study, the authors found that Verreaux's sifaka exhibited anti-terrestrial predator responses after playbacks of growls in the population with a higher threat of predation by terrestrial predators, whereas Coquerel's sifaka living in a raptor-dominated habitat seemed to associate growls with a threat by raptors. The authors interpreted this differential comprehension and usage of alarm calls as the result of social learning processes enhanced by different predation pressures. Even if this is not a case of innovation, it can be considered as an interesting instance of behavioural variation (*sensu* Fichtel and Kappeler, 2010), which might be one of the possible precursors of innovation.

Innovation in primate social groups can occur more or less rarely (Kummer and Goodall, 1985; Huffman, 1996; Leca *et al.*, 2007; 2010) and an interesting aspect is to check if other members of the group or community are able to acquire and maintain novel abilities. The most famous examples in this respect are the spread of the food-washing behaviour in Japanese macaques (Galef, 1992), stone play and stone handling in the same species (Huffman, 1984; Huffman and Quiatt, 1986) or the experiments on tool use learning in capuchin monkeys (Visalberghi and Fragaszy, 2002 for review). More or less anecdotally, 'innovation' has been reported in *Lemur catta*. In the Berenty forest, Alison Jolly (2004) reports that after being expelled from the troop and losing the fights with the former group mates for a while, 'the next day the subordinates...developed a tactic – the only innovation I have seen in ringtailed lemurs. Almost always when troops confront each other, the actual aggression is carried out by only a couple of animals. Now, suddenly, Sly and all her friends presented a united front'. The story ends with subordinates eventually gaining the territory back. Although this example may not exactly represent an innovation, it is certainly an example of a rare behaviour that provides insights on the too often neglected behavioural flexibility of lemurs. In a captive group, ring-tailed lemurs were observed performing a behaviour that was not observed in the wild: some individuals would immerge their tail in the water and then lick the tail for drinking (Hosey *et al.*, 1997). The individuals that were allegedly not performing that behaviour, started doing it sometimes expressing also an incomplete motor pattern (the author cautiously suggested that the lemurs might have tried to copy the behaviour). The author could only suggest that social learning was in place because the behaviour was already present in the group when the observations started. Schnoell and Fichtel (2013) reported on a supposedly novel feeding behaviour observed in one out of four followed groups of *Eulemur rufifrons* in the Kirindy forest starting 2009. The behaviour consisted of depleting spider nests of social spiders of the genus *Stegodyphus* by opening each nest, breaking into the central part where the spiders live and store their prey, and removing food items from it. The authors could not rule out either possibility that the behaviour was 'reinvented different times' or transmitted to others in 2011 (when the behaviour was observed in more than one individual of a group).

Kappeler (1987) was the first to experimentally verify the process of acquisition and propagation of a new feeding habit (flipping a dish over to obtain food concealed underneath) in a captive group of *Lemur catta*. Similar experiments were then repeated in *Eulemur* spp. (Fornasieri *et al.*, 1990; Anderson *et al.*, 1992). Kappeler (1987) reported that six of eight animals that acquired the new feeding system were immature, with no effect of rank or kinship. Consistently, young ring-tailed lemurs at the Beza Mahafaly Special Reserve (Madagascar) showed behaviours consistent with basic social learning through behavioural synchrony in feeding with a nearest neighbour (O'Mara and Hickey, 2012). The same authors

(O'Mara and Hickey, 2012) postulated that the basic learning rules of feeding when a close, older individual feeds and of individual exploration may provide the foundation for the more elaborate social learning needed for complex extractive foraging or the transmission of traditions. Some evidence of social learning also comes from experimental trials conducted on the ring-tailed lemurs of Berenty (Madagascar) (Kendal et al., 2010). Lemurs were faced with tubes (puzzle feeders) that contained food and could be opened by flipping the flap or swivelling and opening the disc to the left. One subgroup of lemur demonstrators learned to open the box using the flap and another subgroup by sliding the disc (see the original article for the exact procedure). Due to the highly despotic nature of *Lemur catta* (see Chapter 4), dominant females were the first to access the tubes and were therefore used as demonstrators. After adjusting the experimental protocol so that dominant females could not monopolise all the puzzle feeders, other lemurs were allowed to access the tubes, after watching the demonstrators performing the task. Results show that lemurs socially learned how to open the box for the flipping option but not for the other option (Kendal et al., 2010). Using three types of feeding boxes that could be opened with different techniques (fliptop, round-top, and cylinder), Dean et al. (2011) found no evidence of social learning in captive *Varecia variegata* and *Varecia rubra*, but found that females were faster than males in solving the box in male-biased groups (probably because there were less females competing for dominance) but not in female-biased groups. Conversely, male rates of solving the puzzlebox increased as the groups became more male biased (possibly because there were less females trying to monopolise the box). These results confirm the role of social setting and possibly of group dynamics in the learning process. Contrary to the above mentioned work, Stoinski et al. (2011) found quite strong evidence of social learning in *Varecia variegata*. The basic methodology (two-action foraging task) involves first training a demonstrator to retrieve food from an apparatus using one of two possible techniques (lifting a lightweight aluminium hinge or sliding an opening disc, in the study case). After observing a demonstrator using one technique, subjects are given access to the apparatus with both techniques available for use. If there is a reliable bias in favour of the observed action rather than the alternative method, the observer is said to have acquired the behaviour socially (Caldwell and Whiten, 2007; Stoinski et al., 2011). All the experimental subjects used the demonstrated technique on their very first attempt and showed a significant preference for the demonstrated technique, thus indicating that social learning was in place (Stoinski et al., 2011).

In summary, lemurs can learn newly introduced behaviour, are influenced by the social setting in their learning performances and are able to socially learn from others at least in some cases (or groups). It is clear, though, that the issue of social learning remains puzzling and needs further investigation.

Social learning is a precondition for the occurrence of tradition defined by Fragaszy and Perry (2003) as 'a distinctive behaviour pattern shared by two or more individuals in a social unit, which persists over time and that new practitioners acquire in part through socially aided learning'. Huffman and Quiatt (1986) recognised three phases in the propagation of innovations, starting with the *transmission* of the new behaviour across individuals in social groups, continuing with the *tradition* phase in which the behaviour is passed down to subsequent generations through mother–infant interactions (with the rate of diffusion becoming similar to the birth rate), and eventually, possibly followed by a *transformation phase*, if the behaviour is modified to make it more efficient. The first experimental attempt to ascertain whether wild lemurs (*Eulemur rufifrons*) could develop long-term traditions introduced by humans was performed by Schnoell et al. (2014), in the Kirindy forest (Madagascar). In a previous study, Schnoell and Fichtel (2012) introduced an artificial feeding box that could be opened by two different techniques (pulling or pushing the door to get access to food). The authors found evidence of transmission through social learning because the lemurs developed a group preference for either one opening technique or another, following the technique that each group had been forced to learn in first place. It was also observed that the individuals that observed others performing the task made fewer mistakes when it was their turn to open the box. In the subsequent work (Schnoell et al., 2014), the authors found that red-fronted lemurs were able to open the artificial feeding boxes successfully more quickly over the years, thus suggesting that they remembered the rewarding character of the box over time, although the presentation of the feeding boxes occurred every 9 months. However, on the population level the lemurs did not maintain a preference for one or the other technique over the three consecutive study years, even if six individuals exhibited a stable feeding technique preference over time. The authors suggested that the task may not have been difficult enough to induce long-term preferences in red-fronted lemurs and that the fluctuating preferences may be due to the fact that the two opening options did not differ in their difficulty to learn. Moreover, most red-fronted lemurs also scrounged during the opening event, thus bypassing the need of opening the box on their own and making the opening process more costly for other individuals.

Another example of tradition in the wild is reported by Mertl-Millhollen (2000) in *Lemur catta*, at Berenty. She found that a troop, intermittently observed for 35 years, in 1998 spent 62% of its time budget in the home range recorded in 1975. Lemurs also used the same sleeping trees in 1975 and 1998, and all of the 1998 scent marks deposited in the 1975 home range were placed in the same locations marked in 1975. It is therefore clear how long-term longitudinal studies are important to detect social transmission of information through generations, necessary to define traditions.

Previously in this book we showed that lemurs build society from social interactions through individual recognition (which is not as obvious as it may sound) and multimodal communication, which includes frontal secondary sexual signals, when vertical locomotion is involved. Conflicts and their resolution shape lemur societies as it occurs in other primates; the individuals are able to manage the anxiety that arises from motivational conflict and engage in anxiety buffering behaviours. Lemurs use play and enact biological market strategies to relate with one another in response to social challenges. In this final part we showed that lemur cognitive potential can go much further than previously thought.

Figure 9.6 Exploring new lemurs' potential. Young individual of *Lemur catta* interacting with a tablet at Greensboro Science Center (North Carolina, US), on a task intended for gibbons. Photo: Joanne Altman.

It is clear that the most effective approach to learn the full potential of lemurs is to study them at different levels (Figure 9.6). Lemurs from different populations, groups or individuals of the same species can behave differently, depending on varying ecological, sociobiological and cognitive correlates. A better understanding of lemur behaviour at any level, along with the measures defined in the last Lemur Conservation Action Plan, can help implementing better conservation strategies to preserve what is currently considered as the most threatened mammal group on Earth (Schwitzer *et al.*, 2014).

References

Addessi, E. (2003). Ruolo delle influenze sociali sulla neofobia alimentare nel cebo dai cornetti (*Cebus apella*) e sullo scimpanzé (*Pan troglodytes*). Ph.D. Thesis, University 'La Sapienza', Rome.

Addessi, E. & Rossi, S. (2010). Tokens improve capuchin performance in the reverse–reward contingency task. *Proceedings of the Royal Society of London, Series B: Biological Sciences*, 278, 849–854.

Addessi, E., Galloway, A. T., Visalberghi, E. & Birch, L. L. (2005). Specific social influences on the acceptance of novel foods in 2–5-year-old children. *Appetite*, 45, 264–271.

Albiach-Serrano, A., Guillen-Salazar, F. & Call, J. (2007). Mangabeys (*Cercocebus torquatus lunulatus*) solve the reverse contingency task without a modified procedure. *Animal Cognition*, 10, 387–396.

Anderson, J. R. (2001). Self- and other-control in squirrel monkeys. In: T. Matsuzawa (ed.), *Primate Origins of Human Cognition and Behavior.* Tokyo: Springer-Verlag, pp. 330–347.

Anderson, J. R., Fornasieri, I., Ludes, E., & Roeder, J-J. (1992). Social processes and innovative behaviour in changing groups of *Lemur fulvus. Behavioural Processes,* 27, 101–112.

Anderson, J. R., Awazu, S. & Fujita, K. (2000). Can squirrel monkeys (*Saimiri sciureus*) learn self-control? A study using food array selection tests and reverse reward contingency. *Journal of Experimental Psychology: Animal Behavior Processes,* 26, 87–97.

Anderson, J. R., Hattori, Y. & Fujita, K. (2008). Quality before quantity: rapid learning of reverse-reward contingency by capuchin monkeys (*Cebus apella*). *Journal of Comparative Psychology,* 122, 445–448. http://dx.doi.org/10.1037/a0012624.

Barton, R. A. (1999). The evolutionary ecology of the primate brain. In: P. C. Lee (ed.), *Comparative Primate Socioecology.* Cambridge University Press, pp. 167–203.

Bond, A. B., Kamil, A. C. & Balda, R. P. (2003). Social complexity and transitive inference in corvids. *Animal Behaviour.* 65, 479–487.

Boysen, S. T. & Berntson, G. G. (1995). Response to quantity: perceptual versus cognitive mechanisms in chimpanzees (*Pan troglodytes*). *Journal of Experimental Psychology: Animal Behavior Processes,* 21, 82–86.

Braeuer, J., Call, J. & Tomasello, M. (2005). All great ape species follow gaze to distant locations and around barriers. *Journal of Comparative Psychology,* 119, 145–154.

Brannon, E. M. & Terrace, H. S. (2000). Representation of the numerosities 1–9 by rhesus macaques. *Journal of Experimental Psychology: Animal Behavior Processes.* 26, 31–49.

Burkhart, J. & Heschl, A. (2006). Geometrical gaze following in common marmosets (*Callithrix jacchus*). *Journal of Comparative Psychology,* 120, 120–130.

Caldwell, C. A. & Whiten, A. (2007). Social learning in apes and monkeys: cultural animals? In: C. J. Campbell, A. Fuentes, K. C. MacKinnon, A. Panger & S. K. Bearder (eds), *Primates in Perspective.* New York, NY: Oxford University Press, pp. 652–663.

Cantlon, J. F. & Brannon, E. M. (2006). Shared system for ordering small and large numbers in monkeys and humans. *Psychological Science,* 17, 402–407.

Caraco, T. & Lima, S. L. (1985). Foraging juncos – interaction of reward mean and variability. *Animal Behaviour,* 33, 216–224.

Carlson, S. M., Davis, A. C. & Leach, J. G. (2005). Less is more: executive function and symbolic representation in preschool children. *Psychological Science,* 16, 609–616.

Cheney, D. L. & Seyfarth, R. M. (1990). *How Monkeys See the World: Inside the Mind of Another Species.* Chicago: University of Chicago Press, p. 377.

Clayton, D. A. (1978). Socially facilitated behavior. *The Quarterly Review of Biology,* 53, 373–392.

De Petrillo, F., Ventricelli, M., Ponsi, G. & Addessi, E. (2015). Do tufted capuchin monkeys play the odds? Flexible risk preferences in *Sapajus* spp. *Animal Cognition,* 18, 119–130.

Dean, L. G., Hoppitt, W., Laland, K. N. & Kendal, R. L. (2011). Sex ratio affects sex-specific innovation and learning in captive ruffed lemurs (*Varecia variegata* and *Varecia rubra*). *American Journal of Primatology,* 73, 1–12.

Deppe, A. M., Wright, P. C. & Szelistowski, W. A. (2009). Object permanence in lemurs. *Animal Cognition,* 12, 382–388.

Digby, L. J., Haley, M., Schneider, A. C. & Del Valle, I. (2008). Sensorimotor intelligence in aye-ayes and other lemurs: a puzzle box approach. *American Journal of Primatology,* 70 (Suppl 1), 54.

Dunbar, R. I. & Shultz, S. (2007). Understanding primate brain evolution. *Philosophical Transactions of the Royal Society: Biological Sciences,* 362, 649–658.

Emery, N. J., Lorincz, E. N., Perrett, D. I., Oram, M. W. & Baker, C. I. (1997). Gaze following and joint attention in rhesus monkeys (*Macaca mulatta*). *Journal of Comparative Psychology,* 111, 286–293.

Feigenson, L., Carey, S. & Hauser, M. D. (2002). The representations underlying infants' choice of more: object files versus analog magnitudes. *Psychological Science,* 13, 150–156.

Feigenson, L., Carey, S. & Spelke, E. (2002). Infants' discrimination of number vs. continuous extent. *Cognitive Psychology*, 44, 33–66.

Ferrari, P. F., Visalberghi, E., Paukner, A., *et al.* (2006). Neonatal imitation in rhesus macaques. *PLoS Biol.*, 4(9), e302. http://dx.doi.org/10.1371/journal.pbio.0040302.

Fichtel, C. & Kappeler, P. M. (2010). Human universals and primate symplesiomorphies: establishing the lemur baseline. In: P. M. Kappeler & J. Silk (eds), *Mind the Gap: Tracing the origins of human universals*. Heidelberg: Springer, pp. 395–426.

Fichtel, C. & Kappeler, P. M. (2011). Variation in the meaning of alarm calls in Verreaux's and Coquerel's sifakas (*Propithecus verreauxi*, *P. coquereli*). *International Journal of Primatology*, 32, 346–361.

Fornasieri, I., Anderson, J. R. & Roeder, J-J. (1990). Responses to a novel food acquisition task in three species of lemurs. *Behavioural Processes*, 21, 143–156.

Fragaszy, D. M. & Perry, S. (2003). *The Biology of Traditions: Models and Evidence*. Cambridge University Press.

Fragaszy, D., Visalberghi, E. & Fedigan, L. (2004). *The Complete Capuchin. The Biology of the Genus Cebus*. Cambridge University Press, p. 339.

Galef, B. G., Jr. (1992). The question of animal culture. *Human Nature*, 3, 157–178.

Genty, E., Palmier, C. & Roeder, J. J. (2004). Learning to suppress responses to the larger of two rewards in two species of lemurs, *Eulemur fulvus* and *E. macaco*. *Animal Behaviour*, 67, 925–932.

Genty, E., Chung, P. C. & Roeder, J. J. (2011). Testing brown lemurs (*Eulemur fulvus*) on the reverse-reward contingency task without a modified procedure. *Behavioural Processes*, 86, 133–137.

Glady, Y., Genty, É. & Roeder, J-J. (2012). Brown lemurs (*Eulemur fulvus*) can master the qualitative version of the reverse-reward contingency. *PLoS ONE*, 7(10), e48378. http://dx.doi.org/10.1371/journal.pone.0048378.

Hauser, M. & Carey, S. (2003). Spontaneous representations of small numbers of objects by rhesus macaques: examinations of content and format. *Cognitive Psychology*, 47, 367–401.

Hauser, M. D., Carey, S. & Hauser, L. B. (2000). Spontaneous number representation in semi-free-ranging rhesus monkeys. *Proceedings of the Royal Society of London, B: Biological Sciences*, 267, 829–833.

Hauser, M. D., Tsao, F., Garcia, P. & Spelke, E. S. (2003). Evolutionary foundations of number: spontaneous representation of numerical magnitudes by cotton-top tamarins. *Proceedings of the Royal Society of London B: Biological Sciences*, 270, 1441–1446.

Hayden, B. Y. & Platt, M. L. (2007). Temporal discounting predicts risk sensitivity in rhesus macaques. *Current Biology*, 17, 49–53.

Heilbronner, S. F., Rosati, A. G., Stevens, J. R., Hare, B. & Hauser, M. D. (2008). A fruit in the hand or two in the bush? Divergent risk preferences in chimpanzees and bonobos. *Biology Letters*, 23, 246–249.

Hosey, G. R., Jacques, M. & Pitts, A. (1997). Drinking from tails: social learning of a novel behavior in a group of ring-tailed lemurs (*Lemur catta*). *Primates*, 38, 415–422.

Huang, C-T. & Charman, T. (2005). Gradations of emulation learning in infants' imitation of actions on objects. *Journal of Experimental Child Psychology*, 92, 276–302.

Huffman, A. M. (1984). Stone-play of *Macaca fuscata* in Arashiyama B troop: transmission of non-adaptive behaviour. *Journal of Human Evolution*, 13, 725–735.

Huffman, M. A. (1996). Acquisition of innovative cultural behaviors in nonhuman primates: A case study of SH, a socially transmitted behavior in Japanese macaques. In: B. G. Galef, Jr. & C. Heyes (eds), *Social Learning in Animals: Roots of Culture*. San Diego, CA: Academic Press, pp. 267–289.

Huffman, M. A. & Quiatt, D. (1986). Stone handling by Japanese macaques (*Macaca fuscata*): implications for tool use of stones. *Primates*, 27, 427–437.

Jacobs, G. H. (2008). Primate color vision: a comparative perspective. *Visual Neuroscience*, 25, 619–633.

Jolly, A. (1964). Prosimians' manipulation of simple object problems. *Animal Behaviour*, 12, 560–570.

Jolly, A. (1966). Lemur social behavior and primate intelligence. *Science*, 153, 501–506.

Jolly, A. (2004). *Lords and Lemurs: Mad Scientists, Kings wth Spears, and the Survival of Diversity in Madagascar.* Houghton Mifflin Harcourt.

Jones, S. M. & Brannon, E. M. (2012). Prosimian primates show ratio dependence in spontaneous quantity discriminations. *Frontiers in Psychology*, 3, 550.

Jones, S. M., Pearson, J., DeWind, N. K., et al. (2014). Lemurs and macaques show similar numerical sensitivity. *Animal Cognition*, 17, 503–515.

Judge, P. G., Evans, T. A. & Vyas, D. K. (2005). Ordinal representation of numeric quantities by brown capuchin monkeys (*Cebus apella*). *Journal of Experimental Psychology: Animal Behavior Processes*, 31, 79–94.

Kacelnik, A. & Bateson, M. (1996). Risky theories – The effects of variance on foraging decisions. *American Zoologist*, 36, 402–434.

Kalenscher, T. and van Wingerden, M. (2011). Why we should use animals to study economic decision making – a perspective. *Frontiers in Neuroscience*, 5, 82.

Kappeler, P. M. (1987). The acquisition process of a novel behavior pattern in a group of ring-tailed lemurs (*Lemur catta*). *Primates*, 28, 225–228.

Kendal, R. L., Custance, D. M., Kendal, J. R., et al. (2010). Evidence for social learning in wild lemurs (*Lemur catta*). *Learning & Behavior*, 38, 220–234.

Koechlin, E. (1997). Numerical transformations in five-month-old human infants. *Mathematical Cognition*, 3, 89–104.

Kralik, J. D., Hauser, M. D. & Zimlicki, R. (2002). The relationship between problem solving and inhibitory control: Cotton-top tamarins (*Saguinus oedipus*) performance on a reversed contingency task. *Journal of Comparative Psychology*, 116, 39–50.

Kudo, H. & Dunbar, R. I. M. (2001). Neocortex size and social network size in primates. *Animal Behaviour*, 62, 711-722.

Kummer, H. (1995). *In Quest of the Sacred Baboon.* Princeton, NJ: Princeton University Press.

Kummer, H. & Goodall, J. (1985). Conditions of innovative behaviour in primates. *Philosophical Transactions of the Royal Society B: Biological Sciences*, 308, 203–214.

Kuznar, L. (2001). Risk sensitivity and value among Andean pastoralists: Measures, models, and empirical tests. *Current Anthropology*, 42, 432–440.

Leca, J. B., Gunst, N. & Huffman, M. A. (2007). Japanese macaque cultures: inter-and intra-troop behavioural variability of stone handling patterns across 10 troops. *Behaviour*, 144, 251–281.

Leca, J. B., Gunst, N. & Huffman, M. A. (2010). Indirect social influence in the maintenance of the stone-handling tradition in Japanese macaques, *Macaca fuscata*. *Animal Behaviour*, 79, 117–126.

Lewis, K. P., Jaffe, S. & Brannon, E. M. (2005). Analog number representations in mongoose lemurs (*Eulemur mongoz*): evidence from a search task. *Animal Cognition*, 8, 247–252.

Lhota, S., Jůnek, T., Bartoš, L. & Kuběna, A. A. (2008). Specialized use of two fingers in free-ranging aye-ayes (*Daubentonia madagascariensis*). *American Journal of Primatology*, 70, 786–795.

Limongelli, L., Boysen, S. T. & Visalberghi, E. (1995). Comprehension of cause-effect relations in a tool-using task by chimpanzees (*Pan troglodytes*). *Journal of Comparative Psychology*, 109, 18.

MacLean, E. L., Merritt, D. J. & Brannon, E. M. (2008). Social organization predicts transitive reasoning in prosimian primates. *Animal Behaviour*, 76, 479–486.

MacLean, E. L., Matthews, L. J., Hare, B. A., et al. (2012). How does cognition evolve? Phylogenetic comparative psychology. *Animal Cognition*, 15, 223–238.

MacLean, E. L., Sandel, A. A., Bray, J. et al. (2013). Group size predicts social but not nonsocial cognition in lemurs. *PLoS ONE*, 8, e66359.

MacLean, E. L., Hare, B., Nunn, C. L., *et al.* (2014). The evolution of self-control. *PNAS*, 111, E2140–2148.

Mallavarapu, S., Perdue, B. M., Stoinski, T. S. & Maple, T. L. (2013). Can black-and-white ruffed lemurs (*Varecia variegata*) solve object permanence tasks? *American Journal of Primatology*, 75, 376–386.

Merritt, D., MacLean, E. L., Jaffe, S. & Brannon, E. M. (2007). A comparative analysis of serial ordering in ring-tailed lemurs (*Lemur catta*). *Journal of Comparative Psychology*, 121, 363–371.

Merritt, D. J., MacLean, E. L., Crawford, J. C. & Brannon, E. M. (2011). Numerical rule-learning in ring-tailed lemurs (*Lemur catta*). *Frontiers in Psychology*, 2, 23.

Mertl-Millhollen, A. S. (2000). Tradition in *Lemur catta* behavior at Berenty Reserve, Madagascar. *International Journal of Primatology*, 21, 287–297.

Mischel, W., Shoda, Y. & Rodriguez, M. I. (1989). Delay of gratification in children. *Science*, 244, 933–938.

Murray, E. A., Kralik, J. D. & Wise, S. P. (2005). Learning to inhibit prepotent responses: successful performance by rhesus macaques, *Macaca mulatta*, on the reversed-contingency task. *Animal Behaviour*, 69, 991–998.

Neiworth, J. J., Burman, M. A., Basile, B. M. & Lickteig, M. T. (2002). Use of experimenter-given cues in visual co-orienting and in an object-choice task by a New World monkey species, cotton top tamarins (*Saguinus oedipus*). *Journal of Comparative Psychology*, 116, 3–11.

O'Mara, M. T. & Hickey, C. M. (2012). Social influences on the development of ringtailed lemur feeding ecology. *Animal Behaviour*, 84(6), 1547–1555.

Parker, C. E. (1973). Manipulatory behavior and responsiveness. *Gibbon and Siamang*, 2, 185–207.

Read, D. (2004). Intertemporal choice. In: Koehler, D. J. & Harvey, N. (eds), *Blackwell Handbook of Judgement and Decision Making*, pp. 424–443.

Real, L., Ott, J. & Silverfine, E. (1982). On the tradeoff between the mean and the variance in foraging: effect of spatial distribution and color preference. *Ecology*, 63, 1617–1623.

Rieskamp, J., Busemeyer, J. R. & Mellers, B. A. (2006). Extending the bounds of rationality: evidence and theories of preferential choice. *Journal of Economic Literature*, 44, 631–661.

Rosati, A. G. & Hare, B. (2011). Chimpanzees and bonobos distinguish between risk and ambiguity. *Biology Letters*, 7, 15–18.

Rosati, A. G., Rodriguez, K. & Hare, B. (2014). The ecology of spatial memory in four lemur species. *Animal Cognition*, 17, 1–15.

Ruiz, A., Gómez, J. C., Roeder, J. J. & Byrne, R. W. (2009). Gaze following and gaze priming in lemurs. *Animal Cognition*, 12, 427–434.

Sandel, A. A., MacLean, E. & Hare, B. (2011). Evidence from four lemur species that ringtailed lemur social cognition converges with that of haplorhine primates. *Animal Behaviour*, 81, 925–931.

Santos, L. R., Mahajan, N. & Barnes, J. L. (2005a) How prosimian primates represent tools: Experiments with two lemur species (*Eulemur fulvus* and *Lemur catta*). *Journal of Comparative Psychology*, 119, 394–403.

Santos, L. R., Barnes, J. L. & Mahajan, N. (2005b). Expectations about numerical events in four lemur species (*Eulemur fulvus*, *Eulemur mongoz*, *Lemur catta* and *Varecia rubra*). *Animal Cognition*, 8, 253–262.

Schilling, A. (2013). Cognitive capacities of captive gray mouse lemurs as evidenced by object manipulation. In: Masters, J., Gamba, M. & Génin, F. (eds), *Leaping Ahead*. New York: Springer, pp. 331–340.

Schnoell, A. V. & Fichtel, C. (2012). Wild redfronted lemurs (*Eulemur rufifrons*) use social information to learn new foraging techniques. *Animal Cognition*, 15, 505–516.

Schnoell, A. V. & Fichtel, C. (2013). A novel feeding behaviour in wild redfronted lemurs (*Eulemur rufifrons*): depletion of spider nests. *Primates*, 54, 371–375.

Schnoell, A. V., Dittmann, M. T. & Fichtel, C. (2014). Human-introduced long-term traditions in wild redfronted lemurs? *Animal Cognition*, 17, 45-54.

Schwitzer, C., Mittermeier, R. A., Johnson, S. E., et al. (2014). Averting lemur extinctions amid Madagascar's political crisis. *Science*, 343, 842-843.

Shepherd, S. V. & Platt, M. L. (2008). Spontaneous social orienting and gaze following in ringtailed lemurs (*Lemur catta*). *Animal Cognition*, 11, 13-20.

Shumaker, R. W., Palkovich, A. M., Beck, B. B., Guagnano, G. A. & Morowitz, H. (2001). Spontaneous use of magnitude discrimination and ordination by the orangutan (*Pongo pygmaeus*). *Journal of Comparative Psychology*, 115, 385-391.

Siegler, R. S. (1986). *Children's Thinking*. Englewood-Cliffs, NJ: Prentice-Hall.

Silberberg, A. & Fujita, K. (1996). Pointing at smaller food amounts in an analogue of Boysen and Berntson's (1995) procedure. *Journal of the Experimental Analysis of Behavior*, 66, 143-147.

Simon, T. J., Hespos, S. J. & Rochat, P. (1995). Do infants understand simple arithmetic? A replication of Wynn (1992). *Cognitive Development*, 10, 253-269.

Smith, B. R., Piel, A. K. & Candland, D. K. (2002). The numerical abilities of a socially-housed Hamadryas Baboon (*Papio hamadryas*) and Squirrel Monkey (*Saimiri sciureus*). *Abstracts of the Psychonomic Society*, 545, 81.

Spence, K. W. (1937). Experimental studies of learning and higher mental processes in infra-human primates. *Psychological Bulletin*, 34, 806-850.

Stevens, J. R. & Mühlhoff, N. (2012). Intertemporal choice in lemurs. *Behavioural Processes*, 89, 121-127.

Stevens, J. R., Hallinan, E. V. & Hauser, M. D. (2005). The ecology and evolution of patience in two New World monkeys. *Biology Letters*, 1, 223-226.

Stevens, R., De Waegenaere, A. & Melenberg, B. (2010). Longevity risk in pension annuities with exchange options: The effect of product design. *Insurance: Mathematics and Economics*, 46, 222-234.

Stoinski, T. S., Drayton, L. A. & Price, E. E. (2011). Evidence of social learning in black-and-white ruffed lemurs (*Varecia variegata*). *Biology Letters*, 7, 376-379.

Tattersall, I. (1982). *The Primates of Madagascar*. New York: Columbia University Press.

Thorpe, W. H. (1963). *Learning and Instinct in Animals*, 2nd edition. London: Methuen.

Tomasello, M. (1996). Do apes ape? In: C. Heyes & B. Galef (eds), *Social Learning in Animals: The Roots of Culture*. Academic Press, pp. 319-346.

Tomasello, M. & Call, J. (1997). *Primate Cognition*. New York: Oxford University Press.

Tomasello, M., Call, J. & Hare, B. (1998). Five primate species follow the visual gaze of conspecifics. *Animal Behaviour*, 55, 1063-1069.

Uher, J. & Call, J. (2008). How the great apes (*Pan troglodytes*, *Pongo pygmaeus*, *Pan paniscus*, *Gorilla gorilla*) perform on the reversed reward contingency task II: transfer to new quantities, long-term retention, and the impact of quantity ratios. *Journal of Comparative Psychology*, 122, 204-212.

Uller, C., Hauser, M. & Carey, S. (2001). Spontaneous representation of number in cotton-top tamarins (*Saguinus oedipus*). *Journal of Comparative Psychology*, 115, 248-257.

van Schaik, C. P. & Burkart, J. M. (2011). Social learning and evolution: the cultural intelligence hypothesis. *Philosophical Transactions of the Royal Society Series B*, 366, 1008-1016.

Visalberghi, E. & Addessi, E. (2000). Seeing group members eating a familiar food affects the acceptance of novel foods in capuchin monkeys, *Cebus apella*. *Animal Behaviour*, 60, 69-76.

Visalberghi, E. & Addessi, E. (2001). Acceptance of novel foods in *Cebus apella*: do specific social facilitation and visual stimulus enhancement play a role? *Animal Behaviour*, 62, 567-576.

Visalberghi, E. & Fragaszy, D. (1990). Do monkeys ape? In: S. Parker & K. Gibson (eds), *'Language' and Intelligence in Monkeys and Apes*, Cambridge University Press, pp. 247-275.

Visalberghi, E. & Fragaszy, D. (1995). The behavior of capuchin monkeys (*Cebus apella*) with food: the role of social context. *Animal Behaviour*, 49, 1089-1095.

Visalberghi, E. & Fragaszy, D. (2002). Do monkeys ape?' Ten years after. In: K. Dautenhahn and C. L. Nehaniv (eds), *Imitation in Animals and Artifacts*. Cambridge, MA: MIT Press, pp. 471–499.

Visalberghi, E. & Limongelli, L. (1994). Lack of comprehension of cause-effect relations in tool-using capuchin monkeys (*Cebus apella*). *Journal of Comparative Psychology*, 108, 15–22.

Visalberghi, E., Sabbatini, G., Stammati, M. & Addessi, E. (2003). Preferences towards novel foods in *Cebus apella*: the role of nutrients and social influences. *Physiology and Behavior*, 80, 341–349.

Visalberghi, E., Addessi, E., Spagnoletti, N., *et al.* (2009). Selection of effective stone tools by wild capuchin monkeys. *Current Biology*, 19, 213–217.

von Neumann, J. & Morgenstern, O. (1944). *Game Theory and Economic Behavior*. Princeton: Princeton University.

Wright, P. C. (1999). Lemur traits and Madagascar ecology: coping with an island environment. *American Journal of Physical Anthropology*, 110(s29), 31–72.

Looking back to the future

Michael A. Huffman

The quest to understand what makes us humans began with the search for the 'missing link', that enigmatic fossilised hominid that was thought would bridge the gap between 'them', the non-human primates, and us. In more recent times the picture growing from this quest looks more like a chain-link fence, than the envisioned simple link idea, with each passing discovery bringing new examples of extinct species sharing a mosaic of traits shared by humans and great apes (e.g. Templeton, 2005; White, 2012). With the pioneering great ape studies in the 1960s and 1970s, the search began to incorporate studies on our closest living relatives, the great apes, species with which we share many obvious behavioural, physiological and anatomical traits. Over the years primatology has methodically removed the perceived barrier between humans and apes, with the discovery of such behaviours as tool use, tool making and hunting with or without tools, communication, conspecific warfare, even the possible use of medicinal plants to treat illness (e.g. Goodall, 1968, 1986; McGrew, 1979, 1981, 1992; Boesch and Boesch, 1981; Nishida *et al.*, 1985; Huffman and Seifu, 1989; Huffman and Kalunde, 1993; Boesch, 1994; Van Schaik and Knott, 2001; Van Schaik *et al.*, 2003; Mitani *et al.*, 2010). For some, this may seem like the beginning and the end point for looking at the origins of characteristics considered unique to humans. But is it really?

As our discipline advances, the boundaries have been pushed even further, with studies promoting macaques and capuchins as suitable models for human evolution of tool use behaviour, through the discovery of their ability to use stone tools as deftly as chimpanzees (e.g. Fragaszy *et al.*, 2004; Moura and Lee, 2004; Ottoni and Izar, 2008; Gumert *et al.*, 2009, 2011; Visalberghi *et al.*, 2009). In this context, behavioural innovation and the long-term maintenance of stone handling, a non-functional cultural behaviour in macaques, and the artefacts of this behaviour left behind provide thought for socioecological forces behind the emergence of tool use in early humans (Huffman and Quiatt, 1986; Quiatt and Huffman, 1993; Huffman *et al.*, 2010). Clearly, the gap between apes and monkeys is not as large as was once thought.

Unexpectedly, in my quest to understand medicinal plant use in chimpanzees, curiosity and several cross-disciplinary collaborations led me to take a much broader perspective on what is uniquely human, and what is not, about the ability to self-medicate (e.g. Huffman, 2002, 2007; Hardy *et al.*, 2013). Looking at a range of primates and even non-primate animals species' ability to control parasite infection and/or its related symptoms, I came to realise that what was once perceived to be unique about us as a species is firmly grounded in behavioural predispositions shared by other primates and indeed a range of other mammals, birds and invertebrates (Huffman, 1997, 2011; Currie et al., 1999; Singer et al., 2009; Su et al., 2013; Suarez-Rodriquez et al., 2013). These behaviours, which can represent adaptations to similar selective pressures that cross over taxonomic boundaries, are expected, however, to be supported by a variety of different cognitive and innate processes.

In spite of my growing appreciation for the bigger picture and the comparative approach, I assumed that lemurs, the living fossils of Madagascar, were most valuable as a comparative link with our nearest mammalian predecessors, offering a view to our 'humble' beginnings, rather than our 'eminent selves'. Fichtel and Kappeler (2010) argue that looking at the gaps between prosimians and the other primates is important, and that this provides the baseline for understanding the origins of our own social and technical skills. I was aware that a great deal of fascinating discoveries had been made about lemur evolution, behavioural ecology, physiology, parasitology and other fundamental topics highlighting their unique adaptations to the very diverse and equally unique island environment of Madagascar (see Masters et al., 2013). However, with the exception of what I know about lemur self-medication (Birkinshaw, 1999; Valderrama et al., 2000; Carrai et al., 2003) and recent reports of lemurs that hibernate in tree holes likes squirrels or underground like groundhogs (Dausmann et al., 2004; Blanco et al., 2013), I thought little about what lemurs could possibly tell us about ourselves. Now I have to ask myself why haven't we been considering lemurs earlier!

This book has challenged me to think harder about lemurs and how they do tmake good models for understanding human behaviour. The study of these living 'missing links' offers not only a window into our past, but a unique avenue for understanding the path travelled by our closest primate cousins. In the early days of primatology, Jolly (1966) insightfully concluded from her studies of *Lemur catta* that having evolved a society and capacity for social learning similar to monkeys, without the ability to manipulate objects, the rudiments of primate sociality preceded and made possible the emergence of primate intelligence.

As primatology has come of age, so have primatologists. A wealth of information has been gathered over the years since Alison Jolly began her pioneering studies. In this book, Elisabetta Palagi and Ivan Norscia have drawn upon their wide experience with lemurs, monkeys and apes to eloquently review the current state of lemur studies and expand this knowledge with their own cutting edge research. The authors have successfully taken paradigms and methods used in research of our closer primate relatives, and shown us that some lemurs are highly relevant to understand who we are and how we got here.

The gap between monkeys and strepsirrhines is narrowing fast. Lemurs provide new ground to search for insights into the evolution and emergence of human behaviour. Like monkeys, apes and humans, lemurs have the capacity for individual recognition – an important skill for the formation and maintenance of complex social networks. They do this with visual, olfactory and acoustic clues. Lemurs communicate with each other multimodally, combining visual and olfactory cues. The role of female secondary sexual signals to attract mates in humans is found in males of some species of lemur. Verreaux's sifaka females are dominant over males, and the male makes use of self-application of a temporary secondary sexual signal. Males apply body 'decoration' to their chest from brown throat secretions as a temporary 'biomorph' that successfully attracts female mates.

Lemur hierarchy is like that of monkeys and apes. Despite the linear nature of their social hierarchy, lemur groups display a wide array of tolerance ranging from more egalitarian to more despotic, much the variation seen among macaque species. Like chimpanzees and humans, some lemur species form coalitions. They can also reconcile! Stress is an affect of group living. Social lemurs also suffer from anxiety, which can be quantified behaviourally by measuring the intensity of scratching, as seen in humans and other primates. Their mechanisms to buffer stress are also similar to other primates. To relieve anxiety, ring-tailed lemurs groom, brown lemurs reconcile and sifaka play.

The role of social play in adult animals is still a puzzle. With the exception of humans and a handful of other primates (e.g. Palagi and Paoli, 2008), social play with peers disappears late in the juvenile period of a primate's life, yet adult lemurs play with each other. They use play apparently to reduce aggression between in-group and out-group males during the mating season. This resembles in some ways the context and function of non-reproductive sex, female–female and male–male genital rubbing greetings initiated by bonobos at stressful times of intergroup encounters (Savage-Rumbaugh and Wilkerson, 1978; Kano, 1980; Kuroda, 1980; Hohmann and Fruth, 2000)!

Were the leaps and bounds we have made as a species in an evolutionarily short period of time set in motion by lemurs? The take-home message of this book for me is that lemurs are really fascinating and there is still a lot to be learned about them. By revisiting old paradigms with new methodologies, this book shows us the way forward.

References

Birkinshaw, C. R. (1999). Use of millipedes by black lemurs to anoint their bodies. *Folia Primatologica*, 70, 170–171.
Blanco, M. B., Dausmann, K. H., Ranaivoarisoa, J. F. & Yoder, A. D. (2013). Underground hibernation in a primate. *Scientific Reports*, 3, 1768.
Boesch, C. (1994). Cooperative hunting in wild chimpanzees. *Animal Behavior*, 48, 653–667.
Boesch, C. & Boesch, H. (1981). Sex differences in the use of natural hammers by wild chimpanzees: a preliminary report. *Journal of Human Evolution*, 10, 585–593.
Carrai, V., Borgognini-Tarli, S. M., Huffman, M. A. & Bardi, M. (2003). Increase in tannin consumption by sifaka (*Propithecus verreauxi verreauxi*) females during the birth season: a case for self-medication in prosimians? *Primates*, 44, 61–66.
Currie, C. R., Scott, J. A., Summerbell, R. C. & Malloch, D. (1999). Fungus-growing ants use antibiotic-producing bacteria to control garden parasites. *Nature*, 398, 701–703.
Dausmann, K. H., Glost, J., Ganzhorn, J. U. & Heldmaier, G. (2004). Hibernation in a tropical primate. *Nature*, 429, 825–826.

Fichtel, C., & Kappeler, P. M. (2010). Human universals and primate symplesiomorphies: establishing the lemur baseline. In: P. M. Kappeler & J. B. Silk (eds), *Mind the Gap*. Berlin: Springer-Verlag, pp. 395–426.

Fragaszy, D., Izar, P., Visalberghi, E., Ottoni, E. B. & de Oliviera, M. G. (2004). Wild capuchin monkeys (*Cebus libidinosus*) use anvils and stone pounding tools. *American Journal of Primatology*, 64, 359–366.

Goodall, J. (1968). Behaviour of free-living chimpanzees of the Gombe Stream area. *Animal Behavioural Monograph*, 1, 163–311.

Goodall, J. (1986). *The Chimpanzees of Gombe: Patterns of Behavior*. Belknap.

Gumert, M. D., Kluck, M. & Malaivijitnond, S. (2009). The physical characteristics and usage patterns of stone axe and pounding hammers used by long-tailed macaques in the Andaman Sea region of Thailand. *American Journal of Primatology*, 71, 594–608.

Gumert, M. D., Hoong, L. K. & Malaivijitnond, S. (2011). Sex differences in the stone tool-use behavior of a wild population of Burmese long-tailed macaques (*Macaca fascicularis aurea*). *American Journal of Primatology*, 73, 1–11.

Hardy, K., Buckley, S. & Huffman, M. A. (2013). Neanderthal self-medication in context. *Antiquity*, 87, 873–878.

Hohmann, G. & Fruth, B. (2000). Use and function of genital contacts among female bonobos. *Animal Behavior*, 60, 107–120.

Huffman, M. A. (1997). Current evidence for self-medication in primates: a multidisciplinary perspective. *Yearbook of Physical Anthropology*, 40, 171–200.

Huffman, M. A. (2002). Animal origins of herbal medicine. In: J. Fleurentin, J-M. Pelt & G. Mazars (eds), *Des Sources du Savoir aux Medicaments du Future – From the Sources of Knowledge to the Medicines of the Future*. Paris: IRD Editions, pp. 31–42.

Huffman, M. A. (2007). Animals as a source of medicinal wisdom in indigenous societies. In: M. Bekoff (ed.), *Encyclopedia of Human-Animal Relationships*, vol. 2. Westport, Connecticut: Greenwood Publishing Group, pp. 434–441.

Huffman, M. A. (2011). Primate self-medication. In: C. Campbell, A. Fuentes, K. MacKinnon, M. Panger & S. Bearder (eds), *Primates in Perspective*. Oxford: Oxford University Press, pp. 677–690.

Huffman, M. A. & Kalunde, M. S. (1993). Tool-assisted predation on a squirrel by a female chimpanzee in the Mahale Mountains, Tanzania. *Primates*, 34, 93–98.

Huffman, M. A. & Quiatt, D. (1986). Stone handling by Japanese macaques *Macaca fuscata*: Implications for tool use of stone. *Primates*, 27, 427–437.

Huffman, M. A. & Seifu, M. (1989). Observations on the illness and consumption of a possibly medicinal plant *Vernonia amygdalina* by a wild chimpanzee in the Mahale Mountains, Tanzania. *Primates*, 30, 51–63.

Huffman, M. A., Leca, J. B. & Nahallage, C. A. D. (2010). Cultured Japanese macaques – a multidisciplinary approach to stone handling behavior and its implications for the evolution of behavioral traditions in non-human primates. In: Nakagawa, N., Nakamichi, M. & Sugira, H., (eds), *The Japanese Macaques*. Tokyo: Springer, pp. 191–221.

Jolly, A. (1966). Lemur social behavior and primate intelligence. *Science*, 153, 501–506.

Kano, T. (1980). Social behavior of wild pygmy chimpanzees (*Pan paniscus*) of Wamba: A preliminary report. *Journal of Human Evolution*, 9, 243–260.

Kuroda, S. (1980). Social behavior of the pygmy chimpanzees. *Primates*, 21, 181–197.

Masters, J., Gamba, M. & Génin, F. (2013). *Leaping Ahead: Advances in Prosimian Biology*. New York: Springer Press.

McGrew, W. C. (1979). Evolutionary implications of sex differences in chimpanzee predation and tool-use. In: D. A. Hamburg & E. R. McCown (eds), *The Great Apes*. Menlo Park, CA: Benjamin Cummings, pp. 441–463.

McGrew, W. C. (1981). The female chimpanzee as a human evolutionary prototype. In: F. Dahlberg (ed.), *Woman the Gatherer*. New Haven: Yale University Press, pp. 35–73.

McGrew, W. C. (1992). *Chimpanzee Material Culture: Implications for Human Evolution.* Cambridge University Press.

Mitani, J., Watts, D. & Amsler, S. (2010). Lethal inter-group aggression leads to territorial expansion in wild chimpanzees. *Current Biology,* 20, R507–508.

Moura, A. C. D. & Lee, P. C. (2004). Capuchin stone tool use in Caatinga dry forest. *Science,* 306, 1909.

Nishida, T., Hiraiwa-Hasegawa, M., Hasegawa, T. & Takahata, Y. (1985). Group extinction and female transfer in wild chimpanzees in the Mahale National Park, Tanzania. *Z. Tierpsychologie,* 67, 284–301.

Ottoni, E. & Izar, P. (2008). Capuchin monkey tool use: overview and implications. *Evolutionary Anthropology: Issues, News, and Reviews,* 17, 171–178.

Palagi, E. & Paoli, T. (2008). Social play in bonobos: Not only an immature matter. In: T. Furuichi & J. Thompson (eds), *The Bonobos: Behavior, ecology, and conservation.* New York: Springer, pp. 55–74.

Quiatt, D. & Huffman, M. A. (1993). On home bases, nesting sites, activity centers, and new analytic perspectives. *Current Anthropology,* 34, 68–70.

Savage-Rumbaugh, E. S. & Wilkerson, B. J. (1978). Socio-sexual behavior in *Pan paniscus* and *Pan troglodytes*: a comparative study. *Journal of Human Evolution,* 7, 327–344.

Singer, M. S., Mace, K. C. & Bernays, E. A. (2009). Self-medication as adaptive plasticity: increased ingestion of plant toxins by parasitized caterpillars. *PLoS ONE,* 4, e4796. http://dx.doi.org/10.1371/journal.pone.0004796.

Su, H., Su, Y. & Huffman, M. A. (2013). Leaf-swallowing and parasite infection in the Chinese lesser civet (*Viverricula indica*) in northern Taiwan. *Zoological Studies,* 52, 22. http://www.zoologicalstudies.com/content/52/1/22.

Suarez-Rodriguez, M., Lopez-Rull, I. & Garcia, C. M. (2013). Incorporation of cigarette butts into nests reduces nest ectoparasite load in urban birds: new ingredients for an old recipe? *Biology Letters,* 9, http://dx.doi.org/10.1098/rsbl.2012.0931.

Valderrama, X., Robinson, J. G., Attygale, A. B. & Eisner, T. (2000). Seasonal anointment with millipedes in a wild primate: a chemical defense against insects? *Journal of Chemical Ecology,* 26, 2781–2790.

van Schaik, C. P. & Knott, C. D. (2001). Geographic variation in tool use on *Neesia* fruits in orangutans. *American Journal of Physical Anthropology,* 114, 331–342.

van Schaik, C. P., Ancrenaz, M., Borgen, G., *et al.* (2003). Orangutan cultures and the evolution of material culture. *Science,* 299, 102–105.

Visalberghi, E., Addessi, E., Truppa, V., *et al.* (2009). Selection of effective stone tools by wild bearded capuchin monkeys. *Current Biology,* 19, 213–217.

Templeton, A. R. (2005). Haplotype trees and modern human origins. *Yearbook of Physical Anthropology,* 48, 33–59.

White, T. (2012). Paleoanthropology: Five's a crowd in our family tree. *Current Biology,* 23, R112–115. http://www.sciencedirect.com/science/article/pii/S0960982212014418.

Index

action component, 16, 18
adornment, 56, 57
alliance, 3, 18, 112, 113, 208
altruism, 221, 222, 223
ancestor, 8, 61, 79, 99
Andean herders, 260
Andean puna, 260
anxiety, 122, 123, 146, 147, 148, 149, 150, 151, 152, 153, 154, 155, 156, 157, 158, 160, 164, 165, 166, 167, 168, 169, 170, 171, 173, 206, 222, 272, 282
Apodemus sylvaticus, 226
appeasement, 94, 121, 124, 125, 194
Ardipithecus ramidus, 115
arousal, 43, 44, 124, 128, 146, 168, 169, 170, 171, 173
Assam macaque, 123
Ateles geoffroyi, 125
authoritarian society, 78
Avahi meridionalis, 99, 102, 172
Avahi occidentalis, 99, 101
aye-aye, 188, 247, 248

Barbary macaque, 152
Bayesian inference, 80
Berenty, xix, xxxi, 47, 48, 65, 88, 92, 93, 113, 114, 126, 132, 154, 155, 168, 169, 200, 231, 271, 272
Beza Mahafaly, 67, 162, 200, 270
bimorphism, 55, 57, 65, 67, 236
biological market, xxii, 65, 100, 220, 221, 223, 224, 226, 229, 233, 236, 272
black lemur, 262, 268
black-and-white ruffed lemur, 250, 254
black-tufted marmoset, 150
Bofi population, 207
Bombus impatiens, 37
bonobo, 32, 84, 85, 86, 87, 88, 96, 115, 123, 124, 125, 196, 197, 208, 228, 260, 282
bottlenose dolphin, 122
Brachyteles hypoxanthus, 227

capuchin monkeys, 9, 22, 79, 96, 125, 128, 168, 225, 226, 248, 263, 264, 265, 266, 267, 270
cathemerality, 99
Cebus apella, 9, 79, 168, 172, 199, 225, 263
Cebus capucinus, 22, 96

Cercocebus torquatus, 262
chacma baboon, 40, 79, 225, 226
chimpanzee, 9, 32, 59, 61, 79, 85, 86, 93, 96, 114, 115, 116, 121, 123, 124, 152, 167, 169, 171, 172, 196, 197, 208, 219, 224, 228, 237, 248, 260, 263, 280, 282
Chlorocebus aethiops, 69, 149
cichlid fish, 135
coalition, 3, 18, 59, 77, 97, 112, 114, 132, 158, 194, 221, 232, 282
coalitionary support, 120, 123, 132, 157, 223, 235
coercion, 64, 219, 220, 235
cognition, 78, 164, 165, 194, 195, 247, 255, 256, 257, 258, 259, 264, 267, 269
cohesion, 20, 91, 92, 120, 132, 165, 172
Columbian ground squirrel, 100
commodities, xxii, 3, 220, 221, 223, 224, 231
common marmoset, 16, 261
conflict management, 116, 119, 131, 134, 121
conflict of interest, 120, 129, 160
conflict resolution, 116, 118, 120, 122, 124, 129, 135, 136, 167
consolation, 121, 167
cooperation, 115, 117, 120, 123, 132, 133, 134, 135, 188, 191, 193, 207, 220, 221, 226, 227
coping model, 128
Coquerel's sifaka, 269
cortisol, 155, 156, 164
Cost-Asymmetry Hypothesis, 100
cotton-top tamarin, 252
coyote, 191, 192, 200
crested macaque, 35, 95, 131, 197
Crocuta crocuta, 132

dark-eyed juncos, 260
Daubentonia madagascariensis, 98, 247
David's score, 80, 82, 83, 90, 91, 94, 102
decision-making, 146, 155
democratic society, 78
density-aggression model, 128
despotic society, 78, 82, 90, 91, 120, 125, 128, 131, 132, 133, 159, 198, 206, 208, 271, 282
despotism, 90, 131, 132, 133
detectability, 39, 41

detection probability, 37, 39, 45, 48
dimorphism, 54, 56, 60, 64, 67, 99, 100, 115, 129, 231, 235, 236
dog, xxii, 37, 122, 129, 165, 185, 190, 191, 192, 194, 201, 250
domestic goat, 122, 129
dominance hierarchy, 80, 82, 120, 162, 226
Drosophila, 6
Duchenne laughter, 196

E. rufus x collaris, 91
economic theory, 235, 236
egalitarian society, 78, 158, 198, 207, 228, 282
elephant shrew, 100
Elephantus rufescens, 100
elevator effect, 120, 128
Energy Conservation Hypothesis (ECH), 99
Eulemur coronatus, 98
Eulemur fulvus, 98, 131, 229, 230, 249, 251, 262, 268
Eulemur macaco, 98, 131, 254, 261, 262, 268
Eulemur mongoz, 250, 251, 252, 254, 259
Eulemur rubriventer, 98
Eulemur rufifrons, 173, 270, 272
Eulemur rufus, 250
Eulemur rufus x collaris, 88, 90, 91, 92, 98, 126, 131, 133, 134, 154, 168
Evolutionary Disequilibrium Hypothesis (EVDH), 99
Expected Utility Maximisation Hypothesis, 260
expression component, 16, 17

feeding ecology, 37, 267
female dominance, 64, 67, 88, 90, 98, 99, 100, 102, 129, 131, 232, 235, 236
firefly, 30
fission-fusion, 80, 207, 208, 229, 261
fitness, 41, 42, 96, 100, 191, 192, 202, 226
flanged male, 57

galago, 13, 99, 209
Galago demidovii, 210
Gallicolumba luzonica, 56
gaze following, 268, 269
gaze priming, 268
gelada, 33, 80, 169, 172, 197
general adaptation syndrome, 147, 155
giant river otter, 100
gibbon, 13, 226
Gombe, 114
gorilla, 32, 33, 60, 61, 124, 125, 128, 158, 195, 196, 197, 219

Gorilla gorilla, 158, 196, 197, 262
gray wolf, 122, 123
grey-cheeked mangabey, 223
grooming, 35, 85, 86, 91, 93, 120, 125, 126, 128, 148, 150, 152, 165, 166, 167, 200, 209, 210, 220, 221, 222, 223, 224, 225, 226, 229, 230, 231, 232, 233, 234, 235, 237, 269
guppy, 219

Habituation/Discrimination Tests (HDT), 17
Habituation/Dishabituation Tests, 14
hamadryas baboon, 80
hamster, 9
Hanuman langur, 194
Hapalemur griseus, 99, 250
Hapalemur simus, 188
Harrier-hawk, 155
Homo ergaster, 60
Homo sapiens, 32, 58, 67, 146, 148, 149, 164, 166, 167, 171, 197, see also humans
honeybee, 30
hormone, 62, 63, 149, 162, 163, 164, 166, 167
house sparrows, 56
howler monkey, 33
humans, 3, 8, 20, 22, 30, 34, 38, 40, 41, 58, 59, 60, 62, 63, 64, 67, 77, 78, 83, 115, 116, 119, 124, 129, 146, 148, 151, 165, 168, 172, 185, 186, 187, 190, 193, 196, 197, 201, 206, 207, 224, 226, 234, 254, 259, 264, 272, 280, 281, 282, see also *Homo sapiens*
hunter-gatherer societies, 207
Hylobates lar, 226
Hylobates syndactylus, 13
hyrax, 100

I&SI method, 80, 83
imitation, 264, 267
impulsivity, 205
individual recognition, 3, 4, 8, 14, 16, 17, 18, 19, 20, 21, 22, 23, 34, 121, 131, 272, 281
Indri indri, 98, 172
infanticide, 86, 114
inhibitory tasks, 269
innovation, 270, 272, 280
intertemporal choice, 254, 259, 261
intrasexual competition, 55, 58, 61, 65, 66, 67, 147

Jacobson's organ (vomero-nasal organ), 13
Japanese macaque, 83, 84, 95, 131, 198, 207, 270

kin selection, 222, 223
Kirindy, 229, 233, 270, 272

lactation, 11, 60, 99, 100
Landau's index, 79, 81, 90
language, 4, 5, 35, 36, 61, 264
large-or-nothing contingency task, 262
learning facilitation, 263
Lemur catta, 14, 16, 18, 19, 20, 23, 31, 37, 43, 44, 45, 46, 47, 48, 65, 88, 90, 91, 98, 99, 122, 126, 128, 131, 132, 133, 154, 158, 159, 160, 161, 162, 163, 164, 166, 172, 188, 194, 199, 208, 232, 249, 250, 251, 253, 254, 259, 268, 269, 270, 271, 281, *Vedi*
Lepus americanus, 100
leverage, 77, 100, 219, 220, 221, 226, 234, 235
linear, 91, 92
linearity, 18, 78, 79, 80, 81, 82, 83, 88, 90, 92, 93, 94, 95, 123, 156, 157, 158, 257, 282
lipsmacking, 33
long-tailed macaque, 122, 149
looking time paradigm, 251
Lophocebus albigena, 223
loris, 13, 99, 100, 209
Luzon bleeding-heart, 56

Macaca assamensis, 123
Macaca fascicularis, 149, 150, 167, 226
Macaca fuscata, 83, 95, 131, 167, 207, 262
Macaca mulatta, 9, 22, 93, 94, 150, 254, 262
Macaca nemestrina, 13
Macaca nigra, 35, 95, 131, 152, 197
Macaca sylvanus, 149, 152, 167
Macaca tonkeana, 94, 131, 158, 207
Macropus rufogriseus, 122
magpy, xxii, 202
mandrill, 13, 32, 37, 90, 124, 197
Mandrillus sphinx, 13, 32, 37, 197
manipulation, 12, 149, 150, 152, 197, 203, 204, 205, 206, 247, 248, 249, 250
marking, 10, 11, 12, 13, 14, 18, 20, 21, 22, 37, 41, 43, 44, 45, 46, 48, 54, 65, 66, 67, 68, 69, 123, 210, 232, 233, 234
mating, 19, 33, 43, 44, 56, 58, 59, 62, 63, 64, 65, 66, 67, 85, 93, 100, 129, 133, 158, 159, 162, 163, 164, 166, 199, 200, 210, 221, 226, 231, 232, 233, 234, 235, 282
 priority, 210, 232, 233
mating system, 11, 55, 204, 226
mediation, 41, 153, 154
Microcebus murinus, 43, 98, 101, 188, 248
Mind-State Attribution Model, 172
Mirza coquereli, 210
mongoose lemur, 256, 257, 268
motivational conflict, 146, 151, 272
mouse lemur, xxvii, 43, 101, 248

multimodal communication, 9, 32, 33, 34, 36, 37, 38, 39, 41, 42, 43, 44, 272
multimodal signal, 33, 36, 37, 38, 39, 40, 42, 43, 44, 45, 47, 48
multiple signal, 39, 40, 42, 47
muriqui, 227, 228, 229
Myocastor coypus, 100

natural selection, 8, 54, 97, 186
neocortex, 147, 269
neophobia, 265, 266
nepotism, 131, 220
non-Duchenne laughter, 196, 197
numerical abilities, 251, 252, 254
numerical discrimination, 252, 258
nutria, 100

object movement re-enactment, 265
observational zeros, 82, 101
odorant-binding proteins (OBPs), 5, 6, 7
odour, 4, 5, 6, 7, 8, 9, 11, 14, 16, 17, 18, 39, 40, 44, 45, 62, 192, 232, 233, 252, 265
oestrus, 43, 57, 62, 64, 65, 85, 86, 119, 133, 221
olfaction, 4, 5, 6, 7, 9, 10, 14, 19, 21, 35
olive baboon, 40, 148, 166, 167
open mouth display, 194, 197, 199
orangutan, 32, 56, 57, 61, 67, 196, 219
oriental small-clawed otter, 202
Otolemur garnettii, 99, 100
ovulation, 60, 62, 63

Pan paniscus, 96, 115, 172, 196, 197, 262
Pan troglodytes, 9, 93, 96, 148, 149, 171, 196, 197, 262, 273
Papio anubis, 149, 166, 167
Papio hamadryas, 80, 119, 168, 177, 226
Papio ursinus, 79, 226
Paranthropus robustus, 59
Passer domesticus, 56
paternity, 57, 228, 233
Pavo cristatus, 56
peacemaking, 116, 131, 135, 167
peacock, 54, 56
perception component, 16, 17, 23
Perception-Action Model, 172
Perodicticus potto, 210
Phaner furcifer, 188
pig-tailed macaque, 13
play face, 194, 195, 196, 197, 198, 201
play fighting, 128, 187, 188, 190, 193, 194, 196, 199, 200, 203, 204, 205, 207, 209
playback, 12, 20, 21, 269

playful signals, 129, 192, 194, 195, 196, 199, 200, 201
Polyboroides radiatus, 154
Pongo pygmaeus, 67, 196, 262
potto, 99
power, 60, 77, 78, 82, 86, 90, 93, 95, 100, 112, 113, 207, 219, 220, 221, 226, 227, 230, 234, 235, 236
predation, 47, 54, 78, 119, 154, 155, 156, 157, 158, 162, 164, 171, 193, 203, 269
prefrontal cortex, 78, 204, 205
pregnancy, 60, 85, 86
Procavia capensis, 100
Propithecus coquereli, 37, 269
Propithecus edwardsi, 43, 65
Propithecus verreauxi, 43, 58, 63, 65, 88, 90, 91, 93, 122, 126, 131, 133, 134, 154, 158, 162, 172, 188, 206, 208, 209, 230, 231, 269
pruritus, 149
Pteronura brasiliensis, 100

rat, xxii, 119, 167, 185, 203, 204
Rattus lutreolus, 100
reaction time, 39, 48
receptive period, 159, 162
reciprocity, 125, 154, 205, 223, 224, 230
reconciliation, 19, 95, 117, 118, 120, 121, 122, 123, 124, 129, 131, 132, 133, 134, 150, 167, 168, 200
red-fronted lemur, 125, 272
red-necked wallaby, 122
reverse reward contingency task, 259
rhesus macaque, 9, 22, 40, 79, 93, 94, 95, 125, 150, 187, 252, 254, 263
Rhinopithecus bieti, 80
Rhinopithecus roxellana, 80, 225
ring-tailed lemur, 14, 18, 19, 20, 21, 36, 39, 40, 43, 45, 47, 88, 90, 91, 93, 94, 113, 128, 163, 164, 169, 170, 194, 200, 201, 208, 248, 249, 254, 256, 257, 258, 268, 270, 271, 282
ritualized signal, 199
rituals, 38
rough and tumble, 187, 193, 194
ruff, 55

Saguinus fuscicollis, 11
Saguinus mystax, 11
Saguinus oedipus, 262
Saimiri boliviensis, 100
Saimiri sciureus, 9, 93
Sainte Luce, xxxi, 102
scent, 9, 10, 11, 12, 13, 14, 16, 17, 18, 19, 20, 21, 22, 31, 37, 39, 40, 41, 43, 44, 46, 47, 48, 54, 65, 66, 67, 68, 69, 123, 210, 231, 232, 233
scorpion fly, 219

scratching, 122, 148, 149, 150, 151, 152, 154, 155, 156, 157, 158, 159, 160, 161, 166, 167, 168, 169, 170, 171, 173, 282
scratching variation index, 156
secondary sexual signals, 58, 60, 67, 68, 272, 282
self-directed behaviour, 148, 149, 150, 151, 152, 153, 154, 167, 168, 169
selfish behaviour, 135
Semnopithecus entellus, 194
sequence learning, 258
serious fighting, 128, 187, 193, 203, 209
serotonin, 150, 205
sexual selection, 10, 11, 12, 54, 55, 57, 61, 63, 65
sexual swelling, 40, 41, 85
sifaka, 39, 43, 63, 64, 65, 67, 68, 88, 93, 94, 129, 158, 166, 169, 170, 209, 210, 230, 231, 232, 233, 234, 235, 236, 269, 282
silent-bared teeth display, 198
smell, 5, 8, 9, 13, 14, 129, 269
smile, 94, 196, 197
snowshoe hare, 100
snub-nosed monkey, 80, 225
Social Bridge Hypothesis, 208
Social Constraint Hypothesis, 131
social facilitation, xxiii, 266
social insect, 5
Social Intelligence Hypothesis, 256, 257, 269
social learning, 61, 247, 248, 265, 267, 268, 269, 270, 271, 272
social network, 14, 79, 83, 97, 157, 208, 222, 282
social play, 115, 128, 166, 185, 187, 188, 191, 193, 204, 205, 207, 208, 209, 210, 282
social relationships, 91, 92, 120, 188, 205, 206, 207, 208, 234, 236
social status, 78, 84, 85, 123, 161
Social Tolerance Hypothesis, 208
sociality, 30, 60, 117, 161, 164, 165, 257, 281
society, 3, 34, 38, 64, 77, 82, 94, 116, 119, 121, 124, 132, 158, 159, 207, 208, 228, 232, 247, 272, 281
solitary play, 197
sooty mangabeys, 225
sperm competition, 228
Spermophilus columbianus, 100
spider monkey, 125, 153, 225
spotted hyena, 100, 122, 129, 132
squirrel monkey, 9, 13, 79, 93, 100, 167, 254
steepness, 82, 91, 92, 94, 96, 100, 102, 134
stimulus enhancement, 265, 266
stink fight, 40, 44, 232
strepsirrhine, 13, 14, 32, 43, 44, 63, 65, 67, 93, 125, 129, 131, 134, 168, 172, 187, 188, 207, 208, 210,

229, 231, 232, 247, 248, 253, 254, 262, 265, 267, 269, 281
stress, 77, 78, 99, 115, 146, 147, 148, 149, 151, 153, 155, 157, 158, 161, 162, 163, 164, 166, 167, 168, 173, 282
subordination, 82, 90, 93
Substitute for Reconciliation Hypothesis, 124

Taeniopygia guttata, 56
tamarin, 10, 11, 263
Tandroy, 113
tarsier, 9, 13, 99
temporal discounting, 261
tension, 120, 146, 148, 151, 169, 222
theory of mind, 191, 192
Theropithecus gelada, 33, 172
third party, 121, 123
tolerance scale, 95
tolerant society, 91, 96, 131, 198, 205, 207, 228
Tonkean macaque, 94, 198, 207
tool, 60, 61, 247, 248, 249, 250, 264, 265, 267, 270
totalitarian society, 78
trading, 100, 220, 221, 231
tradition, 35, 38, 271, 272
transmission, 30, 41, 48, 271, 272
tree shrew, 9, 10, 256
triadic affiliation, 123, 124

triangle transitivity, 82, 83, 91, 92
Trichosurus vulpecula, 100
turbinates, 13, 14

Uncertainty Reduction Hypothesis, 122
unflanged male, 57
UT-down, 45, 46, 47
UT-up, 45, 46, 47, 48

Valuable Relationship Hypothesis, 123
Varecia rubra, 251, 259, 261, 271
Varecia variegata, 37, 98, 172, 250, 261, 271
variance sensitive choice, 259
vervet monkey, 69, 149, 220
Victim Protection Hypothesis, 123
vocalisation, xvii, 21, 22, 33, 34, 35, 37, 39, 40, 43, 54, 79, 152, 171, 193, 196, 235

war, 59, 113, 114, 115, 116, 136, 206
weaning, 99, 194

xenophobia, 208, 209

yawning, 148, 168, 169, 170, 171, 172, 173

zebra finch, 56
β-endorphins, 165, 222